Early Responses to the Periodic System

Edited by Masanori Kaji, Helge Kragh,
and Gábor Palló

OXFORD
UNIVERSITY PRESS

OXFORD
UNIVERSITY PRESS

Oxford University Press is a department of the University of
Oxford. It furthers the University's objective of excellence in research,
scholarship, and education by publishing worldwide.

Oxford   New York
Auckland   Cape Town   Dar es Salaam   Hong Kong   Karachi
Kuala Lumpur   Madrid   Melbourne   Mexico City   Nairobi
New Delhi   Shanghai   Taipei   Toronto

With offices in
Argentina   Austria   Brazil   Chile   Czech Republic   France   Greece
Guatemala   Hungary   Italy   Japan   Poland   Portugal   Singapore
South Korea   Switzerland   Thailand   Turkey   Ukraine   Vietnam

Oxford is a registered trademark of Oxford University Press
in the UK and certain other countries.

Published in the United States of America by
Oxford University Press
198 Madison Avenue, New York, NY 10016

Library of Congress Cataloging-in-Publication Data
Early responses to the periodic system / edited By Masanori Kaji, Helge Kragh, and Gábor Palló.
        pages cm
Includes bibliographical references and index.
ISBN 978-0-19-020007-7
1. Periodic table of the elements.   2. Periodic law.   3. Chemical elements.
4. Chemistry—History—19th century.   I Kaji, Masanori, 1956– editor.
II. Kragh, Helge, 1944– editor.   III. Palló, Gábor, editor.
QD467.E27 2015
346'.8—dc23
2014028840

1 3 5 7 9 8 6 4 2
Printed in the United States of America
on acid-free paper

CONTENTS

LIST OF FIGURES

LIST OF TABLES

FOREWORD

Although a great number of historians have studied Mendeleev's discovery of the periodic system of chemical elements, few have looked at how the scientific community has perceived and employed this system in various areas of the world. This book fills this gap. In addition, for the evaluation of the periodic system, this book looks not only at scientific communities, but also at the educational sector and local popular culture.

The idea of a comparative project on the early reception of the periodic system occurred to one of the editors (Masanori Kaji) during the 6th Science and Technology in the European Periphery (STEP) meeting in Istanbul in June, 2008.

He engaged historians of chemistry and organized sessions devoted to this project at international conferences: the 7th International Conference on the History of Chemistry (Sopron, Hungary, August 2–5, 2009); the 7th STEP meeting (Galway, Ireland, June 17–20, 2010); and the 4th International Conference of the European Society for the History of Science (Barcelona, Spain, November 18–20, 2010). After these meetings, the following fifteen participants were brought together for this project, accounting for eleven countries and one region:

Gisela Boeck (Germany), Nathan Brooks (Russia), Marco Ciardi (Italy), Antonio García Belmar (France), Masanori Kaji (Russia and Japan), Helge Kragh (Denmark), Anders Lundgren (Sweden), Annette Lykknes (Norway), Isabel Malaquias (Portugal), Rosa Muñoz Bello (Spain), José Ramón Bertomeu Sánchez (Spain), Soňa Štrbáňová (the Czech Lands), Marco Taddia (Italy), Bernadette Bensaude-Vincent (France), and Gordon Woods (Great Britain).

These individuals all agreed to contribute papers to a collective work, and Masanori Kaji, Helge Kragh, and Gábor Palló agreed to serve as the editors. This book is important, not only for the obvious audience of historians of chemistry, but also for the larger community of historians of science and ideas and for the much larger community of chemists. Moreover, it contributes significantly to the history of pedagogy and popularization in science. It reexamines various concepts in reception studies other than "reception," such as "response" and "appropriation." It also offers new arguments in the philosophical debate of the impact of scientific discoveries.

Masanori Kaji
Helge Kragh
Gábor Palló

ACKNOWLEDGMENTS

Gestated under the warm sunny skies in Istanbul, this idea materialized into a book project thanks to many colleagues who are also interested in the history of chemistry. The contributors and editors, especially Masanori Kaji, wish to express special thanks to Brigitte Van Tiggelen, Carsten Reinhardt, William Brock, Michael Gordin, and Eric Scerri, who were of immense help during the various stages of the project.

Masanori Kaji
Helge Kragh
Gábor Palló

LIST OF CONTRIBUTORS

Antonio García Belmar
Universidad de Alicante
belmar@ua.es
Antonio García Belmar teaches history of science at the University of Alicante (Spain). His research interests include history of science teaching and learning and the making scientific expertise in modern Spain.

Bernadette Bensaude-Vincent
Université de Paris 1/IUF
bensaude@club-internet.fr
Bernadette Bensaude Vincent is Professor of Philosophy of Science and Technology at Université Paris 1-Panthéon-Sorbonne. She has published extensively on the history of chemistry and is currently working on more materials science and nanotechnology.

José Ramón Bertomeu-Sánchez
Institut d'Història de la Medicina i de la Ciència "López Piñero"
bertomeu@uv.es
José Ramón Bertomeu-Sánchez is Senior Lecturer at the University of Valencia and member of the Institute for the History of Medicine and Science. His research areas include history of science education, material culture of science, and nineteenth-century forensic medicine. His most recent book is *La verdad sobre el caso Lafarge: Ciencia, justicia y ley durante el siglo XIX* (El Serbal, 2014).

Gisela Boeck
University of Rostock
gisela.boeck@uni-rostock.de
Gisela Boeck is Associate Professor at the Rostock University, Institute of Chemistry. Her research interests include pedagogy, didactics, and history of chemistry. She is the author of several textbooks in General Chemistry.

Nathan Brooks
New Mexico State University
nbrooks@nmsu.edu
Nathan M. Brooks is Associate Professor of History at New Mexico State University, where he teaches courses in Russian and Soviet history, the history of science, and world history. His research interests include the history of chemistry and technology in Russia and the Soviet Union.

Marco Ciardi
University of Bologna, Italy
marco.ciardi@unibo.it
Marco Ciardi is Professor of History of Science at the University of Bologna. He has authored, co-authored, or edited 30 volumes and more than 100 scientific articles. He has written extensively on Amedeo Avogadro and the History of Chemistry in Italy, including *Amedeo Avogadro* (2006), *Reazioni Tricolori. Aspetti della chimica italiana nell'età del Risorgimento* (2010), and *Avogadro 1811* (2011).

Masanori Kaji
Tokyo Institute of Technology
kaji.m.aa@m.titech.ac.jp
Masanori Kaji is Professor of the History of Science at the Tokyo Institute of Technology. His research interests include history of chemistry in Russian and in Japan and environmental history. He is the author of *Mendeleev's Discovery of the Periodic Law of Chemical Elements* (in Japanese), Sapporo: Hokkaido University Press, 1997.

Helge Kragh
Aarhus University
helge.kragh@ivs.au.dk
Helge Kragh is Professor of History of Science at Aarhus University, Denmark, where he works on the history of the physical sciences (physics, chemistry, and astronomy). His most recent book, published by Oxford University Press in 2014, is a fictional history of cosmology entitled *Masters of the Universe*.

Anders Lundgren
Uppsala University
Anders.Lundgren@idehist.uu.se
Anders Lundgren is Professor emeritus in History of Science and Ideas at Uppsala University. His research interest includes history of chemistry since the 18th century, the development of chemical industries, and the significance of smell and taste in chemistry.

Annette Lykknes
Norwegian University of Science and Technology
annette.lykknes@ntnu.no
Annette Lykknes is Associate professor in chemistry education at the Norwegian University of Science and Technology (NTNU). Her research interests include history of chemistry in Norway, history of chemistry teaching, history of scientific instruments and objects, and history of women in science. She is the author (with Joakim Ziegler Gusland) of *the History of 100 Years of Chemistry Teaching and Research at NTNU* (forthcoming) and editor (with Donald L. Opitz and Brigitte Van Tiggelen) of *For Better or For Worse? Collaborative Couples in the Sciences* (2012).

Isabel Malaquias
Universidade de Aveiro, Portugal
imalaquias@ua.pt

Isabel Malaquias is Associate professor and teaches physics and history of science at the Department of Physics, University of Aveiro (Portugal). Her research interests include history of science and science education, and material culture of physics and chemistry.

Rosa Muñoz Bello

Universitat de València- CSIC, Valencia, Spain
maromube@uv.es

Rosa Muñoz Bello teaches Physics and Chemistry at the High School in Valencia. Her research interests include history of science textbooks and teaching methods in the eighteenth and nineteenth century and scientific terminology. At present, she is pursuing her (or: a) Ph.D. in the Department of History of Science at the University of Valencia. Her latest article was about chemistry terminology.

Gábor Palló

gabor.pallo@ella.hu

Gábor Palló is Senior Consultant at the Visual Learning Lab, Budapest University of Technology and Economics. His fields of research include history of chemistry and physics, 20th century history of natural sciences in Hungary, philosophy of science, history of migration of scientists, the relationship between science, politics, and philosophy. He is the author of *Zsenialitás és korszellem* (Genius and Zeitgeist), Budapest: Áron Kiadó, 2004.

Soňa Štrbáňová

Centre for the History of Sciences and Humanities
Institute of Contemporary History of the Academy of Sciences of the Czech Republic
sonast2@gmail.com

Soňa Štrbáňová is Associate Professor, Centre for the History of Sciences and Humanities, Institute of Contemporary History of the Academy of Sciences of the Czech Republic in Prague. Her research interests include history of chemistry and biochemistry and national styles in science. She is co-author or co-editor of several books, among others: (with J. Janko) *Science in Purkinje's Time* (1988); (with I. Stamhuis and K. Mojsejová) *Women Scholars and Institutions* (2004); (with A. Kildebaek Nielsen) *Creating Networks in Chemistry: The Founding and Early History of Chemical Societies in Europe* (2008); and (with A. Kostlán) *One Hundred Czech Scholars in Exile* (2011).

Marco Taddia

University of Bologna, Italy
marco.taddia@unibo.it

Marco Taddia is Associate Professor of Analytical Chemistry at the University of Bologna. His scientific research has been focused on the development of methods for the inorganic trace analysis. He had a long-standing interest in the history of chemistry and published over 70 papers also in this area. He is the current President of the "Gruppo Nazionale di Fondamenti e Storia della Chimica" (2014–2017) and official representative of the Italian Chemical Society to the EuCheMS Working Party on the History of Chemistry.

Gordon Woods
Historical Group, Royal Society of Chemistry
gandp16@talktalk.net
Gordon Woods, now retired, is Fellow of the Royal Society of Chemistry, a member of its historical group. His main research interests are the Periodic Table and chemists, particularly British. His latest article was "Mendeleev, Man and his Matrix" (Springer). He treasures a copy of Mendeleev's *Principles of Chemistry* (1896) and his worldwide collection of Periodic Tables.

Early Responses to
the Periodic System

CHAPTER 1

ojo

Introduction

MASANORI KAJI, HELGE KRAGH, AND GÁBOR PALLÓ

Even though there have already been many studies of the reception of scientific discoveries and theories, only a few discoveries have been systematically examined from a comparative perspective, in particular Darwin's theory of evolution in biology and Einstein's relativity theory in physics.[1] In the field of chemistry, the periodic system of the elements is a good candidate for such comparative reception studies. Although the discovery of the periodic system and its later history have generated numerous inquiries,[2] its reception has received only partial or scanty attention. In his noted paper published in 1996, the American historian of science Stephen G. Brush explored the role that successful predictions and accommodation of known facts played in persuading scientists to accept scientific discoveries.[3] He systematically examined textbooks and comprehensive chemistry reference works, observing that, "[the] number of explicit references to the periodic law to be found in late nineteenth-century chemistry journals is small and fluctuates irregularly."[4] Relying on a survey of textbooks and reference works written between 1871 and 1890 and existing in American libraries, he concluded that the periodic law had been generally accepted in the United States and Britain by 1890.[5] In a footnote to the same paper, he suggested the need to extend this study of texts to other countries, especially Germany and France.[6]

In fact, two years before Brush's paper was published, Ludmilla Nekoval-Chikhaoui had completed her dissertation on the diffusion of Mendeleev's periodic classification in France.[7] She studied this subject as part of a project on the diffusion of scientific knowledge from the second half of the nineteenth century to the early twentieth century. Basing her examination on scientific journals and chemistry books, Nekoval-Chikhaoui analyzed the diffusion of the periodic system in the French scientific community.[8] She also surveyed the introduction of periodic classification in higher and secondary education based on an analysis of chemistry textbooks, higher education courses, and public education programs. At the end of her dissertation, she called to conduct a comparative analysis of the diffusion of Mendeleev's discovery in different European countries.[9] However, Nekoval-Chikhaoui's call and Brush's suggestion went unmet; no comparative studies on the reception of the periodic system followed. Therefore, this book constitutes the first major step toward a response.

When we started our comparative work, we listed the following topics for study:

What were the first or earliest journal papers, chemistry textbooks, or reference books that mentioned the periodic law in the country examined? Were they written by local authors or translated from other languages? Who were these authors? Wherever possible, we provide quantitative data concerning the number of textbooks for the period between 1870 and 1920 in which the periodic law was mentioned.

How did local actors perceive the periodic law? Did they regard the new discovery as a law, as a classification, or as a theoretical interpretation? How did this treatment and perception change over time?

Why was the new classification appropriated and employed? Was it used for teaching or research? What happened to the old chemical classifications? Were they abandoned after the introduction of the periodic classification? Did some authors criticize or reject the new classification? If so, why?

Were there any arguments about the implications of the periodic law for the structure of matter? Was the periodic law related to controversies concerning atomic theory? Was it associated with other ideas about the nature of matter, such as elements vs. elementary substances, or more general theories of the universe, such as evolutionism and cosmology?

For the periodic law in the public sphere, was the discovery reported in popular books, lectures, and periodicals? In which journals and for what audience was the discovery reported? Who were the journalists interested in the subject? What was the perception of the discovery and its consequences in the popular media? Did this change during the period under study?

Were the papers on the periodic law translated into the local languages? From which papers were they translated? In which journals/publications did they appear? Who was their intended audience? Who were the translators?

Who was the discoverer of the periodic law/system according to the local actors? Did priority controversies take place?

What was the impact of new discoveries like those of gallium, scandium, germanium, the rare earths, and noble gases or the discovery of radioactivity on the perception of the periodic law? Did they have any influence on the contents of textbooks and reference books? What were the consequences of the periodic law as perceived by the historical actors?

Were there any further research studies to develop the periodic law by local researchers? New classifications? New interpretations? New predictions?

The fifteen authors in this book survey all or some of the questions mentioned previously as they apply to ten countries and one region in Europe, as well as Japan. They describe the various circumstances concerning responses to the periodic system in these locations. The study covers the period spanning the 1870s to the 1920s, before the advent of quantum mechanics. In a few cases, the authors refer to earlier periods for reasons of comparison and to later periods for analysis of secondary education textbooks. Some authors have examined journal articles, reference books, and chemistry textbooks for both higher and secondary education, while others have reviewed popular books and magazines and educational curricula or examinations prepared by local or national governments.

We are aware that the categories of *reception* and *diffusion* have been criticized in recent years for their static conception of scientific theories, which are then *transferred* to passive recipients. Critics have proposed other terms, such as "response" or "appropriation."[10] We allowed the individual authors to decide which term would be appropriate for their paper.

AN OVERVIEW OF THE BOOK

Part I deals with the two countries from which the periodic system originated, namely, Russia and Germany. Chapter 2 on Russia (by Masanori Kaji and Nathan Brooks) discusses Dmitrii Ivanovich Mendeleev's compilation of the first table of elements. By the autumn of 1870, Mendeleev had completed a refined version of the periodic system, with detailed predictions of undiscovered elements in Russian, which was translated into German and published in the German journal "Annalen der Chemie und Pharmacie" the following year. The Russian response to the periodic system was different from that elsewhere because of Mendeleev's presence. As the main figure of the new Russian Chemical Society, founded in 1868, he succeeded in persuading the leading chemists in his home country of the validity of the periodic law. German-speaking chemists in the Russian Empire (our authors call this "German mediation") and Mendeleev's famous chemistry textbook *Osnovy khimii* (*The Principles of Chemistry*) played an important role in disseminating the periodic system both inside and outside Russia. Since Russia did not have a strong educational tradition, as France and Spain had, a new approach based on the periodic system was smoothly implemented in Russian secondary education in the 1880s.

Chapter 3 on Germany (by Gisela Boeck) features Lothar Meyer, another discoverer of the periodic system, as recognized by the Royal Society of London, which awarded the Davy Medal in 1882 to Meyer and Mendeleev jointly. However, in contrast to Mendeleev's case, Meyer's role in the discovery was considered less important, even by his colleagues, and became more or less forgotten in his home country. The periodic system was used only on a limited scale in research in Germany. The German educational tradition was well established, and the periodic system was not used as a novel didactic approach. Instead, popular journals mentioned the periodic system in connection with the origin of the elements, the evolution of inorganic matter, or the theory of descent of biological species.

Part II deals with two countries taking center stage in chemical research: Great Britain and France. Chapter 4 on Great Britain (by Gordon Woods) starts with preliminary research on the terms *periodic system, periodic law, periodic classification*, and *periodic table*. After a brief explanation of British contributions to the periodic system, such as those of J. A. R. Newlands, W. Odling, W. Ramsay, J. J. Thomson, F. Soddy, and H. Moseley, along with an examination of academic books, the author concludes (as Brush did in the case of the United States) that by the late 1880s, the periodic law had been generally accepted by academic chemists. He also broadly analyzes educational scenarios in Britain, using not only textbooks for universities and schools but also curricula, syllabi, and examinations, which reveals that the periodic system was not a central theme of inorganic chemistry until approximately 1920.

Chapter 5 on France (by Bernadette Bensaude-Vincent and Antonio García Belmar) reveals a conspicuous silence regarding the periodic system or classification. A small but significant group of authors adopted Mendeleev's views, including Adolphe Wurtz and Édouard Grimaux. However, they introduced the periodic system not as the final solution to the long-standing quest for a natural classification of elements, but as support for the atomic theory. Their argument was that the system depended crucially on atomic weights and could not be deduced from equivalents. In France, the traditional and unsolved problem of classification in chemistry led educators to consider classificatory issues to be a subject reserved for advanced students.

Part III includes Chapter 6 (by Soňa Štrbáňová), dealing with the central European periphery, that is, the Czech Lands. The Czech chemist Bohuslav Brauner played a crucial role in the reception of the periodic system. He initiated chemical research when European chemists started to pay attention to Mendeleev's system with the discovery of gallium in 1875. From that point, Brauner became an enthusiastic promoter of the periodic system and endeavored to perfect it, especially with regard to the position of the rare earths in the periodic table. The time span of the chapter covers the period when Czech-German antagonism reached Czech scientific institutions. The Society of Czech Chemists, founded in 1866, had an almost exclusively Czech membership, while a specialized German chemical association had never been created in the Czech Lands. Universities split into their Czech and German counterparts. Even though Brauner himself had a cosmopolitan background, Mendeleev and his works, including the periodic system, were celebrated as a brilliant representation of Slavic science.

Part IV deals with the northern European periphery, including the three Scandinavian countries of Sweden, Denmark, and Norway. The chapters in this part describe indifference to the periodic system, much as in France, but for different reasons. Chapter 7 (by Anders Lundgren) explains that a long-standing practical and atheoretical tradition of Swedish chemistry was unaffected by the periodic system, with many new elements discovered by Swedish chemists independently of the system. Because Swedish chemists at the time had little interest in theory, they did not require any explanation of the periodicity of the elements. Nor was the periodic system used as a pedagogical tool for textbooks. Lundgren contends that Mendeleev's periodic system might not have been as important as historians of chemistry have traditionally believed.

Chapter 8 (by Helge Kragh) notes that when the Danish Chemical Society was founded in 1879, the first of its kind in Scandinavia, there were only two full professors and three associate professors in Denmark, and most of the country's chemists had a practical orientation. Kragh argues that although the first published recognition of the periodic system by a Danish chemist dates back to 1880, many chemists were aware of the classification even earlier. The first chemistry university textbook, which included a detailed account of the periodic system, was published in 1902, but the system did not function as an organizing principle of the text. In the early twentieth century, only a few textbooks for secondary education referred to or used the periodic system. Julius Thomsen, a pioneer of thermochemistry, was the only Danish scientist who actively sought to understand the periodic system. Kragh further argues that some of Thomsen's ideas inspired Niels Bohr's develop-ment of his atomic theory in 1913.

Chapter 9 (by Annette Lykknes) discusses the condition of chemistry in Norway, which resembled that of other Scandinavian countries. The Norwegian chemistry community was very small: only one professor in chemistry until 1872, three chairpersons in Christiania and four in Trondheim in the 1930s. Most chemists were employed in industry and had a very practical orientation. The periodic system was first mentioned in a chemistry textbook published in 1888, not in a scientific journal. Norwegian chemists started citing the periodic system in research journals in the 1910s, when problems of radioactivity and atomic theory renewed chemists' interest in the system. However, the periodic system was introduced as a pedagogical tool in the universities only in the 1940s. Remarkably, the inclusion of the periodic system in Norwegian textbooks used in gymnasia (the equivalent of high schools) did not begin until 1970 because of the monopoly of one chemistry textbook, which happened to deny the usefulness of the periodic system.

Part V deals with countries of the southern European periphery, namely, Spain, Portugal, and Italy. Chapter 10 (by José Ramón Bertomeu-Sánchez and Rosa Muñoz-Bello) explains the limited role of the periodic system in the teaching of chemistry in Spain between 1870 and 1920 by the existing tradition of chemistry textbooks. As in France, there were long-standing debates on the classification of chemical elements and compounds into *artificial, natural,* and *hybrid* classifi-cations. Consequently, Spanish textbook authors were not that impressed by the possible classification offered by the periodic system. After the first successful prediction of new elements and the publication of Mendeleev's paper in French journals, the periodic table was disseminated in Spanish textbooks. However, the textbook authors did not consider the system to be the basis of classification of the elements, only as a way of introducing theoretical topics, such as the existence of atoms, Prout's hypothesis, the evolution of inorganic matter, and even the origin of the universe.

Chapter 11 (by Isabel Malaquias) explores traces of Mendeleev's periodic system in Portuguese higher education and secondary textbooks, some popular books, and several booklets published at the end of the nineteenth and early twentieth centu-ries. Although periodic classification was adopted officially as a topic to be taught in Portuguese secondary education only in 1948, some professors had already made reference to the system early in the twentieth century. Malaquias has found a letter by Mendeleev, dated February 4, 1904, to Balthazar Ozorio, a zoology professor at the Lisbon Polytechnic School, where Mendeleev mentioned Ozorio's discovery of possible impurities of iodine.

Chapter 12 (by Marco Ciardi and Marco Taddia) is the first attempt to study the reception of the periodic system in Italy. In 1879 an Italian translation of Wurtz's *La Théorie Atomique* was published, which explained Mendeleev's periodic system in detail. In Florence, Augusto Piccini, an assistant of Stanislao Cannizzaro, played an important role in promulgating Mendeleev's system and received a letter from Mendeleev, dated January 29, 1903. Giacomo Ciamician, another of Cannizzaro's assistants and a chemistry professor in Bologna, also taught general chemistry based on the periodic system from at least the end of the nineteenth century and contributed to work on the system in Italy. The authors have examined university chemistry textbooks and a well-known secondary education textbook by Fausto Sestini and Angelo Funaro in the nineteenth century.

Part VI includes Chapter 13 (by Masanori Kaji) about Japan, one of the few countries outside the Western world that participated in modern scientific research in the nineteenth century. The discovery of the periodic law in 1869–1871 and its dissemination in the 1880s coincided with the institutionalization of chemistry in Japan after the Meiji Restoration in 1868. This factor helped facilitate the appreciation of the periodic system as a basis for chemistry there. Most of the first-generation Japanese chemistry professors accepted without much skepticism the periodic law as one of the recent developments in chemistry in Europe. Furthermore, around this time, Japanese chemists began to contribute to the study of the periodic system. For instance, Ogawa Masataka announced the discovery of a new element called nipponium in 1908, which much later turned out to be rhenium.

The papers in this book thus shed light on a multitude of responses to the periodic system. The smallness of the chemical community, for example, played a role in the Scandinavian countries' reaction to the system. Consequently, even among chemists who had a practical orientation and who did not pay much attention to theory in general, one particular researcher with an interest in theory—such as Julius Thomsen, a pioneer of thermochemistry in Denmark—could change the situation. Thomsen offered a neo-Proutean speculation of internally structured atoms, which Mendeleev denied, but his ideas inspired Niels Bohr's development of an atomic theory in 1913. In Norway, by contrast, one chemistry textbook that happened to deny the periodic system and that was dominant in secondary education, delayed the system's reception there until as late as 1970.

In Sweden, where chemistry remained an atheoretical science, the periodic system did not bring about any change in education or research. The periodic system also did not impress chemists in France and Spain, where there was a long tradition of and debate about the classification of matter. Some research considered the system to be the worst kind of natural classification, which did not show chemical analogies clearly. Before the advent of atomic structural theories and quantum chemistry, the periodic law faced the difficulty of breaking with the tradition, and German teachers within the established practice of chemistry education did not find any novelty in it as a didactic tool.

On the other hand, in countries such as Russia and Japan, where there was no strong tradition in education, the periodic system was readily accepted. In places where there were devoted researchers, such as Mendeleev himself or Bohuslav Brauner, his influential coworker, the acceptance of the periodic system was a momentous event, as in the case of Russia or the Czech Lands. The coincidence of the institutionalization of science, including chemistry, and the discovery of the periodic law, helped Japanese chemists to accept the law without much problem.

These comparative studies reveal the relative insignificance of the periodic system in research and teaching in many countries, which is contrary to the understanding of most historians of chemistry. Of even greater interest is the fact that in the late nineteenth and early twentieth centuries, many metaphysical and philosophical reflections on nature based on the periodic system appeared outside of chemistry. Even in Russia, where there was an exceptional impact on both research chemists and chemical education, some intellectuals were trying to speculate about the reasons for the periodic law at the end of the nineteenth and early twentieth centuries. In Germany, where the periodic system was used only on a limited scale in research and was not employed as didactically, the popular journals mentioned the system in connection with the origin of the elements, the evolution of inorganic matter, or the theory of descent. In Spain, Mendeleev's system offered good opportunities for popularizers of science to speculate on various theoretical topics, such as the existence of atoms, Prout's hypothesis, the evolution of inorganic matter, and the origins of the universe. However, neither the practice nor the strategy changed in research and teaching of chemistry because of the periodic system, probably because chemists were practice oriented, as was clearly the case in the Scandinavian countries.

It may be wondered why the papers in this book are categorized by nation-states.[11] Besides aiding with the manageability of the research, the time span coincides with the age of the nation-states, so taking the nation-state as a unit of study makes sense historically. At the same time, this framework can highlight some features that lie outside its scope, including the German-speaking scientists working outside Germany, such as in Russia and the Czech Lands. In the former Russian Empire, German nationals and subjects of the empire played an important role in promulgating the periodic system beyond the Russian border. The chapter on the Czech Lands poses a question about the expatriate German chemical community when Czech-German antagonism reached the Czech scientific institutions and universities split into their Czech and German counterparts. In addition, French influence played a certain role in the southern European periphery, and in Japan, British chemists, who were employed as teachers in higher education soon after the Meiji Restoration in 1868, played a positive role in the early introduction of the periodic system there.

Our collective work has several limitations that suggest the direction of further studies. The first is the obvious restriction in terms of countries and regions, neglecting other potentially interesting cases, such as Ireland, Canada, Mexico, Brazil, Argentina, China, India, and Australia.

A second limitation pertains to the time frame. The scope of our reviews more or less ends in 1920, before the advent of quantum chemistry. For this later phase, where a quite different response to the periodic system may be seen, we need another book like this one.

A third limitation relates to the lack of graphic representations of the periodic system. None of our papers discusses this component, even though some topics, such as Thomsen's periodic system in Kragh's chapter, imply its significance. We are well aware that a large number of graphic representations were created, with their own histories and visual accounts closely connected to chemistry.[12]

Notwithstanding our awareness of these and other limitations, we hope that this collective work of historians of chemistry might provide inspiration for new scholarly research on the reception of the periodic system.

NOTES

1. See *The Comparative Reception of Darwinism*, ed. Thomas F. Glick (Chicago and London: The University of Chicago Press, first published in 1974, reissued in 1988 with a new preface, "Reception Studies Since 1974"); *The Comparative Reception of Relativity*, ed. Thomas F. Glick (Dordrecht, Boston, Lancaster, and Tokyo: D. Reidel Publishing Company, 1987); *Disseminating Darwinism: The Role of Place, Race, Religion, and Gender*, ed. Ronald L. Numbers and John Stenhouse (New York: Cambridge University Press, 1999); *The Reception of Darwinism in the Iberian World: Spain, Spanish America and Brazil*, ed. Thomas F. Glick, Miguel Angel Puig-Samper, and Rosaura Ruiz (Dordrecht, Boston, and London: Kluwer Academic Publishers, 2001).
2. See, for example, J. W. van Spronsen, *The Periodic System of Chemical Elements: A History of the First Hundred Years* (Amsterdam: Elsevier, 1969); Eric Scerri, *The Periodic Table: Its Story and Its Significance* (Oxford, New York, etc.: Oxford University Press, 2007).
3. Stephen G. Brush, "The Reception of Mendeleev's Periodic Law in America and Britain," *Isis* 87 (1996): 595–628.
4. Brush (note 3), 600.
5. Ibid., 601.
6. Ibid., 601, n. 12.
7. Ludmilla Nekoval-Chikhaoui, "Diffusion de la classification périodique de Mendéléïev en France entre 1869 et 1934" (PhD Diss., Univ. Paris-Sud U.F.R. Scientifique d'Orsay, 1994).
8. Nekoval-Chikhaoui used "la classification périodique de Mendéléïev" [Mendeleev's periodic classification] more often than "la loi périodique" [the periodic law]. Brush (note 3) used the term "law."
9. Ibid., 140.
10. See, for example, Kostas Gavroglu, Manolis Patiniotis, Faidra Papanelopolou, Ana Simões, Ana Carneiro, Maia Paula Diogo, Joé Ramón Bertomeu Sánchez, Antonio García Belmar, and Agustí Nieto-Galan, "Science and Technology in the European Periphery: Some Historiographical Reflections," *History of Science* 46 (2008): 153–175.

11. For information on the limitations of comparative studies based on national boundaries, see James A. Secord, "Knowledge in Transit," *Isis* 95(4) (2004): 654–672, p. 669.
12. For graphical representations, see F. P. Venable, *The Development of the Periodic Law* (Easton, Pa.: Chemical Publishing Co., 1896); E. Mazurs, *The Graphic Representation of the Periodic System During 100 Years* (Tuscaloosa: University of Alabama Press, 1974).

PART I

Discovery and Early Work on the Periodic System

CHAPTER 2

༄

The Early Response to Mendeleev's Periodic System in Russia

MASANORI KAJI AND NATHAN BROOKS

1. INTRODUCTION: DISCOVERY OF PERIODIC LAW OF CHEMICAL ELEMENTS BY MENDELEEV

Mendeleev's first table of elements, entitled "An Attempt at a System of the Elements Based on Their Atomic Weight and Chemical Analogies,"[1] was dated February 17, 1869.[2] His first paper on the discovery, "The Correlation of the Properties and Atomic Weights of the Elements"[3] (hereafter referred to as Paper I), was read in the meeting of the newly established Russian Chemical Society (RCS)[4] on March 6 by Nikolai Aleksandrovich Menshutkin (1842–1907), the secretary of the Society for Dmitrii Ivanovich Mendeleev (1834–1907), who was not able to attend the meeting since he was visiting various cheese-making factories outside the city.[5] The members of the Society who attended this meeting decided not to discuss Mendeleev's paper, and it was tabled until the next meeting. The paper was published in the second/third combined issue of the first volume of the Society's journal in May (Paper I).

Paper I was the first public announcement of one of the most important discoveries in nineteenth-century chemistry: what would soon be called Mendeleev's "periodic law." However, the paper did not draw immediate attention from the chemists at the meeting. While this muted response was similar to the initial reception of the periodic system in other countries,[6] the reception of the periodic system in Russia was distinctive for various reasons. This paper will examine some of these factors first.

The most obvious difference between the reception of the periodic system in Russia as compared to other countries is simply due to the fact that Mendeleev was Russian. Thus, the reception of the periodic system in Russia needs to be considered in the context of Mendeleev's place in the Russian chemical community, as well as in intellectual terms.

In 1869, at the time of Mendeleev's first announcement about the periodic system, the Russian chemical community was in the process of formation. During the first half of the century, the number of chemists with advanced training was small and nearly all of them were employed at higher educational institutions and the Imperial Academy of Sciences. In addition, these chemists exhibited a local orientation in which they concentrated their attentions on matters of concern for their local communities, not the larger community of chemists in Russia or in Western Europe. Some of these chemists did conduct research, but they did not establish sustained research schools or construct nationwide contacts among other chemists in Russia. This situation began to change with the death of Nicholas I (1796–1855, emperor of Russia 1825–55) in 1855 and the beginning of the Great Reform Era (1855–81) under Tsar Alexander II (1818–81, emperor of Russia 1855–81). Russians were again sent abroad at state expense to receive training in preparation for employment in higher educational institutions, something that had been prohibited during the years of 1848 and 1855. The employment opportunities for chemists and other educators also greatly expanded at this time, as the state increased the ranks of the professoriate in both the universities and the technical schools of higher education. The numbers of students studying at higher educational institutions also rapidly expanded at this time. By the late 1850s, there was a growing number of young chemists in St. Petersburg and elsewhere in Russia with advanced training, many of whom had studied abroad. These chemists began to plan the formation of a chemical society that would hold meetings to discuss their research and that would publish the results of this research, using as models la Société Chimique de Paris and the Chemical Society of London. For various reasons, this society was established only in 1868 in St. Petersburg, where a large number of higher educational institutions were located and where the majority of young chemists lived. Mendeleev was one of the organizers of this new society and he played an important role in its operation for many years.[7]

In 1869 Mendeleev was professor of chemistry at St. Petersburg University and had been there since 1865 when he first moved to the chair of technical chemistry.[8] Mendeleev had been a student at the Main Pedagogical Institute (Teacher's College), where many of the teaching staff also taught at St. Petersburg University. He graduated and worked as a secondary school teacher for less than a year before he returned to St. Petersburg to defend his master's dissertation. After working as a lecturer at various schools in the capital, Mendeleev was able to take advantage of study abroad, funded by the Ministry of Public Enlightenment. He then went to Heidelberg, where several other Russians were studying. Mendeleev was slightly older than many of his fellow Russian students, but managed to develop close friendships with other Russians. He did not spend much time in lectures, since he had solid theoretical training (this was the case for many Russian chemistry students who went abroad at this time and later). After spending most of his time in experimental work using equipment purchased in Europe, Mendeleev reluctantly returned to Russia, since he could not afford to stay abroad any longer. Following his return in 1861, Mendeleev worked at various part-time teaching positions until he secured an appointment to the St. Petersburg Technological Institute in 1864. Following the defense of his doctor's dissertation on the densities of alcohol-water combinations early in 1865, he moved to St. Petersburg University as an extraordinary professor of technical chemistry in April. He was promoted to full professorship at the end of the same year, while keeping his position at the Technological Institute until 1871. Upon replacing his teacher A. A. Voskresenskii (1808–80), who had retired from teaching and had moved to a high-level administrative position outside of St. Petersburg, Mendeleev began to teach general chemistry in 1867. It was then that he started to write a textbook of chemistry, which directly led to the formulation of the periodic system.[9]

2. MENDELEEV'S WORK ON THE PERIODIC SYSTEM AND ITS EARLY RECEPTION BY RUSSIAN CHEMISTS

As we have already noted, the reception of Mendeleev's work on the periodic system by chemists in Russia differed in some significant ways from its reception in other countries. Since Mendeleev was a member of the rapidly emerging chemical community in Russia, his work was discussed in the same manner as other chemistry research done by Russian chemists. While Mendeleev's initial communication about the periodic system did not evoke much attention from Russian chemists, Mendeleev continued working on the topic and his presentations and papers soon drew consideration from the Russian chemical community as part of the general discussion of new chemistry research.

As noted previously, the first communication about the periodic system was delivered to a session of the RCS on March 6, 1869, by N. A. Menshutkin in Mendeleev's absence. The protocols of the meeting noted that discussion of the communication would be postponed until the next session of the society, when Mendeleev presumably would be in attendance. Nevertheless, there is no indication in the protocols of the society that any discussion of Mendeleev's initial communication occurred during the next six months or more, even though Mendeleev apparently attended most of these sessions. According to later recollections of B. N. Menshutkin (1874–1938), his father (N. A. Menshutkin) once told him that Mendeleev's communication about the periodic system "did not evoke particular interest or an exchange of opinions."[10]

However, the published version of Mendeleev's first paper on the periodic system (Paper I) included a footnote dated April 5 stating that Fedor Nikolaevich Savchenkov (1831–1900) had informed Mendeleev about a table of elements in William Odling's *Practical Chemistry* (1865) that was similar to the one in Mendeleev's paper. Savchenkov, a chemist in the state mining agency, had recently (1867) translated Odling's textbook into Russian.[11] In the footnote, Mendeleev said that Savchenkov had told him about Odling's work at the April 3 meeting of the Russian Chemical Society. As there is no mention of this in the published protocols, Savchenkov must have stated this to Mendeleev outside of the formal proceedings of the society. Mendeleev downplayed the significance of Odling's table by noting that Odling "did not expand on the meaning of the table and . . . has not mentioned it elsewhere." In addition, Mendeleev emphasized that he had not been aware of Odling's table before this time and argued that had Odling placed any theoretical value on the table, he would have pursued further research on it, but apparently had not published anything more along this vein.

Mendeleev's first paper on the periodic system (Paper I) presented not only what he discovered, such as periodicity, correlation between the chemical properties and the position in the table, but also the following five problems or research proposals from his new classification:

(1) determining the positions in the table for certain elements
(2) other possible forms of the periodic law of the elements
(3) the relationship between the chemical properties of elements and their groups
(4) the correction of some elements' atomic weights
(5) undiscovered elements

Mendeleev's research until the end of 1871 (See Table 2.1) can be regarded as the development of problems proposed in Paper I.

As soon as Mendeleev had finished his original paper on the periodic system, he began to work on determining several periodic functions of atomic weights, including atomic volumes and the composition of higher salt forming oxides. He presented the results of this work in August 1869 at the Second Congress of Russian Naturalists meeting in Moscow. Briefly summarizing this work, he wrote that "the comparison of specific weights and specific volumes of elements belonging to different rows shows to some extent the naturalness of the system."[12]

Several other chemists presented papers at this conference that concerned topics related to the periodic system and perhaps were responses to Mendeleev's initial work on the topic. Nikolai Erastovich Lyaskovskii (1816–71), professor of chemistry at Moscow University, presented a paper in which he proposed a "law that determines the relative energy belonging to different members of natural groups of elements."[13] In addition, Nikolai Nikolaevich Beketov (1827–1911), professor of chemistry at Kharkov University, presented a paper on "The Atomicity [valence] of Elements" in which he examined the limiting values for the valences of various elements.[14] The protocols of the sessions noted that these communications "resulted in lively discussion." Mendeleev himself expressed disagreement with the comments of one respondent to Lyaskovskii's paper.

In the fall of 1869, Mendeleev attempted to determine some chemical properties of the elements as a function of their atomic weights. In early October, he presented a paper on "The Quantity of Oxygen in Saline Oxides and the Atomicity of Elements" at a meeting of the RCS. In this paper he showed that the quantity of oxygen in saline oxides varies in a periodic function according to the atomic weight of the element.[15]

While Mendeleev continued to pursue work on the periodic system, this work was either ignored or discounted by some Russian chemists. For example, in the

---

Table 2.1 MENDELEEV'S STUDY ON THE PERIODIC SYSTEM OF ELEMENTS
AFTER 1869 (IMENDELEEV (NOTE 1) 1958, 753–759 AND NOTE 5)

Ann. = *Annalen der Chemie und Pharmacie*

AN = *Akademiya nauk* (Academy of Sciences at St. Petersburg)

Ber. =*Berichite der Deutschen chemischen Gesellschaft*

RCS = *Russkoe khimicheskoe obshchestvo*

SRE =*Sezd russkikh estelltvoispitatelei* (Congress of Russian Scientists)

TSRE = *Trudy sezda russkikh estestsvoispitatelei* (Proceedings of the Congress of Russian Scientists)

JRCS = *Zhurnal russkogo khimicheskogo obshchstva*. The numbers indicate the volume.

OR = Oral Report P = Pamphlet [ ]: Only in manuscript

[1868 May-Jun. *The Principles of Chemistry*, 1st pt., 1st vol.]

1869 Feb.17 "Attempt of System of Elements on Atomic Weights and Chemical Similarity" (P)

Mar.1 Preface for the 1st pt. of *The Principles of Chemistry*.

Mar.6 "The Relationship between Atomic Weight of Elements and Properties" (OR at RCS; JRCS vol. 1)

Mar.(second half) *The Principles of Chemistry*, 1st pt., 2nd vol.

Aug.23 "On the Atomic Volume of Simple Bodies" (OR at SRE; TSRE)

Oct.2 "On the Quantity of Oxygen in Salified Oxycides and Valency of Elements" (OR at RCS; JRCS vol. 2)

Nov.6 "On the Law of Heat Capacity and Complexity of the Carbon Molecule" (OR at RCS; JRCS vol. 2)

1870 Mar.5 "On Metal Ammonia Compounds" (OR at RCS)

Feb.(end) or Mar. (early). *The Principles of Chemistry*, 2nd pt., 3rd vol.

Oct.8 "On Thionic Acid" (OR at RCS; JRCS vol. 2)

Nov.5 "On Compounds with NO2 group" (OR at RCS; JRCS vol. 3)

Nov.24 "Über die Stellung des Ceriums im System der Elemente" (OR at AN; Bulletin de l'Académie
impériale des sciences de St.-Pétersbourg 1871)

Dec.3 "The Natural System of Elements and Its Application to Show Properties of Undiscovered
Elements" (OR at RCS; JRCS vol. 3)

1871 Feb. (end) *The Principles of Chemistry*, 2nd pt., 4/5th vol.

Mar. "Zur Frage über das System der Element" (Ber.)

Jul. Finished writing "Die periodische Gesetzmässigkeit der chemische Elemente" (Ann.) and
published in Germany on Nov.6 (Gregorian calendar)

Aug.21 "On Specific Volume of Chlorine Compounds" (OR at SRE)

Aug.24 "On Crystal Water" (OR at SRE)

Oct. [On Some So-Called Molecular Compounds]

Nov. [On Polymerization in Mineral Compounds]

Nov. "Note on Peroxides" (JRCS vol. 3)

Dec.2 "On the Atomic Weight of Yttrium" (OR al RCS)

---

fall of 1869, Nikolai Nikolaevich Zinin (1812–80), a leading organic chemist and the first president of the Russian Chemical Society (RCS), advised Mendeleev to do "[real] work," meaning do something experimental, preferably on organic chemistry, which was the mainstream research discipline in Russia at that time. Mendeleev drafted a letter to Zinin in response, although apparently he did not send it.[16] In this letter, Mendeleev defended his current research activities, obviously stung by

Zinin's criticisms. It is interesting to note that Mendeleev did not explicitly mention the periodic system by name, nor did he discuss his work on the textbook *The Principles of Chemistry*. Instead, Mendeleev tried to downplay the importance of organic chemistry, arguing that at present, organic chemists were only concerned with "petty facts," not larger issues, as was the case fifteen years prior. Mendeleev then related an incident from 1856 where he had done some research but had not immediately published the results. Soon after, a German chemist had published the same results. What Mendeleev obviously was trying to argue in this letter was that his instincts about research topics were valid since influential—foreign—chemists also had worked on these topics. It is clear from the tone of this letter that Mendeleev felt defensive about his current work on the periodic system but did not intend to give it up. It also appears that Zinin wanted Mendeleev to work on organic chemistry rather than the periodic system, which involved inorganic and physical chemistry. This implies that Zinin felt that organic chemistry—not inorganic or general chemistry—was the most important field in chemistry at that time and that Mendeleev was wasting his time working on any other field.

Although it appears that organic chemists in Russia did not take interest in Mendeleev's research on the periodic law at this time, chemists working in areas other than organic chemistry did. One indication of this that we had already noted is the lively discussion of topics related to the periodic law at the August 1869 meeting of the Second Congress of Russian Naturalists. In addition, as Mendeleev worked to expand his research on the quantity of oxygen in various oxides as a periodic function in 1870, he needed some mineral specimens for his experiments. He asked the rector of St. Petersburg University to provide him with the needed specimens. Only twelve days after the initial request, Mendeleev received permission to obtain these samples from the State Mint from F. N. Savchenkov, the Mining Department official who had directed Mendeleev's attention to Odling's work.[17] Later in 1870 or in early 1871, P. A. Kochubei (1825–92), who worked at the Mineralogy Museum of the Mining Institute, supported a similar request from Mendeleev about the release of minerals for Mendeleev's search for the proposed eka-silicon.[18]

During this time, Mendeleev conducted research on aspects of the periodic law in order to place certain elements in their correct positions in the table. For example, in the fall of 1870, he investigated the heat capacity of indium and cerium in order to correct their atomic weights.[19] On the basis of this work, Mendeleev proposed changes in the atomic weights of these two elements. Similarly, he also determined that the atomic weights of uranium and thorium should be doubled. In November 1870, Mendeleev reworked his table into a new short table of the elements that divided all of the elements into eight groups.

Mendeleev also began to direct his attention to describing the properties of undiscovered elements that he proposed could fit into his periodic table. He predicted the properties of three elements in great detail, naming them eka-boron, eka-aluminum, and eka-silicon, arguing that they were analogs of boron, aluminum, and silicon, respectively. Mendeleev used the term *eka*, which means numeral one in Sanskrit. These three elements turned out to be scandium, gallium, and germanium. Once discovered between 1875 and 1886, these elements drew worldwide attention to Mendeleev's periodic law, since their properties turned out to be almost identical to those he had predicted. Stephen Brush has argued that Mendeleev's periodic law began to become accepted in the West only after 1875, in direct consequence of his accurate predictions of these three new elements.[20]

In a paper dated November 29, 1870, and at a meeting of the RCS on December 3, Mendeleev proposed a series of modifications to his periodic system that "satisfies the condition of a natural system" as well as predicting the properties of several undiscovered elements. This paper and presentation appear to be a turning point in the recognition—if not necessarily the full acceptance—of Mendeleev's periodic law in Russia. In contrast to the silence that met Mendeleev's first communication about the periodic law in 1869, this work at the end of 1870 evoked widespread attention both at the society's meeting and afterward. A striking example of this new recognition can be seen in the dramatically changed attitude of N. N. Zinin, the respected elder chemist who we saw had earlier harshly chastised Mendeleev in late 1869 for wasting time working on topics such as the periodic law. In early 1871 Mendeleev called on Zinin at his laboratory in an attempt to obtain some mineral specimens to use in his work. Not finding Zinin there, Mendeleev departed, leaving a copy of his recently published paper. Soon after, Zinin sent a note to Mendeleev promising to dispatch the mineral specimens, as well as effusively praising Mendeleev's newly published article: "With great attention I read your paper on the natural system of the elements, etc. It is very, very good."[21]

This presentation and paper also greatly impressed other chemists. For instance, V. Yu. Rikhter (Victor von Richter) (1841–91), who reported on the meetings of the Russian Chemical Society for the *Berichte* of the German Chemical Society, wrote a detailed summary of Mendeleev's work, noting that "The most interesting would be the discovery of eka-silicon, Es=72, which forms the transition from silicon to tin; it would have a specific volume of about 13, specific weight—5.5. The atomic volume of the oxide would be 22, specific weight = 4.7 . . . These would be interesting predictions if one could succeed in actually discovering this element!"[22]

Similarly, in early 1871, Savchenkov published a long review of Mendeleev's work on the periodic system up to that point in time. In this review, Savchenkov stated, "In our time when the work of chemists each day uncovers new analytical properties separating particular bodies and when spectral analysis shows us paths to discovering new elements, it is very natural to see science striving to group together phenomena and the properties of bodies, as well as to generalize our scientific views, all of which help us to find that red thread in the mass of chemical facts that is increasing every day." Here, Savchenkov appears to appraise highly the organizing value of the periodic system. Later in this article, Savchenkov argues, "Relying on successive changes in the differences in the quantities of atomic weights . . . it is possible to theoretically correct atomic weights of those elements which have been determined with little precision at the present time, and to uncover some conclusions regarding both chemical as well as physical properties of those elements which remain in the system and which are still not discovered but of which discovery is very certain." Savchenkov also emphasized the importance of prediction, but accorded it less significance than accommodation of known facts: "The future will show how true is the proposition of D. I. Mendeleev about the existence of various simple bodies in our time and about their expected properties; but his proposition about a natural system of elements, in my opinion, has a great importance at the present time for grouping chemical facts that are known up to now according to sequential changes in properties, linked with changes in the atomic weight of simple bodies."[23]

Shortly after Savchenkov's review was published, Konon Ivanovich Lisenko (1837–1903), professor at the Mining Institute, wrote a highly laudatory review of Mendeleev's *The Principles of Chemistry* in the same journal. Lisenko stated that Mendeleev's textbook was distinguished from other treatments of general chemistry, since "Mendeleev places his newly proposed classification of the elements as the foundation of his treatment," and referred readers to Savchenko's earlier review of Mendeleev's work on the periodic law.[24]

Another example of the widening interest of Russian chemists in topics related to the periodic law is shown by the discussions at the chemistry section of the Third Congress of Russian Naturalists in Kiev in August 1871. There, Mendeleev had presented several short communications in which he attempted to expand the idea of the periodic law to wider areas of chemistry. In one of these presentations, Mendeleev tried to link the idea of periodicity to aspects of the crystallization of water. The protocols of this session noted that Mendeleev's ideas were "met with general agreement from the members of the section. [Mendeleev's presentation] evoked, at the same time, a prolonged discussion."[25] At this same meeting, V. V. Markovnikov (1837–1904) presented a paper on a similar topic, which also led to extensive discussions by Mendeleev, A. A. Verigo (1837–1905), A. M. Butlerov (1828–86), N. N. Beketov, and others.[26] What makes this of particular significance is that we now see leading Russian organic chemists, such as Markovnikov and Butlerov, fully participating in discussions about various aspects of the periodic law.

By the end of 1871, Mendeleev had finished writing *The Principles of Chemistry* and had summarized his work on the periodic law in a long article written for the German journal *Annalen der Chemie und Pharmacie*. In this article, Mendeleev elaborated on his conception of the periodic law and showed how it could be used to correct the atomic weights of several little-studied elements, as well as predict the properties of some undiscovered elements. Later, Mendeleev looked back at this work and concluded: "This is the best collection of my views and reflections about the periodicity of the elements . . . This is the main reason for my scientific renown, because much was proved correct much later."[27]

At the end of December 1871, Mendeleev abruptly started a new research project on gas expansion. Even though there still remained some problems with the periodic law, such as finding the predicted elements and determining the places of the rare earth elements, Mendeleev did not seem to have sufficient patience to continue the experimental work that these problems necessitated. Instead, he directed his attention to the expansion of gases, explaining that this new research was a search for a true physical foundation of the periodic law.[28]

3. MENDELEEV'S TEXTBOOK *THE PRINCIPLES OF CHEMISTRY* AND THE PERIODIC SYSTEM

Mendeleev's textbook *The Principles of Chemistry* (hereafter referred to as "*Principles*") was a very important factor for the acceptance of the periodic system. The textbook was his main work and he continued to revise it throughout his life, thus staying current with the advancing knowledge in chemistry.

As for the periodic system, it was a key aspect of this textbook from the very start. In the aforementioned Paper I, Mendeleev indicated the close relationship between the discovery of the periodic system and *Principles*:

> In undertaking to prepare a textbook called "Osnovy khimii" [*The Principles of Chemistry*], I wished to establish some sort of system of simple bodies in which their distribution is not guided by chance, as might be thought instinctively, but by some sort of definite and exact principle.[29]

First, let us consider the chronology of the publications of the first edition of *Principles* and the discovery of the periodic system. In May or June 1868, Mendeleev published the first volume, which includes chapters 1 through 11. On February 17, 1869,[30] he compiled the first periodic table.[31] On March 6 Menshutkin read Mendeleev's first paper on his discovery (Paper I) in the meeting of the Society. At almost the same time, he published the second volume of *Principles*, chapters 12 through 22. At the end of February or early March the next year, the third volume, which comprises chapters 1 through 8 of part 2, appeared. Finally, the last volumes (fourth and fifth), which include chapters 9 through 23, were published in February 1871. In July of that year, Mendeleev's most comprehensive paper on the periodic law, "The Periodic Law of the Chemical Elements," was published in the supplemental volume of the *Annalen der Chemie und Pharmacie*.[32] This chronology makes clear that Mendeleev discovered the periodic law in the middle of writing *Principles*. As Kedrov has pointed out, a careful reading of its text reveals exactly when he discovered that law.[33]

Let us examine Paper I and the early chapters of the second part of his textbook, which must have been written around the same time. He organized the first part of *Principles* on the basis of the principle of valence: first he discussed univalent hydrogen, then divalent oxygen, trivalent nitrogen, and tetravalent carbon.[34] After his treatment of the univalent halogens, which concludes the first part of the textbook, Mendeleev began the second part with a description of the univalent alkaline metals. At the end of the chapter on heat capacity, which follows the alkaline metals, he explained that he should treat analogs of copper next, but would write about the divalent alkaline-earth metals instead, even though the former awkwardly exhibits both uni- and divalence.[35] Although he had followed the principle of valence to this point in the textbook, he immediately began the next chapter on a different principle, comparing alkaline-earth metals with alkaline metals on the basis of their atomic weights.

In this connection it should be noted that toward the end of Paper I, Mendeleev stressed that "[t]he purpose of my paper would be entirely attained if I succeed in turning the attention of investigators to the relationships in the size of the atomic weights of *nonsimilar* elements, which have, as far as I know, been almost entirely neglected until now."[36] He emphasized the word "nonsimilar" with italics. Alkaline metals and alkaline-earth metals were obviously such *nonsimilar* groups of elements.

If Kedrov's analysis of Mendeleev's process is followed,[37] then Mendeleev noticed this comparison of nonsimilar groups of elements in the middle of February 1869, and he first compiled the central part of the table based on this principle. With the help of cards of the chemical elements, which he made for this occasion, Mendeleev finally succeeded in organizing a table of all the known elements on the basis of their atomic weights. He completed this on February 17, 1869.[38] Clearly, at that moment, Mendeleev had conceived the idea that atomic weight might be the fundamental numerical property of the elements.

In Paper I Mendeleev wrote:

> No matter how properties of simple bodies may change in the free state, *something* remains constant, and when the element forms compounds, this *something* is material existence and establishes the characteristics of the compounds, which include the given element. In this respect we know only one constant peculiar to an element, namely the atomic weight. The size of the atomic weight, by the very essence of matter, is common to the simple body and all its compounds. Atomic weight belongs not to coal or diamond but to carbon.[39]

This "something," italicized in the quotation above, exactly corresponds to Mendeleev's definition of the elements. In other words, atomic weight belongs to elements.

As a result of this reconceptualization, or discovery, Mendeleev realized that he should use atomic weights, not valence, as the guiding principle for the remainder of his textbook. This was the moment when he started to write the chapter on alkaline-earth metals. However, since he defined the concept of elements without the notion of atoms, he considered atomic weights to be the fundamental property of the elements. They were not necessarily based on atomic theory, which was still speculative in some respects. Thus, the scope of atomic weights would have to be broader than that of definite proportions on which the atomic theory was thought to be based. Mendeleev even once suggested the use of the word "elementary weight" instead of "atomic weight."[40]

4. CHANGES IN THE LATER EDITIONS OF *THE PRINCIPLES OF CHEMISTRY*

Contrary to many statements in the existing literature on the periodic law that Mendeleev kept the original version of *Principles* unchanged through every edition,[41] he actually changed the structure of the textbook significantly in each edition. Much confusion has resulted from this misunderstanding. In all, eights editions were published during Mendeleev's lifetime. Let us look briefly at the changes in each edition.[42]

The first four editions had two type fonts in their text: sections in a larger font for beginning students and sections in a smaller font for advanced learners. In the second edition, published in 1872–73, only one year after the completion of the first, there were only minor changes in the text. Mendeleev moved indium and uranium to the appropriate chapters because of improved values in their atomic weights. He also changed the positions of the rare earths, which remained problematic throughout his life.

The third edition, which appeared in 1877, underwent substantial change, and the chapters were completely reorganized in accord with the periodic law. The textbook was divided into two parts, as were the first two editions, but the chapters were now numbered successively throughout. Only small changes were needed in the first part, which was introductory and devoted to the elements frequently encountered in daily life. Mendeleev placed the chapter on the periodic system, entitled "Similarity of Elements and Their System," in the second part, immediately after the description of the alkaline and alkaline-earth metals. Following these chapters, he described the elements in order of their position in the periodic table: from the second group to the sixth group, ending with the eighth group, iron and platinum analogs. The final chapters were devoted to the noble metals. The third edition also included gallium, the first element predicted by Mendeleev to be discovered.

The fourth edition in 1881–82 was the same as the third in principle. The book was slightly larger, increasing in size from 18 x 11 cm to 20 x 12 cm. Mendeleev first mentioned the discovery of scandium in this edition.

The fifth edition in 1889 underwent the second major change since the third edition. The book became considerably larger in size, and the text was printed in double rather than single columns for the first time. Therefore, the whole work became much shorter, reduced from 1,176 pages in the fourth to 789 pages in the fifth. Some of the material that had previously appeared in the text was moved into footnotes in the smaller font. There were no longer two parts, only one, bound as a single volume (the textbook retained this format through all subsequent editions). The chapters were also completely reordered. Many of them were combined, and the forty-four chapters in the fourth edition became only twenty-four chapters in the fifth. The chapter on the periodic law was expanded to include the history of its discovery and the problem of priority.[43] This fifth edition was translated into English, German, and French.[44]

The sixth edition in 1895 underwent no substantial change in its format from the fifth, but Mendeleev rewrote many of the footnotes. He added notes on argon, the newly discovered gas from the air, at the end of the textbook, and he argued for the possibility of argon being N_3.

In the seventh edition of 1902–03 Mendeleev abandoned N_3 and fully accepted the noble gases, which he incorporated into the chapter on nitrogen and the air. Mendeleev asked the Czech chemist Bohuslav Brauner (1855–1935) to write the section on the rare earths for the seventh and eighth editions, even though they had somewhat different opinions on the positions of the rare earths within the periodic system.[45] They agreed to place scandium, yttrium, and lanthanum in the third group and tantalum in the fifth. However, while Mendeleev believed that future research would reveal sufficient numbers of rare-earth elements with different properties so that they could be placed in different groups to fit neatly in his periodic table, Brauner proposed that the rare earths should be all placed together in Group IV, which was formerly occupied by cerium alone. Effectively, this demonstrates Mendeleev's admission of the difficulties in solving the placement of the rare earths, so many in number and so similar in properties, within his periodic system. He also mentioned the discovery of radium in this edition, but denied the possibility of the transformation of the elements. He suggested other possible explanations of radioactivity, such as a "state" like a magnetic property or absorbency and the projection of the "ether" in the vicinity of the radioactive atom.

The eighth edition in 1906 was the last edition published before Mendeleev's death. All the notes were separated from the main text and placed in the second half of the book. He argued for the possibility of "chemical ether" as an extremely light element in the noble-gas group, which he thought could explain radioactivity.[46]

As shown in his textbook, Mendeleev's concept of the chemical elements demonstrates his persistent and firm belief in their conceptual priority. His clear understanding of the elements is evident from the very first edition. In his concept of elements, Mendeleev clearly departed from Lavoisier, who had offered a negative definition of an element as an undecomposed substance. For Mendeleev, the concept was defined positively as something abstracted from the diverse properties of simple bodies and their compounds. Therefore, elements were strictly distinguished from simple bodies.

Beginning with the first edition of *Principles*, Mendeleev carefully denied the speculative connotations of the atomic hypothesis. Although it is tempting to say that his "element" is a substitute for "atom," Mendeleev resisted the use of the hypothetical atom. He was also opposed to any suggestion that tried to reduce simple substances to a single substance or a few substances called "primary matter."[47] This attitude was in sharp contrast to those of other individuals who also sought a system of the elements during the 1860s such as Lothar Meyer (1830–95).[48]

5. GERMAN MEDIATION AND RUSSIAN RECEPTION

Even though the priority debates between Lothar Meyer and Mendeleev have been well-documented,[49] these can shed light on the role of German mediation for the Russian reception of the periodic law. First, one should pay attention to Germans in the Russian Empire. Soon after the publication of Paper I, Mendeleev asked Friedrich Konrad Beilstein (1838–1906), then professor of chemistry at the St. Petersburg Technological Institute,[50] to translate the summary of his first paper on the periodic law for publication in a German journal. One of Beilstein's students translated it into German. However, the translator mistakenly interpreted the word "periodical (periodichnyi)" as "stepwise (stufenweise)." The translated sum-mary was sent to *Zeitschrift für Chemie*,[51] one of the editors of which was Beilstein himself. According to the student who translated it, the summary was sent to the journal through Lothar Meyer.[52]

At the end of the same year, Lothar Meyer submitted his famous paper, "The Nature of the Chemical Elements as a Function of Their Atomic Weights," which was published early in 1870.[53] Although Meyer admitted in the paper that his table was essentially the same as Mendeleev's,[54] his table of elements was more refined than Mendeleev's first table, especially in showing clearly the so-called transition metals. Meyer also had the correct atomic weight of indium, which was incorrect in Mendeleev's first table. Meyer had succeeded in vividly conveying a periodic dependency of properties of the elements on their atomic weights by plotting the solid-state atomic volumes of the simple bodies of their elements against their atomic weights. However, the conclusion of Meyer's paper was very tentative and cautious.

Mendeleev's work on the periodic system was conveyed through German-speaking subjects in Imperial Russia. Beilstein was such an example, even though his relationship with Mendeleev was delicate at best.

Felix Wreden (Feliks Romanovich Vreden, 1841–78) was another German living in Russia, but one who was more sympathetic toward Mendeleev than Beilstein. Wreden was born in Riga and his father was a teacher of German lan-guage. Soon after Wreden's birth, the family moved to St. Petersburg, where Felix Wreden received his education. After graduation from the Physico-mathematical faculty of St. Petersburg University in 1863, he first became custodian of the mineral cabinet of the university and then an assistant of the analytical chem-istry laboratory. He also studied organic chemistry, especially hydrocarbon components of coal and petroleum. After receiving a master's degree he became docent of chemistry in Warsaw University in 1873, then an adjunct professor at the Mining Institute in St. Petersburg in 1876.[55] Wreden was one of the found-ing members of the RCS as well as a correspondent for the German Chemical Society, sending information to the German organization about the activity of the RCS. Wreden also translated Mendeleev's Russian text of the 1871 paper into German for *Annalen der Chemie und Pharmacie*.[56] After the appearance of this paper on the periodic system in German in 1871, Mendeleev started to be recognized in Western Europe. Meyer himself became assured of the correctness of the periodic law and tried to apply it fully to systematize inorganic chemistry. Meyer's paper "For Systemization of Inorganic Chemistry" in 1873 was one of its results.[57]

Another factor for the change in attitudes toward Mendeleev in Russia in the late 1870s through the early 1880s was Mendeleev's failure to be elected to full membership in the Academy of Sciences in 1880. The event became a huge scandal because it was seen as rejecting a worthy Russian, and instead electing a foreigner (German) to continue a trend in which the Academy of Sciences was dominated by foreigners. Mendeleev became very well known to the entire country, not just among chemists or scientists, developing into a national icon.[58]

The priority dispute between Mendeleev and Meyer was one of several that Russian chemists had with foreign chemists in the years after the 1860s. These priority disputes stoked Russians' feelings of nationalism, which were growing very strong at this time in many other aspects of Russian life. Even some Russians who did not particularly like Mendeleev appeared to defend him in this priority dispute.

In 1880 Meyer[59] and Mendeleev[60] had a disagreement over the priority and contributions toward the discovery and the development of the periodic law in the journal of the German Chemical Society. Here, Meyer suggested that there was an unfavorable atmosphere toward theoretical work in the German chemical community for "a paper without any new data."[61] However, Mendeleev was bold enough to send a very long paper "without any new data" to the same journal. For one thing, as B. Brauner, a Czech chemist and Mendeleev's friend[62] wrote later, the younger members of the editorial board of the journal[63] strongly supported the publication of Mendeleev's paper. With powerful argument in the paper itself, such as detailed prediction of undiscovered elements, which had persuaded Russian chemists to take Mendeleev's side, these favorable conditions in the German editorial board helped to promulgate Mendeleev's discovery.

This "German mediation" as well as the support and encouragement of the Russian chemical community helped Mendeleev to continue concentrating his study on the periodic system during this important period.

6. SOME WORKS ON THE PERIODIC SYSTEM BY RUSSIAN CHEMISTS AFTER 1871

In the years following Mendeleev's long paper in 1871 summarizing his research and thoughts about the periodic law, Russian chemists published only a small number of articles and comments relating to the periodic law in the journal of the Russian Chemical Society. The Russian Chemical Society was the sole nationwide professional organization for Russian chemists, so it is reasonable to view publications in the society's journal as reflecting the attitudes of professional chemists in Russia to the periodic law.

Several of these articles investigated the valences of different elements and how variable valences could impact the periodic law. At the Third Congress of Russian Naturalists in August 1871, N. N. Beketov presented a paper concerning the valences of chlorine and fluorine, which drew a comment from Mendeleev, who illustrated his point with reference to work done by a foreign chemist.[64] More substantial was a paper published in 1873 by Aleksandr Ivanovich Bazarov (1845–1907) that examined the theories of valence and structure. In this paper, Bazarov criticized Mendeleev's use of his "theory of limits" to help understand the variation of valences and how this theory could be employed to help place elements in their proper position in the periodic table. While Bazarov obliquely criticized Mendeleev's conception of the periodic law, he concluded that the "principle of periodicity requires further elaboration."[65]

One of Mendeleev's students, Aleksei Lavrent'evich Potylitsyn (1845–1905), published an article in 1876 that explicitly used the periodic law, by name, as the framework for studying the force of attraction in halogen compounds. Potylitsyn tried to relate the atomic weights of the halogens to their force of attraction and how easily the halogens were expelled from different compounds. He concluded that the force of attraction varied according to the place of the particular halogen in the periodic table and was proportionate to the atomic weight of the halogen.[66]

The Journal of the Russian Chemical Society also published reviews of chemistry textbooks as well as abstracts of chemistry papers published in foreign countries. In 1878 Petr Petrovich Alekseev (1840–91), professor of chemistry at St. Vladimir's University in Kiev, wrote a highly laudatory review of the third edition of Mendeleev's *Principles of Chemistry*. Alekseev praised the textbook for including several chapters that examined the periodic law in detail, while noting that this was the only chemistry textbook in Russia up to that time that discussed the periodic law at all. Alekseev also noted that because there had been no translation of Mendeleev's textbook into a foreign language up to that time, chemists in other countries had formed a superficial and often incorrect view of the periodic law.[67]

N. A. Menshutkin, Mendeleev's colleague in chemistry at St. Petersburg University, also highly praised Mendeleev's periodic law in an article published in 1885. In this article, Menshutkin compared the theories of substitution and chemical structure for inorganic compounds. He emphasized that Mendeleev's periodic law has "contributed much to the development of chemistry in the short time of its existence."[68]

7. RESPONSE TO THE PERIODIC SYSTEM IN CHEMISTRY TEXTBOOKS IN RUSSIAN HIGHER EDUCATION

Other than Mendeleev's *Principles*, the chemistry textbook of Victor von Richter (1841–91), another German subject of the Russian Empire, played an important role in the early promulgation of the periodic system both in Russia and in Germany.

A Baltic German born in Dobele, not far from Riga, Richter was the son of a Protestant priest. He was educated in a secondary school in St. Petersburg and graduated from Dorpat (now Tartu) University, a German-speaking Imperial University in the Russian Empire.[69] From 1864 to 1872 he worked at the St. Petersburg Technological Institute as a colleague of Mendeleev. Richter, a native speaker of German also fluent in Russian, became one of the first correspondent-chemists of the German Chemical Society, He sent many articles to Germany about the activities of the Russian Chemical Society, including Mendeleev's research, as did Wreden. In 1873 Richter was appointed as professor of general and analytical chemistry at the Institute of Agriculture and Forestry in Novoaleksandriia (now Puławy, part of Poland). Due to his ill health (he suffered from tuberculosis), he moved to Germany for medical treatment and became privatdocent of the University of Breslau in 1875. He then became extraordinary professor in 1879, but was never promoted to full professor due to his premature death and the lack of a vacancy of the post, which was occupied by an elderly professor.[70]

In 1874 he published a textbook of inorganic chemistry, "A Textbook of Inorganic Chemistry Based on the Newest Points of View" in Russian [see Figure 2.1],[71] when he worked as professor of chemistry at the Institute in Novoaleksandriia. It was the first textbook on inorganic chemistry based on the periodic law other than that of Mendeleev.[72] Richter refined the structure of the textbook and added new information in every following edition up to the sixth edition.[73]

After Richter's death, Lyudvig Yul'evich Yavein (1854–1911),[74] chemistry lecturer in St. Petersburg Technological Institute, continued to expand and refine Richter's textbook until the thirteenth edition.[75] The textbook was also translated into other European languages,[76] including German (at least six editions during his lifetime by Richter himself in 1875,[77] 1878, 1881, 1884, 1886, and 1889 and seven more after his death in 1893,[78] 1895, 1897, 1899, 1902, 1910, and 1914), English (at least five editions in 1883, 1885, 1887, 1892, and 1900, with some reprinting of the same edition in Philadelphia, four in 1884, 1886, 1892, and 1896 in London, two in 1893 and 1897 in Tokyo),[79] Dutch (1877), [80] and Italian (1885, 1889, 1895).[81] This extremely popular and successful chemistry textbook played a very important role in promulgating the periodic system not only in Russia but outside Russia as well.

In the 1870s textbooks still followed traditional classification or that of the author's own idea. Ivan Dmitrievich Bokii's chemistry textbook *Osnovaniia khimii* [Foundations of Chemistry]," published in 1873–74,[82] for example, classified the elements based on the forms of compounds with hydrogen, equivalent to a classification based on valence.[83] Its second edition was based on the same principles[84]. Another chemistry textbook (*Vvedeniia k izucheniiu khimii* [The Introduction to study of chemistry]), published in 1876 by an unknown author,[85] was based on the acidity and basicity of oxide compounds.

However, most textbooks on inorganic chemistry for higher education in Russia after the 1880s took the periodic system of elements as the basis of classification, such as those of Aleksei Romanovich Shuliachenko (1841–1903),[86] Grigorii Dmitrievich Volkonskii (1849–?),[87] and Aleksei Lavrent'evich Potylitsyn.[88] All the major inorganic textbooks were based on the periodic systems, often with some extensions, such as those of Nikolai Pavlovich Nechaev (1841–1917) and Nikolai Ivanovich Lavrov (1836–1901),[89] Flavian Mikhailovich Flavitskii (1848–1917),[90] Aleksandr Nikolaevich Reformatskii (1864–1937),[91] Vladimir Nikolaevich Ipat'ev (1867–1952) and Aleksei Vasil'evich Sapozhikov (1868–1935),[92] and Ivan Pavlovich Osipov (1855–1918).[93] Among these textbooks Aleksandr Nikolaevich Shchukarev's introductory textbook of chemistry, *Obshchii kurs khimii* [General course of chemistry], published in 1908,[94] was somewhat different from the other textbooks. Even though he mentioned the periodic law, it seems that he was not so enthusiastic toward the law, writing that "even though the historical contribution of the periodic system was very great, which allowed one to indicate the possibility of some new elements, it is not, however, possible to deny the fact that the basis used for its construction was not stable."[95] His expression was quite indirect and showed, regardless, his somewhat cool attitude toward the periodic law. After a discussion

УЧЕБНИКЪ

НЕОРГАНИЧЕСКОЙ ХИМІИ

ПО НОВѢЙШИМЪ ВОЗЗРѢНІЯМЪ.

В. РИХТЕРА,

ДОКТОРА ХИМІИ.

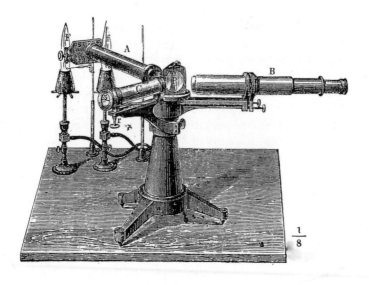

ВАРШАВА

ВЪ ТИПОГРАФІИ ИВАНА ЯВОРСКАГО

Краковское-Предмѣстье № 415

1874.

Figure 2.1 Richter's textbook of inorganic chemistry, *The Textbook of Inorganic Chemistry Based on the Newest Point of View* in Russian (St. Petersburg, 1874).

of hydrogen, oxygen, and water, he wrote about the halogens (Group I), oxygen and sulfur (Group II), nitrogen, arsenic, and antimony (Group III), carbon and organic compounds (Group IV), metals (Group V), and rare gases (Group VI)). This approach was similar to but not exactly the same as the periodic system. It is possible that this approach could be related to the fact that he was a physical chemist.

8. RESPONSE TO THE PERIODIC SYSTEM IN SECONDARY SCHOOL TEXTBOOKS

As for the chemistry textbooks in nineteenth-century Russia, one should pay attention to the fact that chemistry was taught as an independent subject in only a limited area of secondary education. In the gymnasia,[96] the main secondary schools for boys and girls in Imperial Russia, chemistry was taught as a part of physics and natural history. Only in the real schools (real'noe uchilishche), established in 1872 as schools of specialized secondary education, was chemistry taught as an independent subject.[97] However, in 1888, when the graduates of real schools were permitted to enter the physico-mathematical faculty and medical faculty of the universities, chemistry as an independent subject was dropped from the educational plan.[98]

Chemistry textbooks for secondary education in Russia from the 1870s to the early 1920s were surveyed at the Russian National Library (Rossijskaja natsional'naia biblioteka) in St. Petersburg and the Slavonic Library of the National Library of Finland in Helsinki. Eighteen chemistry textbooks for secondary education were found without counting their different editions: (1) Kovalevskii 1874; (2) Al'medingen 1885; (3) Al'medingen 1910; (4) Drentel'n 1886; (5) Fedrov 1892; (6) Kurilov 1896; (7) Kurilov 1915; (8) Boglanovskaia 1897; (9) Jakovlev 1902; (10) Barotoshevich 1902; (11) Dukel'skii 1910; (12) Filosofov 1911; (13) Kukulesko 1912; (14) Tumskii 1912; (15) Nikolaev 1913; (16) Sozonov and Verkhovskii 1915; (17) Jaroshevskii 1900; (18) Ravinskii 1913.[99]

Most chemistry textbooks for real schools were based on the periodic law ((1), (2), (5), (15)). Among them, Kovalevskii's chemistry textbooks were especially interesting. Stephan Ivanovich Kovalevskii (?–1907) was a teacher of chemistry and physics at the First Real School in St. Petersburg. His first textbook of chemistry for real schools in 1874, two years after the start of the real school system, already mentioned Mendeleev's system of elements.[100] However, he employed an original grouping of elements, eleven groups of elements loosely based on the similarity of elements: I (P, As, Sb), II (Cl, Br, I, F), III (S, O), IV (C, Si, Sn, Pb), V (Na, K, NH_3), VI (Mg, Zn), VII (Ca, Ba), VIII (Cu, Ag, Hg), IX (B, Al), X (Fe, Mn), and XI (Au, Pt). He continued to use this grouping of elements in the second edition in 1878. However, the third edition, published in 1880, was thoroughly organized by the periodic law. During his lifetime he published seven more editions in 1882, 1884, 1886, 1889, 1894, 1898, and 1903,[101] with refinement in every edition, based on the periodic law. After his death, two more editions[102] were published in 1907 and 1910 with major revision in the part for organic chemistry, but with only minor changes in the inorganic chemistry part. All the other chemistry textbooks for real schools ((2),

(5), (15)) were more or less based on the periodic law. The authors described most simple metalloid elements and ended the first part with the periodic law. They then explained all the other elements as components of a group of elements, based on the periodic law. This is the same order of description as in Mendeleev's *Principles*. One textbook for commercial and technical schools (10) mentioned the periodic law in the last page, but employed a classification of elements for analytical chemistry, developed by N. A. Menshutkin.

However, the books for a short course on chemistry, which presented only a minimum knowledge of chemistry, described the basic laws of chemistry and typical elements of metalloids and metals. None of them mentioned the periodic law, perhaps because they were designed for workers in night schools (9), for a dental course (18), or for mineralogy (17), although two of them were written for gymnasia ((6), (8)). Even Kovalevskii did not employ the periodic law as a guiding principle when he wrote a short textbook of chemistry for physics students.[103]

In the early twentieth century almost all of the textbooks for secondary education ((7), (11), (12), (13), (14), (16)) not specifically designed for real schools mentioned the periodic law but did not rely on the law. One author pointed out the shortcoming in the order of tellurium and iodine, which seemed reversed from atomic weight ordering.[104] One can therefore imagine that for some authors the periodic law remained controversial as the basic organizing principle for chemistry for secondary school students.

One must note that many popular foreign chemistry textbooks also were translated into Russian during this period, such as those of Henry E. Roscoe (1833–1915),[105] Ira Remsen (1846–1927),[106] and William Ramsay (1852–1916).[107] They all mentioned the periodic law and were also translated into other languages, even Japanese.

9. POPULAR JOURNALS AND SOME RESPONSE BY INTELLECTUALS TO THE PERIODIC SYSTEM

Also in the mid 1870s, there appeared one of the first articles about the periodic law in a popular journal. In 1876 Aleksandr Pavlovich Sabaneev (1843–1923), a chemist in Moscow, wrote an article on the classification of chemical elements in "Priroda," a short-lived (1873–77) popular journal of science and history published in Moscow. He surveyed the history of the concepts of elements and classification from the seventeenth century until the mid-1870s,[108] praising Mendeleev's recent classification and mentioning the discovery of gallium by Lecoq de Boiboudron in 1875, one year earlier.

As shown in the previous sections, by the end of the 1880s the majority of introductory chemistry textbooks in Russia adopted the periodic system, or at least mentioned the periodic law. By the end of the 1890s most of the chemistry textbooks for secondary schools were based on the periodic law. One can also notice some interest in the periodic system beyond the academic and educational spheres.

For example, Boris Nikolaevich Chicherin (1828–1904),[109] a famous jurist and political philosopher, wrote a long article on the system of chemical elements for publication in the journal of the RCS at the end of the 1880s [see Figure 2.2].[110] After his dismissal from the Moscow mayor's post, he devoted his time to original research in various fields, including that of chemistry, while residing at his estates in Tambov and Yalta. According to his memoir, when he decided to study natural sciences, he encountered Wurtz's book on atomic theory,[111] which introduced him to Mendeleev's periodic system.[112] Attempting to find some mathematical law behind the periodic system, he discovered some regularity in atomic volume and mass within the group of alkali metal elements. He then tried to expand these kinds of mathematical relations to other elements, but with difficulties. After a year or so of lonely attempts, he sought expert guidance and sent his manuscript to Mendeleev on the advice of a friend at Moscow University. Mendeleev, who received the manuscript just before he was to travel from St. Petersburg to southern Russia, went straight to Chicherin's estate on his way to the south. After receiving Chicherin's explanations to his questions, Mendeleev sent the paper to the editor of the journal of RCS. Mendeleev also recommended Chicherin for membership in the Society. Chicherin joined the RCS in 1888 and remained a member until 1896.[113] Mendeleev even mentioned Chicherin's work in his famous paper for the Faraday Lecture in 1889.[114] However, Mendeleev eventually lost interest in Chicherin's work, as was often the case with him.

Even though Chicherin's attempts were not successful, since his arguments were based on limited numerical data such as density, he presented many interesting ideas on the structure of the material world. For example, he classified elements into four groups based on their density:[115] formal elements (elements with low densities, such as alkali metals and hydrogen), material elements (with high densities, such as most other metals), central elements, and peripheral elements.[116] With analogy to the solar system, he imagined an atom as a central nucleus with circulating forces and vibrational forces.[117] As an intellectual with a high level of Western culture, including that of natural sciences, the idea of a system of chemical elements stimulated his fantasy on the basis of the material world as a whole.[118]

The attempt to speculate on the possible structure of atoms by Nikolai Aleksandrovich Morozov (1854–1946)[119] was another example of popular interests about the periodic system outside of chemical circles. Morozov was the son of a landlord, but after a home education and a few years of schooling in a gymnasium and two years at Moscow University, he joined a revolutionary group in approximately 1874, at the age of twenty. He was arrested in 1881, after illegally returning to Russia from exile. He was imprisoned in 1882, first in the Peter and Paul fortress at St. Petersburg and then, from 1884 to 1905, in the fortress at Shlisselburg on Lake Ladoga, 35 kilometers east of St. Petersburg. During his twenty-five-year imprisonment, he was allowed to independently study various subjects, including the natural sciences. In 1901 he submitted a long manuscript, entitled "Periodicheskie sistemy stroeniia veshchestva: Teoriia obrazovaniia khimichskikh elementov [The Periodic System of the Structure of Substances: A Theory of the Formation of Chemical Elements]" to the Ministry of Internal Affairs and

СИСТЕМА

ХИМИЧЕСКИХЪ

ЭЛЕМЕНТОВЪ.

Б. Н. Чичерина.

С.-ПЕТЕРБУРГЪ.

Тип. В. Демакова, Новый пер., 7.

1888.

Figure 2.2 The title page of Chicherin's offprint "System of Chemical Elements" (St. Petersburg, 1888).

asked the Ministry to send the manuscript to D. I. Mendeleev or N. N. Beketov. The manuscript was sent to Dmitrii Petrovich Konovalov (1856–1929),[120] Mendeleev's successor at St. Petersburg University. Konovalov dismissed it as pure speculation, with little experimental base.[121] Morozov argued for the complexity of atoms based on an analogy between the homologs of organic compounds and the periodicity of chemical elements in the periodic system. In fact, these arguments were not unusual or original. Many chemists before him developed similar arguments. Mendeleev himself was against these kinds of speculations about the complexity of chemical elements.

These two cases showed that the periodic system stimulated broad speculative imagination on the structure of nature for intellectuals outside of natural sciences in Russia, as in other countries.

Articles on subjects related to the periodic law only began to appear in popular scientific journals[122] at the end of the nineteenth century or early twentieth century in Russia. Most of these articles contained new information or explanations about recent development in the natural sciences.

10. CONCLUSION

The case of the Russian response to the periodic system was different from other cases, because of the presence of Mendeleev. He not only discovered the periodic law, but also within two years showed its possible consequences, firmly grounding his conclusions on his concept of elements. As the main figure of the newly founded Russian Chemical Society, he succeeded in persuading most of the leading chemists in Russia of the validity of the periodic law by 1871. His famous chemistry textbook *Osnovy khimii* (The Principles of Chemistry), which continued to be revised for the rest of his life, including translations into the major European languages, played a significant role in the dissemination of the periodic system.

German subjects of the Russian Empire played an important role in promulgating the periodic system both inside and outside of Russia. This German mediation was another important factor for the acceptance of the periodic system. Victor von Richter, for example, conveyed new developments of the Russian chemical community to Germany as a correspondent for the German Chemical Society. He wrote extremely popular chemistry textbooks, one of the first ones based on the periodic system. His books both in Russian and German were not only published in many editions, but also translated into major European languages.

Since Russia did not have a strong educational tradition like that in France or Spain, a new approach, based on the periodic system—both in education and research—was smoothly implemented. Thus, the periodic system was rather easily accepted in Russia in the academic chemical communities in the 1870s and in secondary education in the 1880s. However, the periodic law was considered to be a somewhat advanced part of chemistry knowledge, so most of the purely elementary textbooks employed for a quick understanding of chemistry did not use or even mention the periodic law.

Figure 2.3 The Bronze Monument dedicated to Mendeleev with the Periodic Table in the wall located in the front yard of the former Chief Bureau of Weight and Measures in St. Petersburg. The statue was made by I. Ya. Gintsburg (1859–1939), based on Mendeleev's portrait in 1890 as a university professor, and erected in 1932. The periodic table on the wall was placed there in 1935[123]. Photographed by Masanori Kaji.

At the end of the nineteenth and the early twentieth centuries, some intellectuals outside of the chemical communities, such as Chicherin and Morozov, published detailed speculations on the basis of the periodic law or on the structure of atoms.

NOTES

1. D. I. Mendeleev, Opyt sistemy elementov, osnovannoi na ikh atomnom vese i khimicheskom skhodstve [1869]. This table was self-published then and reproduced in D. I. Mendeleev, *Periodicheskii Zakon* [Periodic Law], ed. B. M. Kedrov (Moscow: Izd-vo AN SSSR, 1958), 9.

2. The dates for events in Russia are given in the Julian calendar, which lags twelve days behind the Gregorian calendar in the nineteenth century and thirteen days behind in the twentieth century.

3. D. Mendeleev, "Sootnoshenie svoistv s atomnym vesom elementov," *Zh. Russ. Khim. Obshch* [Journal of Russian Chemical Society, thereafter *JRCS*]. **1** (2/3) (1869): 60–77. This paper was reproduced in Mendeleev (note 1) 1958, 10–31.

4. The inaugural meeting of the Russian Chemical Society was held on November 6, 1868. The first paper on Mendeleev's discovery was read in the fifth meeting, only four months after its foundation.

5. R. B. Dobrotin, I. G. Karpilo, L. S. Kerova, and D. N. Trifonov, *Letopis' zhizni i deyatel'nosti D. I. Mendeleeva* [The Chronicle of D. I. Mendeleev's Life and Works] (Leningrad: Nauka, 1984), 110.

6. See the cases of other countries in this book.

7. Nathan M. Brooks, Masanori Kaji, and Elena Zaitseva, "The Formation of the Russian Chemical Society and Its Development until 1914" In *European Chemical Societies: Comparative Analysis of Demarcation*, ed. Anita Kildebaek Nielson and Soňa Štrbáňová (London: Royal College of Chemistry, 2008), 281–304.

8. For Mendeleev's biography, see Michael D. Gordin, *A Well-Ordered Thing: Drimtrii Mendeleev and the Shadow of the Periodic Table* (New York: Basic Books, 2004); B. M. Kedrov, "Mendeleev, Dmitrii Ivanovich," in *Dictionary of Scientific Biography* (Detroit: Charles Scribner's Sons, 1981), vol. 9, 286–295, and its critical supplement by Nathan M. Brooks in *Complete Dictionary of Scientific Biography* (Detroit: Charles Scribner's Sons, 2008), vol. 23, 105–110, with supplementary bibliography; Igor S. Dmitriev, *Chelovek epokhi peremen: Ocherki o D. I. Mendeleeve i ego vremeni* [A Man in a Changing Epoch: A Treatise on D. I. Mendeleev and his Times] (St. Petersburg: Khimizdat, 2004).

9. For the process of discovery, see B. M. Kedrov, *Den' odnogo velikogo otkrytiia* [A Day one Great Discovery] (Moscow: Izd. Sotsial'no-ekonomicheskoi Literatury, 1958, this book was reprinted by Editorial URSS in Moscow in 2001); Igor S. Dmitriev, "Nauchnoe otkrytie *in statu nascendi*: periodicheskii zakon D. I. Mendeleeva [Scientific Discovery in *statu nascendi*: Mendeleev's Periodic Law]," *Voprosy istorii estestvoznaniia i tekhniki* (1) (2001): 31–82; Igor S. Dmitriev, "Scientific Discovery *in statu nascendi*: The Case of Dmitrii Mendeleev's Periodic Law," *Historical Studies in the Physical and Biological Sciences* **34** (2004): 233–275; Masanori Kaji, "Mendeleev's Discovery of the Periodic Law: the Origin and the Reception," *Foundation of Chemistry* 5 (2003): 189–214.

10. B. N. Menshutkin, *Khimiia i puti ee razvitiia* [Chemistry and the Way of Its Development] (Moscow: Izd. AN SSSR, 1937), 229.

11. Vill'yam Odling, *Kurs prakticheskoi khimii* (St. Petersburg: O. I. Bakst, 1867), translated by F. Savchenkov. Its original is William Odling's *A Course of Practical Chemistry, arranged for the use of medical students*, 2nd. ed. (London: Longmans, Green, 1865).

12. Mendeleev (note 1) 1958, 606.

13. N. E. Lyaskovskii, *JRCS* **1** (1869): 232–234. Lyaskovskii was the son of a Polish aristocratic father and a German Protestant mother. After training in pharmacy he studied pharmacology in the medical department of Moscow University. He studied further in Europe in 1843–46, including under Liebig. He became extraordinary professor of chemistry in Moscow University in 1858 and full professor in 1862. His main area of study was in protein chemistry. See V. A. Volkov and M. V. Kulikova, *Rossiiskaia professura XVIII-nachalo XX v.: Khimicheskie nauki, Biograficheskii slovar'* [Professors from 18th to Early 20th Centuries in Russia: Chemistry, Biographical Dictionary](St. Petersburg: Izd. Russkogo Khistianskogo gumanitarnogo instituta, 2004), 146–147.

14. N. N. Beketov, *JRCS* **1** (1869): 235. Beketov studied chemistry under N. N. Zinin in Kazan University. He taught in Kharkov University from 1855 to 1887 (extraordinary professor in 1860 and full professor in 1865). After election to membership in the St. Petersburg Academy of Sciences in 1886, he moved to St. Petersburg. He was one of the pioneers in physical chemistry in Russia. Under his initiative, thermochemistry laboratories were created in Kharkov and in St. Petersburg. See Volkov and Kulikova (note 13), 22–23.

15. Mendeleev (note 1), 1958, 50–58 or *JRCS* **2** (1870): 14–21.

16. Musei-arkhiv D. I. Mendeleeva, Arkhiv [D. I. Mendeleev's Museum-Archive, Archive], 44—1—A—6. See B. M. Kedrov, 1959. *Filosofskii analiz pervykh trudov D. I. Mendeleeva o periodicheskom zakone (1869-1871)* [Philosophical Analysis of Mendeleev's Work on Periodic Law (1869–1871)] (Moscow: Izd-vo AN SSSR, 1959), 243–244. For the text of this letter, one can find its partial English translation in Gordin (note 8), 44.

17. D. I. Mendeleev, *Nauchyi arkhiv. T. 1. Periodicheskii zakon. Estestvennaia sistema elemenlov. Rukopisi i tablitsy. 1869–1871* [Scientific Archives. Vol. 1 Periodic Law. Natural System of Elements: Manuscripts and Tables, 1869–1871] (Moscow: Izd-vo AN SSSR, 1953), 649.

18. Mendeleev (note 17), 187.

19. I. S. Dmitriev, "Problema razmeshcheniia indiia v periodicheskoi sisteme elementov [Problem of Placement of Indium in Periodic System of Elements]," *Voprosy istoriia estestvoznaniia i tekhniki* (2) (1984): 3–14.

20. Stephen Brush, "The Reception of Mendeleev's Periodic Law in America and Britain," *Isis* 87 (1996): 595–628.

21. Muzei-arkhiv D. I. Mendeleeva, arkhiv [D. I. Mendeleev's Museum-Archive, Archive], 13–39–133, ll. 1–2.

22. Mendeleev (note 17), 191.

23. F. Savchenkov, "Otnosheniia mezhdu atomnymi vesami elementov [Relationship between atomic weights of elements]," *Gornyi zhurnal,* ch. II, no. 3, 234–251; reprinted in Mendeleev (note 17), 749–761.

24. Mendeleev (note 17), 762–763.

25. Mendeleev (note 17), 297.

26. Mendeleev (note 17), 672.

27. *Arkhiv Mendeleeva: Avtobiograficheskie materialy* [Mendeleev's Archive: autobiographical materials], vol. 1. (Leningrad: Leningrad State University, 1951), 54.

28. Mendeleev (note 17), 226; D. I. Mendeleev, *Periodicheskii Zakon. Dopolnitel'nye materialy* [Periodic Law. Additional Materials], ed. B. M. Kedrov (Moscow: Izd-vo AN SSSR, 1960), 671. Also see Gordin (note 8), chap. 3.

29. Mendeleev (note 3), 65.

30. See the clarification of dates in note 2.

31. Mendeleev (note 1).

32. D. Mendelejeff, "Die periodische Gesetzmässigkeit der chemischen Elemente," *Annalen der Chemie und Pharmacie Supplementband* **8** (1871): 133–229.

33. Kedrov (note 9), 32, 138–145.

34. D. I. Mendeleev, *Osnovy khimii.* ch. 1, v. 2, gl. 12–22 [The Principles of Chemistry, part 1, volume 2, chapters 12-22] (St. Petersburg: Tip. tov-va "Obshehestv. pol'za," 1869), chap.19. The first edition of Mendeleev's textbook was reprinted as vol. 13 and 14 in D. I. Mendeleev, *Sochineniya* [Collected Works]. 25 vols. (Leningrad: Izd. AN SSSR, 1934–1954). See vol. 13, 650–652.

Mendeleev argued that their compounds could be types for all the other compounds. Obviously, Gerhardt's "type theory" could be seen as influential here since Mendeleev was familiar with it ever since his student days. However, he did not mention Gerhardt and went directly to the concept of valence, for which he used the word atomnost' (atomicity).

35. D. I. Mendeleev, *Osnovy khimii*. ch. 2, v. 3, gl. 1–8 [The Principles of Chemistry, part 2, volume 3, chapters 1-8] (St. Petersburg: Tip. tov-va "Obshehestv. pol'za," 1870), chap.3; Mendeleev, Collected Works (note 34), vol. 14, 120–121.

36. Mendeleev (note 3), 77.

37. Kedrov (note 9), 39–91.

38. D. N. Trifonov criticized Kedrov's version on several minor points (D. N. Trifonov, "Versiya-2: k islorii otkrytiya periodicheskogo zakona D. I. Mendeleevym [Version Two: Toward History of Discovery of Periodic Law by D. I. Mendeleev]" *Voprosy istorii estestvoznaniya i tekhnika* (2) (1990): 25–36; (3) (1990): 20–32). I. S. Dmitriev has offered an alternative version of Mendeleev's discovery (Dmitriev, note 9), including doubt on the use of cards.

39. Mendeleev (note 3), 66.

40. Mendelejeff (note 32), 136n. On Medeleev's concept of elements, see Kaji (note 9).

41. It is often said that Mendeleev kept the text of *Principles* unchanged through all the subsequent editions, but with addition of footnotes that became longer and longer. As shown here, this interpretation is a misunderstanding or at least is inaccurate. This may originate partly from the fact that most Western literature refers to the translations of later editions of *Principles* and partly from the rather vague description of the textbook in Henry M. Leicester's paper on Mendeleev's discovery (M. Leicester, "Factors which Led Mendeleev to the Periodic Law," *Chymia*, **1** (1948): 67–74).

42. For a more detailed study, see Masanori Kaji, "D. I. Mendeleev's Concept of Chemical Elements and *The Principles of Chemistry*," *Bulletin for the History of Chemistry* **27** (1) (2002): 4–16.

43. D. Mendeleev, *Osnovy Khimii* [The Principles of Chemistry], 5th ed. (St. Petersburg: tip. V. Demakova, 1889), chap.15, 448–472.

44. Because the formats of the fifth and subsequent editions were completely different from that of the preceding editions, they were bound as a single volume. However, the English and French translations of these later editions were issued in multiple volumes, which has given rise to the incorrect ideas about the formats of Mendeleev's textbook. The English translation appeared in two volumes (Mendeléeff, D. 1891. *The Principles of Chemistry*. Translated from the Russian (fifth edition) by George Kamensky. Edited by A. J. Greenway in two volumes. London and New York: Longmans, Green & Co., vol. I, xvi+611 pp. & vol. II, vi+487 pp.). Later, the sixth and seventh Russian editions were also translated into English and published as the second and third English editions. These English editions also appeared in two volumes.

The German translation was issued in one volume like the Russian edition (D. Mendeleeff, *Grundlagen der Chemie*. Aus dem russischen übersetzt von L. (St.-Petersburg: Jawein und A. Thillot. Verlag von C. Ricker, [1890]–91), [4], 1127 S.).

Both the fifth and the sixth editions were used for the French translation since the sixth edition appeared while the French edition was being prepared. The French translation should have been published in three volumes, but the third volume was not published for some unknown reason (Dimitri Mendéléeff, *Principes de chimie*. Traduit du russe par E. Achkinasi [et. H. Carrion, avec préface de M. le professeur Armand Gautier., vol.1–2. (Paris: Éditeur B. Tignol, 1895–96) vol. I, [4], iv+585 pp. & vol. II, [4], 499 pp.).

45. On Brauner, see the chapter on the Czech Land of this book.

46. D. Mendeleev, *Popytka khimicheskogo ponimaniia mirovogo efira* (St. Petersburg, 1903). An English translation appeared as *An Attempt Toward a Chemical Conception*

of the Ether, translated by George Kamensky (New York: Longmans, Green & Co. Mendeléeff, 1904). See also Bernadette Bensaude-Vincent, "L'éther, élément chimique: un essai malheureux de Mendéléev (1902)?" *British Journal for the History of Science* **15** (1982): 183–188.

47. Mendeleev (note 35), Part 2, Chapter 6; reprinted in Collected Works, vol. 24, 247.
48. Bensaude-Vincent has pointed out that the logical consequence of Lavoisier's definition was the hypothesis of a primordial matter (Bensaude-Vincent (note 45), 12).
49. See the chapter on Germany of this book for a detailed analysis of the priority debates.
50. Beilstein, famous for the editorship of the most comprehensive handbooks of organic compounds, was born in St. Petersburg as the first son of a German family. Even though he was appointed extraordinary professor of chemistry at Göttingen University at the age of twenty-seven in 1865, the sudden death of his father made him return to St. Petersburg for the family and become a subject of the Russian Empire, later to be appointed as professor at the St. Petersburg Technological Institute. For the details of his biography, including his delicate relationship with Mendeleev, see Michael D. Gordin, "Beilstein Unbound: The Pedagogical Unraveling of a Man and His Handbuch" in *Pedagogy and the Practice of Science*, ed. David Kaiser (Cambridge, Massachusetts, and London, England: The MIT Press, 2005), 11–39.
51. D. I. Mendelejeff, "Über die Beziehungen der Eigenschaften zu den Atomgewichten der Elemente." *Zeitschrift für Chemie* **12**(1869), 405–406.
52. V. A. Krotikov, "Dve oshibki v pervykh publikatsiyakh o periodicheskom zakone D. I. Mendeleeva [Two Mistakes in the Early Publications on Mendeleev's Periodic Law]," *Voprosy istorii estestvoznaniya i tekhniki*, no. 29 (1969): 129–131, p. 129.
53. Lothar Meyer, "Die Natur der chemischen Elemente als Function ihrer Atomgewichte," *Annalen der Chemie und Pharmarcie Supplemente band* **7** (1870): 354–364.
54. Meyer quoted the abstract in *Zeitschrift für Chemie* (Mendelejeff note 50).
55. V. A. Volkov and M.V. Kulikova, *Rossiiskaia professura XVIII-nachalo XX v.: Khimicheskie nauki, Biograficheskii slovar'* [Professors from 18th to Early 20th Centuries in Russia: Chemistry, Biographical Dictionary](St. Petersburg: Izd. Russkogo Khistianskogo gumanitarnogo instituta, 2004), 51; V. Rykhlyakov "Vreden Feliks Romanovich." in the site "Nemtsy Rossii [Germans of Russia]," http://www.rusdeutsch-panorama.ru/jencik_statja.php?mode=view&site_id=34&own_menu_id=4159.   Accessed September 7, 2014.
56. See Mendeleev (note 17), 344–438. Here, one can find both the Russian original and Wreden's German translation shown side by side, the latter of which was published in *Annalen der Chemie und Pharmacie*.
57. Lothar Meyer, "Zur Systematik der anorganischen Chemie," *Berichte der Deutschen Chemischen Gesellschaft* **6** (1873): 101–106.
58. I. S. Dmitriev, "Skuchnaia istoriia (o neizbrannii D. I. Mendeleeva v Imperatorskuyu akademiyu nauk v 1880 g) [A Boring Story (on Failure of Mendeleev's Election in Imperial Academy of Sciences in 1880)]," *Voprosy istorii estestvoznaniya i tekhniki* (2) (2002): 231–280; Gordin (note 8), 113–141.
59. Lothar Meyer, "Zur Geschichte des periodischen Atomistik." *Berichte der Deutschen Chemischen Gesellschaft* **13** (1880): 259–265; "Zur Geschichte des periodischen Atomistik." *Berichte der Deutschen Chemischen Gesellschaft* **13** (1880): 2043–2044.
60. D. Mendelejeff, "Zur Geschichte des periodischen Gesetzes," *Berichte der Deutschen Chemischen Gesellschaft* **13** (1880): 1796–1804.
61. Meyer (note 59), 263.
62. See section 3 of this chapter.

63. They must be Emil Erlenmeyer (1825–1909) and Jacob Volhard (1834–1910) (Mendeleev (note 17), 710). After the retirement of Kopp from editorship in March 1871, Erlenmeyer, Volhard, and Liebig were the editors of *Annalen* (Alan J. Rocke, *The Quiet Revolution* (Berkeley, Los Angeles, London: U. of California Press, 1993), 288). Erlenmeyer was very much a Russophile.

64. JRCS **3** (1871): 249.

65. JRCS **5** (1873): 144–174, p. 171. Bazarov was educated as a chemist under A. M. Butlerov and studied organic chemistry, but after 1887 he started to work in the field of fruit growing and became famous for his work on wines (http://dic.academic. ru/dic.nsf/enc_biography/7298/Базаров, accessed October 20, 2013).

66. JRCS **8** (1876): 199–210. Potylitsyn became professor at Nowa Aleksandria Institute of Agriculture and Forestry (1881–83), then extraordinary professor at Warsaw University (1883), and later full professor (1886–95). After 1895 he became director of Nowa Aleksandria Institute (Volkov and Kulikova (note 13), 181–182).

67. JRCS **10** (1878): 261–262. After graduation from St. Petersburg University, Alekseev studied in Europe (1860–64), including under C. A. Wurtz in Paris. He became extraordinary professor (1868) and full professor (1869–91) at St. Vladimir's University in Kiev (Volkov and Kulikova (note 13), 14–15).

68. JRCS **17** (1885): 303-340, p. 333. Menshutkin graduated from St. Petersburg University in 1862. After further study in Europe in 1863–65, he taught at St. Petersburg University from 1866 (privatdocent in 1867, docent in 1867, extraordinary professor in 1869, full professor in 1876). His main work was in organic chemistry, but his textbook on analytical chemistry was very famous (*Analiticheskaia khimiia* (St. Petersburg, 1871); its sixteenth ed. was published in 1931). He was one of the founding members of the RCS and served as the first editor of its journal (JRCS) for many years (1869–1900). See Volkov and Kulikova (note 13), 153–154.

69. G. Prausnitz, "Nekrolog: Victor von Richter," *Berichte der Deutschen Chemischen Gesellschaft* **24** (3) (1891): 1123–1130; Volkov and Kulikova (note 13), 190–191.

70. Prausnitz (note 69), 1124.

71. V. Yu. Richter, *Uchebnik neorganicheskoj khimii po novejshim vozzreniyam* (Warsaw: tipografiya Ivana Yavorskogo, 1874).

72. He used the natural system of elements for the title of section on the system. He also utilized the periodic system and periodic law (zakon periodichnosti) following Mendeleev's wording. He refereed to Mendeleev's first paper on the periodic system in the Journal of RCS (Paper I) and the German paper (Mendelejeff (note 32)) as well as Lothar Meyer's paper (Meyer (note 53)) for further reading (Richter (note 71), p. 218n).

73. V. Yu. Richter, *Uchebnik neorganicheskoj khimii po novejshim vozzreniyam* (Leipzig: tipografiya V. Lrugukina, 1876); 3rd ed. (Warsaw: tipografiya K. Kovalevskogo, 1878); 4th ed. (St. Petersburg: tipografiya Tovarishchestva "Obshchestvennaya Pol'za," 1880); 5th ed. (St. Petersburg: tipografiya Tovarishchestva "Obshchestvennaya Pol'za," 1883); 6th ed. (St. Petersburg: tipografiya Demakova, 1887). He first mentioned the discovery of Mendeleev's predicted new elements (gallium and scandium) in the fourth edition (p. 248) and all three predicted elements (with germanium) in the sixth edition (p. 281).

74. Yavein translated Mendeleev's *Principles of Chemistry* into German with A. Tillo, assistant of chemical laboratory of St. Petersburg Technological Institute (Mendeleeff (note 44)). Here is another "German mediation."

75. V. Yu. Richter, *Uchebnik neorganicheskoj khimii po novejshim vozzreniyam*, 7th ed. (St. Petersburg: Knizhnyi magazine A. F. Tsinzurkinga, 1893); 8th ed. (St. Petersburg:

Knizhnyi magazine A. F. Tsinzurkinga, 1895); 9th ed. (St. Petersburg: Knizhnyi magazine A. F. Tsinzurkinga, 1897); 10th ed. (St. Petersburg: Knizhnyi magazine A. F. Tsinzurkinga, 1899); 11th ed. (St. Petersburg: Knizhnyi magazine A. F. Tsinzurkinga, 1902); 12th ed. (St. Petersburg: Izdanie A. F. Tsinzurkinga, 1906); 13th ed. (St. Petersburg: Izdanie A. F. Tsinzurkinga, 1910).

76. Prausnitz (note 69), 1128.
77. V. von Richter, *Kurzes Lehrbuch der anorganischen Chemie* (Bonn: Verlag von Max Cohen & Sohn, 1875).
78. V. von Richter and Heinrich Konrad Klinger, *Kurzes Lehrbuch der anorganischen Chemie* (Bonn: Verlag von Max Cohen & Sohn, 1893).
79. Victor von Richter, *Text-book of Inorganic Chemistry*, transl. Edgar Fachs Smith (Philadelphia: Blakiston Son & Co., 1883); 2nd American ed. (Philadelphia: Blakiston Son & Co., 1885); 3rd American ed. (Philadelphia: Blakiston Son & Co., 1887, 1888, 1889, 1890); 4th American ed. (Philadelphia: Blakiston Son & Co., 1892, 1893, 1894, 1896, 1898, 1899); 5th American ed. (Philadelphia: Blakiston Son & Co., 1900, 1901, 1902, 1903, 1904, 1905, 1909 (with W. T. Taggart)).

Victor von Richter, *Text-book of Inorganic Chemistry*, transl. Edgar Fachs Smith (London: Henry Kimpton, 1884); 2nd ed. (London: Henry Kimpton, 1886); 3rd ed. (London: Henry Kimpton, 1892); 4th ed. (London: Kegan Paul, Trench, Trubner & Co., 1896).

Victor von Richter, *Text-book of Inorganic Chemistry*, transl. Edgar Fachs Smith (3rd American ed.) (Tokyo: S. Saito, 1893); 2nd ed. (4th American ed.) (Tokyo: Shueisha, 1897).
80. V. von Richter, *Beknopt leerboek der anorganische scheikunde*, transl. L. Aronstein (Breda: Broese, 1877).
81. Victor von Richter, *Trattato di chimica inorganica*, transl. Augusto Piccini (Torino etc.: Loescher, Ermanno, 1885); 2nd ed. (Torino: Loescher, 1889); 3rd ed. (Torino: Loescher, 1895).
82. I. D. Bokii, *Osnovaniia khimii*. Two volumes (Tiflis: tip. A. Enfiadziantsa i K°, 1873–74).
83. Ivan Dmitrievich Bokii (1845–97) was a long-time gymnasium instructor, first in Tiflis, Georgia, and then in St. Petersburg. His son Gleb Bokii (1879–1937) became a revolutionary and a leading member of the Soviet secret police.
84. I. D. Bokii, *Osnovaniia khimii*, 2nd ed. (St. Petersburg: A. G. Muchnik, 1881).
85. *Vvedeniia k izucheniiu khimii*. Two volumes (St. Petersburg: tip. i khromolit. A. Transhelya, 1876).
86. Shuliachenko was a military engineer officer and professor at Nikolai military engineering academy in St. Petersburg and specialized in chemical technology, especially that of cement (Volkov and Kulikova (note 13), 255–256). A. R. Shuliachenko, 1881. *Neorganicheskaia khimiia* [Inorganic Chemistry] (St. Petersburg: tip. Mor. m-va, 1881).
87. Prince Grigorii Dmitrievich Volkonskii was a laboratory assistant at Moscow University. G. D. Volkonskii, *Kratkii uchebnik neorganicheskoi khimii* [A Short Textbook of Inorganic Chemistry](Moscow: Izd. tip. S. P. Arkhipova i Ko., 1881).
88. On Potylitsyn, see note 66. His textbook numbered eight editions. A. L. Potylitsyn, *Nachal'nyi kurs khimii* [Elementary Course of Chemistry](St. Petersburg: tip. V. F. Demakova, 1881); 2nd ed. (St. Petersburg: tip. V. F. Demakova, 1883); 3rd ed. (St. Petersburg: tip. V. F. Demakova, 1887); 4th ed. (Warsaw: tip. varsh. Ucheb. okr., 1892); 5th ed. (Warsaw: tip. varsh. Ucheb. okr., 1895); 6th ed. (Warsaw: tip. varsh. Ucheb. okr., 1897); 7th ed. (Warsaw: tip. varsh. Ucheb. okr., 1900); 8th ed. (St. Petersburg: tipo-lit. M. P. Frolovoi, 1904).

89. Nechaev, N. P. and N. I. Lavrov, *Kurs khimii: v ob'eme programmy voennykh uchilisch* [Course of Chemistry: for the program of military schools] (Moscow: izdanie knizhn. magaz. V. Dumonova, 1893). Nechaev was a teacher of military schools and general of artillery. Lavrov was a teacher of chemistry at the Imperial Academy of Art, the Second Military Konstantin School, and the St. Petersburg Commercial School.

90. F. M. Flavitskii, *Obshchaia ili neoganicheskaia khimiia* [General Chemistry or Inorganic Chemistry] (Kazan': tipo-lit. Universiteta, 1893–94); 2nd ed. (Kazan': tipo-lit. Universiteta, 1898); 3rd ed. (Kazan': tipo-lit. Universiteta, 1907). Flavitskii was professor of chemistry in Kazan University (1884) and a corresponding member of the St. Petersburg Academy of Sciences (1907) (Volkov and Kulikova (note 13), 230–231).

91. A. N. Reformatskii, *Lektsii neorganicheskoi khimii* [Lectures of Inorganic Chemistry] (Moscow: t-vo tip. A. I. Mamontova, 1902). Aleksandr Nikolaevich Reformatskii (1864–1937) was a privatdocent of Moscow University. After resignation from the post as a protest against the government's reactionary policies in 1911, he taught in various private schools in Moscow, including at Shaniavskii Moscow People's University, one of the first private universities in Russia (Volkov and Kulikova (note 13), 189–190).

92. This textbook was first published in handwritten lithographical version in 1900. After the first printed edition was published in 1902, five more editions were published in 1903, 1907, 1909, 1912, and 1918. V. N. Ipat'ev and A. V. Sapozhikov, *Kurs neorganicheskoi khimii* [Course of Inorganic Chemistry] (St. Petersburg: tip. V. Demakova, 1902); 2nd ed. (St. Petersburg: tip. V. Demakova, 1903); 3rd ed. (St. Petersburg: tip. V. Demakova, 1907); 4th ed. (St. Petersburg: tip-ki. M. P. Frolovoi, 1909); 5th ed. (St. Petersburg: tip-ki. M. P. Frolovoi, 1912); 6th ed. (Moscow: Mosk. nauch. izd-vo pri Mosk. nauch. in-te, 1918). Ipat'ev was a military officer and professor of Mikhail Artillery Academy and from 1916 an Academician of St. Petersburg Academy of Sciences. He stayed in the Soviet Union, but in 1930 fled to the United States (Volkov and Kulikova (note 13), 97–99). Sapozhikov was also professor of chemistry at Mikhail Artillery Academy (Ibid., 198).

93. I. P. Osipov, 1911. *Lektsii neorganicheskoi khimii* [Lectures of Inorganic Chemistry] (Kharkov: tip. "Pechatnik," 1911); 2nd ed. (Khar'kov: Izdanii T-va A. S. Suvorina, 1913).

94. A. N. Shchukarev, *Obshchii kurs khimii* (Moscow: Tipografiya O. L. Somovoi, 1908); 2nd ed. (Kharkov: Tipo-lit. M. Zil'berberg i S-v'ya, 1915). Osipov was one of the students of N. N. Beketov and professor of chemistry at Kharkov University (1894–1906) and at Kharkov Institute of Technology (1906–18) (Volkov and Kulikova (note 13), 165–166). Shchukarev (1864–1936) graduated from Moscow University in 1889 under the supervision of V. V. Markovnikov, a famous organic chemist. In 1893 he started to work as an assistant in the thermochemistry laboratory at Moscow University, created by V. F. Luginin (1834–1911), a famous thermochemist. He was privatdocent in Moscow University from 1906 to 1910, after the defense of his master's dissertation on "the inner energy of gas-forming liquid" and became professor of chemistry at Kharkov Institute of Technology after receiving a doctor's degree following the dissertation "The property of solution under critical temperature of mixture" (Volkov and Kulikova (note 13), 257–258).

95. Shchukarev (note 94, 1908), 125.

96. The first gymnasium was founded in 1726, attached to the Academy of Science. However, the system of gymnasia throughout the Empire was started only in 1803, under the regime of Alexander I. The gymnasia for girls were founded in 1862.

See V. Rudakov, "Gimnaziya," *Entsikropedicheskii slovar' Brokgauza i Efrona*, vol. VIIIa (1893), pp. 694–707.

97. Real gymnasia were established in 1864 and reorganized into real schools in 1872.

98. See "Real'noe uchilishche" on the site http://www.wikiznanie.ru/ru-wz/index. php/Реальное_училище, accessed October 20, 2013.

99. (1) S. I. Kovalevskii, *Uchebnik khimii* [A Textbook of Chemistry](St. Petersburg: Tipografiya M. Stasyulevich, 1874); 2nd ed. (St. Petersburg: Tipografiya M. Stasyulevich, 1878); 3rd ed. (St. Petersburg: Tipografiya M. Stasyulevich, 1880); 4th ed. (St. Petersburg: Tipografiya M. Stasyulevich, 1882); 5th ed. (St. Petersburg: Tipografiya M. Stasyulevich, 1884); 6th ed. (St. Petersburg: Tipografiya M. Stasyulevich, 1886); 7th ed. (St. Petersburg: Tipografiya A. S. Suvorina, 1889); 8th ed. (St. Petersburg: Tipografiya A. S. Suvorina, 1894); 9th ed. (St. Petersburg: Tipografiya A. S. Suvorina, 1898); 10th ed. (St. Petersburg: Tipografiya A. S. Suvorina, 1903) (last edition of Kovalevskii's life time); 11th ed. (St. Petersburg: Tipografiya A. S. Suvorina, 1907); 12th ed. (St. Petersburg: Tipografiya A. S. Suvorina, 1903); 11th ed. (St. Petersburg: Tipografiya A. S. Suvorina, 1910).

(2) A. N. Al'medingen, *Uchibnik khimii* [A Textbook of Chemistry] (St. Petersburg: Tipografiya A. M. Kotomina, 1885).

(3) A. N. Al'medingen, *Nachal'nyi uchebnik khimii* [Elementary Textbook of Chemistry] (St. Petersburg: Izdanie Knigoizdatel'stava, Knizhnogo Skada I Tipo-Litografii A. E. Vineke, 1910).

(4) N. S. Drentel'n, *Nachal'nyi kurs khimii* [Elementary Course of Chemistry] (St. Petersburg: Tipografiya Yu. N. Erlikh, 1886).

(5) A. I. Fedrov, *Uchebnik khimii dlia tekhnicheskikh uchilishch* [A Textbook of Chemistry for Technical Schools] (Kiev: Tip. S. V. Kul'zhenko, 1892).

(6) V. V. Kurilov, *Kratkii uchebnik khimii dlya gimnazii i real'nykh uchilishch* [A Short Textbook of Chemistry for Gymnasia and Real Schools] (St. Petersburg: Tipo-litorgrafiya S. M. Nikolaeva, 1896); 2nd ed. (Ekaterinoslav [now Dnipropetrovsk in Ukraina]: Tipografiya L. M. Rotenberga, 1901); 3rd ed. (Ekaterinoslav: Tipografiya L. M. Rotenberga, 1904).

(7) V. V. Kurilov, *Uchebnik khimii dlya srednikh uchebnykh zavedenii* [A Textbook of Chemistry for Secondary Schools] (Moscow & Petrograd: Izdanie T-va V. V. Dumnov—nasl. Br. Salaevykh, 1915).

(8) V. E. Boglanovskaia, *Nachal'nyi uchebnik khimii* [Elementary Textbook of Chemistry] (St. Petersburg: Tipografiya V. Kirshbauma, 1897); 2nd ed. (St. Petersburg: Tipografiya zhurn. "Stroitel", 1901). Vera Evstaf'evna Bogdanovskaya-Popova (1867–96) got her PhD in chemistry at the University of Geneva in 1892 and taught stereochemistry in Higher Women's Courses at St. Petersburg. She was killed by an accident in a home chemistry laboratory (http://ru.wikipedia.org/wiki/Попова,_Вера_Евстафьевна, accessed October 20, 2013).

(9) N. N. Jakovlev, *Uchebnik khimii (Nach. kurs)* [A Textbook of Chemistry (Elementary course)] (St. Petersburg: Izdanie Postoyannoi Kommissii po Tekhnicheskomu Obrazovaniyu pri Imperatorskom Russkom Tekhincheskom Obshchestve, 1902); 2nd ed. (St. Petersburg: Izdanie Postoyannoi Kommissii po Tekhnicheskomu Obrazovaniyu pri Imperatorskom Russkom Tekhincheskom Obshchestve, 1907).

(10) S. T. Barotoshevich, *Kratkii uchebnik khimii, neorganicheskoi, analiticheskoi i organicheskoi. Dlia vospitannikov sred. ucheb. zavedenii* [A Short Textbook of Chemistry, inorganic, analytical and organic chemistry] (Warsaw: Tipogra. P. Kanevskii i Tipogra. P. Kanevskii i V. Vatslavovich, 1902); 2nd ed. (St. Petersburg: Izdanie G. V. Gol'stena, 1907).

(11) M. P. Dukel'skii, *Uchebnik khimii dlia srednikh uchebnykh zavedenii* [A Textbook of Chemistry for Middle Schools] (Kiev: Izdanie Pirogovskogo T-va, 1910).

(12) P. S. Filosofov, *Uchebnik khimii dlya srednikh shkol* [A Textbook of Chemistry for Middle Schools] (St. Petersburg: Tipografiya V. Bezobrazov i Ko., 1911).

(13) I. M. Kukulesko, *Elementarnyi kurs khimii dlya srednykh uchebnykh zavedenii* [An Elementary Course for Secondary Schools] (St. Petersburg & Kiev: Knigoizdatel'stvo "Sotrudnik," 1912).

(14) Konstantin Ivanovich Tumskii, *Nachala khimii kratkoe obschchedostupnoe izlozhenie osnovnykh ponyatii khimii dlya obshcheobrazovatel'nykh i spetsial'nykh uchilishch* [Basis of Chemistry: Short Popular Explanation of Concepts of Chemistry for General and Special Schools] (Moscow: Tipografiya Imperatoskogo Moskovskogo Universiteta, 1912).

(15) V. I. Nikolaev, *Uchebnik khimii (neorganicheskoi i organicheskoi) dlia srednei shkoly i samoobrazovaniia* [A Textbook of Chemistry (inorganic and organic) for Secondary Schools and Self-Education] (Samara: Gubernskaya Tipografiya, 1913).

(16) S. I. Sozonov and V. Verkhovskii, *Uchebnik khimii. Kurs sred. shk.* [A Textbook of Chemistry. A Course of Secondary Schools] (Petrograd: Izdanie T-va I. D. Sytina, 1915); 2nd ed. (Petrograd: t-vo I. D. Sytina, 1918); 3rd ed. (Mosovw: Gos. izd., 1921, 1923); 4th ed. (Moscow & Petrograd: Gos. izd., 1924); 5th ed. (Leningrad: Gos. izd., 1925); 6th ed. (Leningrad: Gos. izd., 1925); 7th ed. (Leningrad: Gos. izd., 1926); 8th ed. (Leningrad: Gos. izd., 1926); 9th ed. (Moscow & Leningrad: Gos. izd., 1926); 10th ed. (Moscow & Leningrad: Gos. izd., 1927 [1926]); [11th ed.] (Moscow & Leningrad: Gos. izd. tip. im. N. Bukharina v Leningrade, 1928).

(17) K. F. Jaroshevskii, *Uchebnik mineralogii i nachal'nye svedeniia iz khimii* [A Textbook of Mineralogy and Elementary Knowledge of Chemistry] (Moscow: Tipografiya G. Lissnera i A. Geshelya, 1900).

(18) I. Ya. Ravinskii, *Uchebnik khimii (sostavil po programme zubovrachebnoi shkol)* [A Textbook of Chemistry, compiled following program of dental school] (Odessa: Tsentral'naya tipografiya S. Rozunshtraukha i N. Lemberga, 1913).

100. Kovalevskii (note 99(1), 1874), x.

101. See Kovalevskii (note 99(1)).

102. Ibid.

103. S. I. Kovalevskii, *Elementarnye svedeniia iz khimii (Ucheb. Posobie k kersu fiziki)* [Elementary Knowledge of Chemistry: A textbook for a course of physics] (St. Petersburg: Tipografiya M. Stasyulevicha, 1873); 6th ed. (St. Petersburg: tipografiya A. S. Suvorina, 1896); 8th ed. (St. Petersburg: tipografiya A. S. Suvorina, 1903); 9th ed. (St. Petersburg: tipografiya A. S. Suvorina, 1909).

104. Dukel'skii (note 99 (11)), 178.

105. Rosko [Roscoe], *Kratkii uchebnik mineral'noi i organicheskoi khimii* [*Kurzes Lehrbuch der Chemie nach den neuesten Ansichten d. Wissenschaft* (Braunschweig, 1867)], transl. by G. G. Gustavson, M. Ya. Kagustin, and N. M. Popov (St. Petersburg: Izdanie Tovarishchestva "Obshchevennaya pol'za," 1868). This was the translation from the German translation by Carl Schorlemmer (1834–92) and accompanied by a preface from Mendeleev, who recommended this textbook for middle schools. 2nd ed. (St. Petersburg: Izdanie Tovarishchestva "Obshchevennaya pol'za," 1873); 3rd ed. (St. Petersburg: Izdanie Tovarishchestva "Obshchevennaya pol'za," 1876); 4th ed. (St. Petersburg: Izdanie Tovarishchestva "Obshchevennaya pol'za," 1880); 5th ed. (St. Petersburg: Izdanie Tovarishchestva "Obshchevennaya pol'za," 1885); 6th ed. (St. Petersburg: Izdanie Tovarishchestva "Obshchevennaya pol'za," 1893).

106. A. Remsen [Ira Remsen], *Vvedenie k izucheniiu khimii* [*An Introduction to the Study of Chemistry* (London: Macmillan, 1886)] transl. M. I. Konovalov (Moscow: Tipografiya Tovarishchestva I. D. Sytina, 1901); 2nd ed. (Moscow: Tipografiya Tovarishchestva I. D. Sytina, 1906).

107. Vil'yav Ramzai [William Ramsay], *Sovremennaia khimiia* [*Modern Chemistry*, 1900?], transl. by S. V. Lebedev and E. P. Ostroumova (St. Petersburg: K. L. Rikker, 1906–07). Vil'yav Ramzai [William Ramsay], *Noveishaya khimiya* [*Modern Chemistry*, 1907–09], transl. by Lev Aleksandrovich Chugaev, S. Severinov and E. Menshikov (Moscow: t-vo I. D. Sytina, 1910).

108. A. N. Sabaneev, "O sootnoshenii i klassifikatsii elementov [On Correlation and Classification of Elements]," *Priroda*, (3) (1876): 186–197.

109. For his life, see Shuichi Sugiura, 1999. *Roshia jiyu-shugi no seiji-shiso* [in Japanese, Political Philosophy of Russian Liberalism] (Tokyo: Mirai-sha, 1999) and Gary M. Hamburg, *Boris Chicherin and Early Russian Liberalism, 1826–1866* (Stanford, Calif.: Stanford University Press, 1992).

110. B. N. Chicherin, "Sistema khimicheskikh elementov I," JRCS **20** (1888); "Sistema khimicheskikh elementov II," JRCS 21 (1889); "Sistema khimicheskikh elementov III," JRCS 24 (1892). These articles were published as a supplement for the journal and later published in an independent booklet: B. N. Chicherin, *I. Sistema khimicheskikh elementov; II. Zakony obrazovaniia khimicheskikh elementov* (Moscow: "Pechatiya S. P. Yakovleva", 1911).

111. The Russian translation of the second edition (Charles Adolphe Wurtz, *La théorie atomique* (Paris: Baillière, 1879)) was published in 1882 in Kiev, However, it is possible that Chicherin read the French original.

112. B. N. Chicherin, *Vospominaniia* [in 2 vols] (Moscow: Izd. Im. Sobashnikovykh, 2010).

113. One can see his name in the list of the membership published dated January 15th every year from 1889 until 1896 in the journal of the Society.

114. D. Mendeléeff, "The Periodic Law of the Chemical Elements (Faraday Lecture delivered before the fellows of the Chemical Society in the theatre of the Royal Institution on Tuesday, June 4, 1889)" in *The Principles of Chemistry*, the third English Edition, translated by George Kamensky, vol. II (London, New York and Bambay: Longmans, Green, and Co., 1905, reprinted in 1969 by Kraus Reprint Co, New York), 489–508, pp. 495–496.

115. Actually they are the densities of simple bodies. Thus, Chicherin did not distinguish simple bodies from elements, unlike Mendeleev.

116. Chicherin (note 110, 1911), 218–219.

117. Ibid., 226–227.

118. Ibid., 359.

119. On his biography and his major work, see *Nikolai Aleksanrovich Morozov: uchenyi-entsiklopedist* [Morozov: scholar-encyclopedist] (Moscow: Izdatel'stvo "Nauka," 1982); *Nikolai Alkecandrovich Morozov (1854–1946): Materialy k biobibliografii uchenykh SSSR* [Morozov (1854–1946): Materials for a biobliography of scholars in Soviet Union] (Moscow: Izdatel'stvo "Nauka," 1981); http://ru.wikipedia.org/wiki/Морозов,_Николай_Александрович_(революционер), accessed October 24, 2013.

120. After graduation from the Mining Institute in St. Petersburg, Konovalov attended lectures in St. Petersburg and studied under Butlerov and Mendeleev, while later further studying at Strasbourg University. After his return from Europe he became an assistant of analytical chemistry for Menshutkin. He taught at St. Petersburg University until 1907 (privatdocent in 1884, extraordinary professor in 1886, full

professor in 1893). His main field was physical chemistry, especially chemical thermodynamics and kinetics. See Volkov and Kulikova (note 13), 109–110.

121. On Konovalov's assessment, see *Nikolai Aleksanrovich Morozov* (note 119), 62–63. This manuscript was published after Morozov's release from prison. *Periodicheskie sistemy stroeniia veshchestva: Teoriia obrazovaniia khimichskikh elementov* (Moscow: Sytin, 1907).

122. The following journals were surveyed: *Fizicheskoe obozrenie* [Physical Review] (Warsaw and Kiev, 1900–18), *Priroda* [Nature] (Moscow, 1912–18), *Gornyi zhurnal* [Mining journal] (St. Petersburg, 1886–1916), *Nauchnoe obozrenie* [Scientific Review] (St. Petersburg, 1894–1903, 1911–12), *Priroda i lyudi* [Nature and People] (1891–1918), *Vestnik znaniia (Ezhemesyachyi illyustrirovanyi literaturnyi i populyarno-nauchyi zhurnal s prilozheniyami dlia samoobrazovanniia)* [Bulletin of Knowledge: Monthly Illustrated Literary and Popular Scientific Journal with supplements for Self-Teaching] (1903–16), *Niva: ill. Zhurn. Literatury, politiki i sovrem. zhini [Niva: Illustrated Journal of Literature, Politics and Modern Life]* (St. Petersburg: T-vo A. F. Marks, 1870–99), *Russkoe bogatstvo* [Russian Wealth] (1893–1911).

123. V. G. Isachenko, *Pamyatniki Sankt-Peterburga. Spravochnik* (St. Petersburg: Paritet, 2004), 172.

CHAPTER 3

cᴠᴑ

The Periodic System and Its Influence on Research and Education in Germany between 1870 and 1910

GISELA BOECK

1. INTRODUCTION

In 1895 Karl Seubert (1851–1942) published some of the most important papers by Lothar Meyer (1830–1895) and Dmitrii I. Mendeleev (1834–1907) on the so-called natural system of elements. He wrote:

> At first it seems incomprehensible to today's reader of these essays that the general reception of the system was delayed for many years even though it was presented in a final form and its benefit for theoretical, practical and pedagogical purposes had been explained in detail.[1]

Seubert discovered a lack of interest in the field of inorganic chemistry, but also an inadequate description of the system. He remarked that Meyer's explanations were too short, and Mendeleev's too circuitous.

> The system became a resounding success when the deductions which were drawn from it were confirmed by experiments in rapid succession: the selection of the atomic weight with respect to the known number of equivalents, as in the case of indium and uranium; the change in the order, regardless of the valid atomic weights, such as the platinum group; and, last but not least, the prediction of new elements and their chemical properties which were proved true with the

discoveries of scandium, gallium and germanium quickly one after the other. The brilliant vision and the boldness of Mendelejeff led the system to its unquestioned victory.[2]

Seubert[3] was Meyer's colleague for many years. From 1878 to 1895, they worked together on the redetermination of atomic weights and published several papers on this topic. Seubert was the first biographer to write about Meyer and was responsible for publishing his most important papers. Nevertheless, Seubert regarded Mendeleev's role in the discovery of the periodic system to be of greater importance. This is shown by the last sentence of the previously quoted passage.

Seubert's remark elicits two questions: First, why did Seubert consider Meyer's role in the discovery of the periodic system as less important? Second, was its reception in Germany truly delayed? These questions are connected to several different factors: politics within German chemistry; didactic approaches to teaching chemistry in schools and universities; and the role of the periodic system in the public sphere. At first glance, it seems that these factors are independent of one another. However, the development of a science is strongly connected to educational questions, the material itself, and the ideational support of a science by the society, which has a direct link to the image of the science. This is linked to questions of popularization.[4] The first part of this chapter discusses the setting for the reception of the periodic system in Germany between 1870 and 1910. In the first half of the nineteenth century, German chemists were already part of the search for a classification of the elements. Nevertheless, neither Meyer's nor Mendeleev's system was celebrated in the 1870s as significant progress even though both systems, and the priority debate that followed, were published in German journals. It was only Mendeleev's successful predictions, and later the placement of the noble gases, that accelerated the reception process. On the other hand, Meyer's contributions to the periodic system became more and more forgotten. This can be attributed to Meyer's disposition, his reticence concerning predictions, and the fact that Meyer was a representative of the field of physical chemistry, whereas the German chemical scene was dominated by organic chemistry.

Furthermore, German chemists provided significant contributions to the redetermination of atomic weights and the definition of an international standard. The question of the reference for atomic weight (H or O) divided German chemists. Meyer was a member of the H-minority. This fact negatively influenced the acceptance of his ideas.

The second part of this chapter discusses the importance of the periodic system as a didactic tool. Both Meyer and Mendeleev established a periodic system as a two-dimensional illustration while writing their textbooks and looking for a new didactic introduction to the elements. An immediate effect on the content of new textbooks was expected, but preventing this was, first, the fact that older chemical classifications were well established and, second, that new didactic approaches— especially in school instruction—in the second half of the nineteenth century favored an introduction to chemistry based on reactions and/or compounds that were already well known to students.

The last part of this chapter shows that in the public sphere, the periodic system was connected to questions concerning the understanding of nature, its internal connections, and the descent theory. After 1910 the situation regarding the periodic system was changing rapidly as a result of the new understanding of atomic theory and the law of Henry Moseley (1887–1915). This time period, however, is not discussed in this chapter.

2. THE PERIODIC SYSTEM IN THE GERMAN ACADEMIC FIELD

2.1 A Short Prehistory

At the end of the eighteenth century, the great amount of knowledge in chemistry required a system that first explained the internal connection between the chemical objects, and second improved the teachability of the existing chemical knowledge. Jeremias Benjamin Richter (1762–1807) discovered the first quantitative connection between elements. His law of progression was the basis for the determination of several atomic weights using the masses of oxides and hydroxides. Johann Wolfgang Döbereiner (1780–1849) found a connection between the atomic weights of groups of three elements (named triads) with comparable properties; he failed, however, to establish a connection between these triads.

Leopold Gmelin (1788–1853) expanded on such connections, moving from "triads" to maximal "hexads." He arranged the elements in the shape of a parabola: on the left side the more electronegative elements, on the right side the more electropositive elements. Mathematical relationships between elements were discovered by Max Pettenkofer (1818–1901), who showed that the difference between the atomic weights of similar elements could be divided by eight. Peter Kremers (1827–?) wrote about "conjugated triads," including the comparison of dissimilar elements. A key advancement came in the display of a bidirectional system. In addition, Ernst Lenssen (1837–?) arranged fifty-eight elements in twenty triads.[5]

These examples clearly show that the question of the taxonomy of elements was one of the major topics on the agenda in German chemistry. Aside from the quantitative trials in classification, there were other classifications that were only concerned with the physical and chemical properties of substances. The atomic weight, however, was not a classification criterion. Artificial and natural classifications[6] could be found in several chemistry textbooks in the first half of the nineteenth century.[7] For example, Rudolf Arendt (1828–1902) wrote the following:

> Elements can be divided into either non-metals (or metalloids) and metals, or light and heavy metals. Non-metals can be further divided into the oxygen group, the group of the halogens, the azote group, or the carbon group. Light metals can be divided into the potassium, the magnesium, the calcium, or the aluminium group. Heavy metals can be split into the ferric, the tin, and the lead groups.[8]

2.2 Lothar Meyer and His System of Elements

Julius Lothar Meyer[9] studied medicine at the University of Zurich and natural sciences at the University of Würzburg, where he received his doctoral degree in medicine in 1854. He focused his studies on chemistry and mathematical physics at Heidelberg University and was tutored by Robert Bunsen (1811–1899) and Gustav Kirchhoff (1824–1887). He enrolled as a medical student because enrollment was a precondition for attending lectures and gaining access to the laboratory. He wrote his second medical dissertation on the gases of the blood. In Heidelberg, Meyer became acquainted with Friedrich Beilstein (1838–1906), Henry Roscoe (1833–1915), Hans Landolt (1831–1910), and August Kekulé (1829–1896). As a result of his interest in theoretical physics, Meyer, together with Beilstein, followed Landolt to Königsberg in 1856. It was there that Meyer attended lectures by Franz Ernst Neumann (1798–1895). He was also a physiology student of Gustav Werther (1815–1869). In his laboratory, he investigated the influence of carbonic oxide gas on blood. Based on these results, Meyer was awarded a Doctor of Philosophy degree from the University of Breslau. Meyer's first papers placed him in the field of physiological chemistry. With his Habilitation on the chemical theories of Claude Louis Berthollet (1748–1822) and Jöns Jacob Berzelius (1779–1848), he entered the field of general chemistry. In 1859 he started his academic career as a docent at Breslau.

Meyer attended the Karlsruhe Conference in 1860. Deeply impressed[10] by the paper of Stanislao Cannizzaro (1826–1910), Meyer started to write his textbook *Die modernen Theorien der Chemie und ihre Bedeutung für die chemische Statik* [Modern Chemical Theory and Its Meaning for the Chemical Statics, hereafter *Moderne Theorien*].[11] His goal was to summarize modern chemical theory[12] and to eliminate contradictions in the theories on atomic weight, the atom, the equivalent, and the molecule. Section 91 discusses the "peculiar regularities"[13] that were found among the atomic weights. Meyer used the notion of there being an arithmetic relationship between atomic weights. He suspected that the regular relationships were responsible for the idea that atoms are an aggregate of smaller units. This explanation was adopted from the homologous series in organic chemistry. These series are characterized by the repeated addition of constant fragments, something Meyer also found when comparing atomic weights. He thought these fragments should be based on smaller units of the atom, like those in the homologous series, which are smaller units of a molecule.[14]

In the book *Moderne Theorien*, Meyer arranged fifty elements in two tables. The first table included twenty-eight elements, which were grouped with respect to their increasing atomic weights and valence. The second (not very well organized) table contained the remaining twenty-two elements. In the first table, differences between the atomic weights of elements arranged on top of each other could be found. This difference was the same if you were to compare a vertical pair with the neighboring vertical pair, also creating horizontal relationships. Today, the elements in the first table are known as the main group elements and those in the second table as the transition group elements. The order of increasing atomic weight was interrupted where tellurium and iodine were transposed. These elements were arranged only with respect to their chemical properties, and not their atomic weights. Meyer assumed that incorrect atomic weights were the source of this irregularity and called for them to be more exactly determined. In 1866 Meyer started to correct the weights, but because he left for Karlsruhe, this work was interrupted for about fifteen years. It was later continued together with Seubert.[15]

The first table published in 1864 also contained gaps. An example of such a gap concerns the precautionary prediction of the atomic weight. The element following silicon should have had an atomic weight 44.55 higher than silicon (28.5). The book *Moderne Theorien*, in which Mendeleev's ideas were anticipated,[16] was published in 1864, even though Meyer had already largely completed the manuscript in 1862[17]— only two years after the Karlsruhe Conference!

In 1866 Meyer moved to Neustadt-Eberswalde, where he became professor at the Forestry Academy. It was there that he worked on a new edition of his book *Moderne Theorien* and, in 1868, constructed an extended arrangement with fifty-two elements in fifteen vertical rows, leaving the sixteenth row empty (Figure 3.1). This system was not published at that time. Meyer left the draft to his successor, geologist Adolf Remelé (1839–1915). It is not clear why Remelé only returned the manuscript to Meyer in 1893, and not in the 1880s during the priority dispute. It is also unknown why Meyer did not ask Remelé for the draft. The system from 1868 was finally published in 1895.[18] Seubert wrote of a "system" in the headline of his 1895 publication, but there is no evidence that Meyer used this term.

To summarize, Lothar Meyer developed two two-dimensional illustrations of the arithmetical connections between the atomic weights and the properties of the elements for use in his textbooks. He arranged the elements in order of increasing atomic weight, and horizontal relationships could be found. But until 1870, he spoke about a regular change in the chemical character, and did not use the terms "periodic dependence" or "law." He found consistent differences in the atomic weights, which he discussed as proof of the complex characteristics of atoms.

2.3 Mendeleev's Publications Reach Germany

It is well known that Mendeleev developed his first system of elements in February of 1869. Nikolai A. Menshutkin (1842–1907) presented the paper on this to the Russian Chemical Society, who then published it in the first volume of their journal. The paper and journal were in Russian, which created a linguistic barrier. It is often said that Mendeleev himself sent copies of his system to other chemists in Russia and several other countries, but there is no useful information about the recipients.

As a result of the installation of correspondents for the German Chemical Society in St. Petersburg, London, Paris, and other European cities, who spoke German and the corresponding foreign language, the means for the fast transmittance of information on new developments in chemistry was established. The reports were available to the members of the German Chemical Society, founded in 1867,[19] through its journal, the *Berichte der Deutschen Chemischen Gesellschaft* [Reports from the German Chemical Society]. At the end of 1869, correspondent Viktor von

Anhang.

6

Entwurf eines Systems der Elemente von Lothar Meyer. 1868.

§ 91   Nicht gedruckt. Wiedergabe nach dem Manuscript.

1	2	3	4	5	6	7	8
		Al — 27,3⁴) $\frac{28,7}{2}=14,3$	Al = 27,3 *)				C = 12,00 / 16,5 / Si = 28,5 $\frac{89,1}{2}=44,55$
Cr = 52,6	Mn = 55,1 / 49,2	Fe = 56,0 / 48,3	Co = 58,7 / 47,3	Ni = 58,7	Cu = 63,5 / 44,4	Zn = 65,0 / 46,9	$\frac{89,1}{2}=44,55$
	Ru = 104,3 / 92,8 = 2·46,4	Rh = 104,3 / 92,8 = 2·46,4	Pd = 106,0 / 93 = 2·46,5		Ag = 107,94 / 88,8 = 2·44,1	Cd = 111,9 / 88,3 = 2·44,15	Sn = 117,6 / 89,4 = 2·44,7
	Pt = 197,1	Ir = 197,1	Os = 199,0		Au = 196,7	Hg = 200,2	Pb = 207,0

(right margin: Lothar Meyer)

*) Im Original durchstrichen und durch daruntergesetzte Punkte wieder gültig gemacht. K. S.

9	10	11	12	13	14	15	16
			Li = 7,03 / 16,02	Be = 9,3 / 14,7			
N = 14,04 / 16,96	O = 16,00 / 16,07	Fl = 19,0 / 16,46	Na = 23,05 / 16,08	Mg = 24,0 / 16,0			
P = 31,0 / 44,0	S = 32,07 / 46,7	Cl = 35,46 / 44,51	K = 39,13 / 46,3	Ca = 40,0 / 47,6	Ti = 48 / 42	Mo = 92 / 45	
As = 75,0 / 45,6	Se = 78,8 / 49,5	Br = 79,97 / 46,8	Rb = 85,4 / 47,6	Sr = 87,6 / 49,5	Zr = 90 / 47,6	Vd = 137 / 47	
Sb = 120,6 / 87,4 = 2·43,7 / Bi = 208,0	Te = 128,3	J = 126,8	Cs = 133,0 / 71 = 2·35,5 / ? Tl = 204 ?	Ba = 137,1	Ta = 137,6	W = 184	

(right margin: Natur der Atome: Gründe gegen ihre Einfachheit.)

S. L. *Gmelin*, Hdb. 5. Aufl. I, 47 ff.; Münch. gel. Anz. 1850 Bd. 30, S. 261, 272, abgedr. Ann. Chem. Pharm. 1858. 105, 187; *J. Dumas*, C. r. 1857, t. 45, p. 709; auch Ann. Chem. Pharm. 105, S. 74 u. a.

Seite 7 hat man sich in der Weise seitlich an Seite 6 angereiht zu denken, dass N = 14,04 in Spalte 9 neben C = 12,00 in Spalte 8 zu stehen kommt, P neben Si, Sb neben Sn, Bi neben Pb. K. S.

7

Figure 3.1 Draft of Meyer's system from 1868, first published by Karl Seubert *in Das natürliche System der chemischen Elemente*, Ostwald's Klassiker der exakten Wissenschaften, no. 68, (Leipzig: W. Engelmann, 1895) (no permission needed).

Richter (1841–1891) reported on the interesting relationships in the system of elements that Mendeleev had developed.[20] He noted that Mendeleev believed the atomic weights were the "really consistent" property of the elements. He also gave a short excerpt of the system and mentioned the review in the *Zeitschrift für Chemie* [Journal of Chemistry], which describes the contents of Mendeleev's paper on the relationships between the properties of the elements and their atomic weights.[21] In the review, Mendeleev's main idea was incorrectly translated: instead of the *periodicity* of the properties, the translator wrote about a *stepwise* change in the properties. This mistake was already discovered during the lifetimes of Mendeleev and

Meyer. It is known that Mendeleev had first asked Beilstein to do the translation, but it was then translated by A. A. Ferman (or Fehrmann).[22] Beilstein later sent the translation to Meyer with the request to publish it in the *Zeitschrift für Chemie*. It is still unclear who was responsible for the incorrect translation[23] and if Ferman or Beilstein understood the meaning of Mendeleev's discovery. In this case they were unaware of the consequences of a loose translation.

2.4 Further Publications on the Periodic Law by Meyer and Mendeleev in Germany

In 1870 Meyer published the paper entitled "Die Natur der chemischen Elemente als Funktion ihrer Atomgewichte [The Nature of the Chemical Elements as a Function of Their Atomic Weights]."[24] A plot of the atomic volume versus the atomic weight showed a periodic dependence, which Meyer then explicitly spoke of for the first time. Meyer saw in this periodicity the key to understanding the nature of atoms. He stated that his table, for the most part, was similar to Mendeleev's. This led to the assumption that Meyer would not have published his own ideas. Mendeleev answered with two publications in 1871: "Die periodische Gesetzmäßigkeit der Elemente [The Periodic Regularity of the Elements]"[25] and "Zur Frage über das System der Elemente [On the Question Concerning the System of Elements]."[26] The second publication focused on the questions of priority and already noted the incorrect translation.

After 1871 Mendeleev and Meyer focused on other scientific problems. It was only in 1873 that Meyer published a paper on the system of inorganic chemistry[27] in connection with discussions on atomicity resulting from a paper by Julius Thomsen (1826–1909) on the basicity of "*Überjodsäure*" (periodic acid).[28] He used the system to make his argument. After 1878 Meyer published several papers on the determination of atomic weights.

2.5 The Priority Dispute in the *Berichte der Deutschen Chemischen Gesellschaft*

In 1879 Adolphe Wurtz (1817–1884)[29] sent a letter to the German Chemical Society[30] complaining about the incorrect changes in the translation of his book, *La théorie atomique*. He believed that the translated text favored Meyer's work.[31] Wurtz stressed that only Mendeleev had the idea to arrange the elements according to their atomic weight in two rows, and that Meyer just completed this idea.

In response to the letter, Meyer published what he believed to be the history of the periodic system, comparing his tables to those of Mendeleev. First, he noted that his earliest two tables of elements in 1864 were not just simple arrangements of similar elements.[32] At that time, he was not able to arrange one single table using incorrect atomic weights. However, with the knowledge of the exact atomic weights, he constructed such a table.[33] But—as we already know—it was not published then.

Second, Meyer acknowledged that Mendeleev was the first to mention the *periodic* change in the properties in connection with the atomic weights, the need to correct atomic weights,[34] and the possibility of predicting elements.[35] Third, Meyer complained that Mendeleev failed to construct *one* row with increasing atomic weights. Meyer was the only one to do this and published it in 1870. In the original paper this fact is not stated explicitly.[36] Meyer also spoke of double periodicity in 1880, but described it only indirectly in his 1870 paper.[37] Meyer regretted that he could only submit a short paper in 1870, making it impossible for him to compare the different tables in detail. On the other hand, the priority dispute required a new view on the ten-year-old results. Meyer accepted Mendeleev's priority in the prediction of elements, even though, as mentioned earlier, a cautious prediction can be found in Meyer's *Moderne Theorien*. This is not, however, comparable to Mendeleev's predictions.

2.6 After the Priority Dispute—M eyer and the German Chemical Community

The priority dispute took place in the *Berichte der Deutschen Chemischen Gesellschaft* and could be easily read by the chemical community, which at the time was dominated by organic chemists. How did the chemical community respond to the priority dispute? At first, no other papers could be found that explicitly discussed the priority dispute. It cannot be excluded that the correspondences reflected the dispute.[41] But it is obvious that the authors who dealt directly or indirectly with the questions of atomism, elements, and their classification or the determination of atomic weights often mentioned the successful predictions of Mendeleev and showed some reticence concerning Meyer's work. It seems that the success of (some)[42] predictions of new elements influenced the process of forgetting about Meyer's precautionary contribution and overemphasizing the role of the predictions in the process of acceptance.[43]

Several examples can be given to underline and explain these facts. The following paragraphs focus on Meyer's standing in the physicochemical community.

Meyer was a representative of physical chemistry, a specialization that, because of the efforts of Wilhelm Ostwald (1853–1932), started its institutionalization as a discipline, but did not play a dominant role in the chemical community then. Among physical chemists there was a type of "trench warfare" going on over the existence of atoms and the theory of solutions. There were several conflicts between Meyer and Ostwald. Meyer had reservations about the theory of dissociation and the establishment of the *Zeitschrift für physikalische Chemie* [Journal of Physical Chemistry], and later about the theory on osmotic pressure. Ostwald wrote to Svante Arrhenius (1859–1927) saying that, "Lothar wrote a very foolish article on osmotic pressure."[44] Meyer warned Ostwald on several occasions against being too critical in his journal. The repeated differences of opinion intensified with time.[45] Meyer and Seubert's book from 1883, *Die Atomgewichte der Elemente aus den Originalzahlen neu berechnet* [The Atomic Weights of the Elements Newly

Calculated from the Original Numbers],[46] was based on the ratio of atomic weights H:O=1:15,96. This was also a moot point.[47]

How did Meyer value his own contribution to the periodic system after 1882, when he and Mendeleev were awarded with the Davy Medal? Two years before his death, Meyer published a paper on the use of the periodic system in the study of inorganic chemistry.[48] He perceived his contribution to the periodic system as a modification of the Döbereiner system and not as a new qualitative step; he spoke neither of a new theory nor a new law. He also touched on questions of didactics in his paper, emphasizing that this system would be well suited to give students an overview. In addition, he stated that regardless, it received only little attention in textbooks in being mentioned or explained; only a small number of textbooks used it as a fundamental part of the arrangement; and it was normally used only in a distorted or randomly changed form. Meyer gave reasons for this: first, the leading chemists had turned to organic chemistry; second, and much more importantly, someone unacquainted with the system required instruction—it was not self-explanatory.[49] He elaborated, noting that he had modified his course several times,[50] and described his system as a didactic tool rather than a law. Meyer mentioned the use of a big plate[51] to illustrate the system in the lecture hall, as well as a model using a rotatable cylinder.[52]

In 1895 Meyer edited *Die Anfänge des natürlichen Systems der chemischen Elemente* [The Beginnings of the Natural System of Chemical Elements] in *Ostwald's Klassiker No. 66,* which included a reprint of Döbereiner's work.[53] Meyer's last lecture, "Über naturwissenschaftliche Weltanschauung [About the Scientific Worldview]," does not mention the periodic system, which is surprising given that thirty years prior to that he tried to determine the nature of atoms.[54]

In contrast to this, Mendeleev clarified his priority in the fifth edition of *Osnovy khimii,*[55] which was translated into several languages. It is unclear whether Meyer's lecture in 1893 and the publication of the most important papers on the periodic system in 1895 and 1896 can be characterized as an answer to Mendeleev's efforts concerning priority.

In the following paragraphs, an evaluation of Meyer's work by other German chemists is presented. Hans Landolt (1831–1910) showed his appreciation of Meyer's work in 1888 by recommending his election to the Berlin Academy. Landolt's letter contained a comparison of the works of Mendeleev and Meyer, in which he pointed out the big differences between them. Mendeleev was interested in the gaps and predictions of new elements. These predicted elements were later discovered, with not only chemists understanding this. Meyer's results were less "popular," but Landolt thought that they were more important from a scientific point of view. Meyer discovered irregularities and, what he thought to be, incorrect atomic weights. Because of this, he started a redetermination project together with his assistant Seubert, which led to the desired corrections. They later improved their calculation method for atomic weights and published them in 1883.[56] Landolt saw an important role in the predictions that led to the acceptance of the periodic system and the high recognition of Mendeleev. The same assessment was given by Seubert, who described the "view of a genius and the boldness" of Mendeleev (and not of his teacher Meyer), which led to the victory of the periodic system.[57]

It is well known that Mendeleev's predictions were also verified by German chemist Clemens Winkler.[58] In 1886 he detected the new element germanium[59] in the mineral argyrodite, which was found by Albin Weisbach (1833–1901) in 1885 at the Himmelsfuerst silver mine near Freiberg. His report dates back to February 6, 1886.[60] Winkler, who was not very well acquainted with the periodic system, assumed that germanium filled the gap between Sb and Bi (eka-stibium). Already on February 14 (26),[61] 1886, Mendeleev wrote to Winkler saying that, with respect to the chemical properties, the new element should be eka-cadmium with an atomic weight of 160–165. On March 7 (19), the new element was presented at a session of the Russian Physico-Chemical Society in St. Petersburg as eka-silicium. However, the first to express this idea was von Richter. He wrote to Winkler on February 13 (25), 1886, and later to Mendeleev, saying that germanium must be the predicted eka-silicium. Meyer wrote the same to Winkler on February 27, 1886.[62] In several papers and lectures, Winkler praised the success of the periodic system based on Mendeleev's predictions. Mendeleev, in turn, praised Winkler as a "verifier" of his system (Figure 3.2).[63] In 1889 Victor Meyer (1848–1897) gave a lecture entitled *Chemische Probleme der Gegenwart* [The Chemical Problems of Today] at a session of the German Scientists and Physicians Conference. Victor Meyer held the work of Mendeleev in much higher esteem than that of Meyer. The successful predictions were also determinative for him as organic chemist.[64] But he was interested in a

Figure 3.2 Mendeleev and Winkler in 1894 (with permission of the archive of the Technical University Mining Academy Freiberg).

more complex explanation. He demanded to know the (physical) reasons for the periodicity of the properties: *"We are missing a clear understanding of the reasons for the internal connection between the elements which is expressed by the system."*[65] German physical chemists, however, did not work in this direction.

Like Nikolai N. Zinin (1812–1880) in St. Petersburg,[66] German organic chemist Herman Kolbe (1818–1884) in Leipzig also rejected the periodic system. He said that it is possible to discuss and philosophize about the periodicity, and that one can "easily construct a trite hypothesis that dazzles, especially if has an ingenious air to it. The splendor of it, however, vanishes when the prosaic experiment exposes the truth."[67]

2.7 The Periodic System in the Berichte—Additional Examples

A review of the *Berichte* shows that from 1870 to 1910 only a few papers were closely connected to the questions concerning element classification. Most of the authors are unknown to today's readers. In 1871 Max Zaengerle (1832–1903)[68] reported on the regularities of the atomic weights,[69] and in 1873, Heinrich Baumhauer (1848–1926)[70] published a natural system of chemical elements with references to Meyer and Mendeleev.[71] In 1878 physicist Friedrich Wächter (1855–1940)[72] tried to improve Mendeleev's system.[73] Hermann Friedrich Wiebe (1852–1912)[74] explained the elongation of inelastic elements as a function of the atomic weights based on Meyer's periodic system.[75] Thomas Carnelley (1857–1890) explained the magnetic properties on the basis of Mendeleev's system.[76] Mendeleev praised this work as the one and only extension of his theory.[77] In 1884 Carnelley tried to find a connection between the atomic weight and the occurrence of elements in nature.[78] In 1885 the dispute over the unit of measurement for atomic weight was reflected and discussed in connection with the hypothesis of William Prout (1786–1850). Prout hypothesized that all elements can be reduced to one smallest unit, hydrogen, the primary matter. In 1894 Isidor Traube (1860–1943)[79] presented the fundamentals of a new system of elements.[80] William Preyer (1841–1897), who at the time was a professor of physiology at the University of Jena and an opponent of Darwinism,[81] tried to determine the correct positioning of argon and helium in the periodic system.[82]

The paper by Clemens Winkler on the discovery of new elements[83] also dealt with the periodic system. Another example is the work of William Ramsay (1852–1916) on the newly discovered gases and their relationship to the periodic system.[84]

There were several trials in using the periodic system to explain element properties and to improve the display of the system, as well as to discuss a connection to a possible origin of the elements from primary matter. Until the end of the nineteenth century, papers mentioning the periodic system made reference to Meyer roughly as often as they did to Mendeleev.

Experimental papers examined the redetermination of atomic weights with regard to their position in the periodic system. But it is doubtful that the periodic system was really the only reason for the redetermination.[85] New analytical methods allowed a higher degree of accuracy. Some examples of this are given below.

In 1872 chemist and mineralogist Karl Rammelsberg (1813–1899) published his results on the atomic weight of uranium, with a clear reference to Mendeleev.[86] In the years that followed, there were discussions on the atomic weights of several elements. In 1875 the discovery of gallium[87] and the connection between the prediction and its real properties were reported.[88] In 1878 Meyer published his results on beryllium.[89] Bohuslav Brauner (1855–1935), Lars Friedrich Nielsen (1840–1899), and Sven Otto Pettersson (1848–1941) also discussed their experimental findings.[90] In 1879 the first information on scandium appeared[91] and in 1880, the connection to Mendeleev's prediction could be found.[92] Results on the determination of the atomic weights were presented by Brauner, Seubert, Gerhard Kruess (1859–1895), and others. At the end of the 1880s there were disagreements on the reference for atomic weight; Brauner and Ostwald pushed for O=16, whereas Seubert and Meyer defended H=1. This debate was revived in 1896/97. Consequently, the German Chemical Society established an atomic weight commission in 1897 with members Ostwald and Seubert, and Landolt as the chairman.[93] The commission determined O=16 to be the reference atomic weight for Germany.[94] The commission invited chemical societies and other similar institutions from other countries to take part in this debate.[95] Later, an international commission with more than fifty members from the United States, Germany, Belgium, Holland, Japan, Italy, Austria-Hungary, Russia, Sweden, and Switzerland was established to create a worldwide standard and to decide how many decimal places to use, what the rule is for the last decimal place, and if it is necessary to constitute a special working division for actualizing the atomic weights on an annual basis.[96]

Generally, it can be said that the *Berichte* provided information regarding successful predictions, problems with the reference of the atomic weight, placement of beryllium and the noble gases, and attempts to position the rare earths. These papers were written by known and lesser-known scientists.

3. THE PERIODIC SYSTEM, GERMAN TEXTBOOKS, AND PEDAGOGICAL JOURNALS

Stephen G. Brush wrote a comprehensive paper on the reception of the periodic law in the United States and Britain.[97] He also studied some German books, including forty-six books from the years 1871 to 1890. These books were used in schools and universities.[98] According to Brush about 45 percent of these texts mention the periodic system. The number is surprisingly high, but the relatively small number of reviewed books should not be overlooked. Some books provided diagrams of the system, others discussed the question of the best way to illustrate the periodic law or deal with the question of primary matter. Brush established that mention of the periodic system increased sharply after the discovery of predicted elements.[99]

Why is a direct influence of the periodic system on the structure and the con-tent of textbooks expected? Meyer and Mendeleev were both writing textbooks when they arranged their periodic systems. Their intentions were to improve the methods for teaching. This provides a simple explanation for the notable influence of their ideas on textbooks after 1870. A survey[100] of German[101] school textbooks was conducted for the time period of 1870[102] to 1910, in which several editions of a book were examined. This allowed for the observation of changes in the didac-tic approaches and in the selection and order of the material. Furthermore, the *Zeitschrift für mathematischen und naturwissenschaftlichen Unterricht* [Journal for Instruction in Mathematics and the Natural Sciences] and the *Zeitschrift für den Physikalischen und Chemischen Unterricht* [Journal for Instruction in Physics and Chemistry] were reviewed in their entireties.

The general conclusion was that neither Meyer's nor Mendeleev's periodic system significantly influenced the textbooks. Some of the reasons for this are explored in the following section. It is shown that an introduction to chemistry with the help of everyday phenomena is possible without a classification of the elements. If a systematic approach was preferred, long-proven classifications based on elemental properties (both physical and chemical) could be used. Some irregularities in the atomic weights and an unusual grouping of dissimilar elements in the periodic sys-tem were not understood, providing reasons for its rejection. And the didactic value of the two-dimensional display in the form of a table was believed to be ill-suited for students.

3.1 Methodical or Systematic Approach?

It should be noted that in the nineteenth century, textbooks were not clearly specified for either school or university use.[103] Their titles were also not a strong indicator. The textbooks offered a great deal of knowledge on substances, describing them and their reactions. Undoubtedly, the order of the substances corresponded to the general didactic requirement needed to proceed from the supposedly simple (the elements) to the more complex (the compounds).[104] The elements were presented using the conventional system of being grouped with respect to their properties, and later with respect to their valence.[105] These types of textbooks were referred to as systematic in the discussions started by German teachers about the methods of teaching chemistry. Arendt and Ferdinand Wilbrand (1840–1914) had a determining influence.[106] They demanded a new approach based on reactions or everyday phenomena, and called the constructed textbooks methodical.

Even in 1890, every second textbook in Prussian high schools used a system-atic approach.[107] It could, therefore, be expected that the periodic system would be used in these textbooks. Consequently, Arendt, a methodologist, implemented the periodic system only in 1894 in the fifth edition of his book, *Grundzüge der Chemie* [The Basic Principles of Chemistry].[108] The author devoted four of the 367 pages at the end of the chapter on inorganic chemistry to the periodic system according to Meyer.

3.2 Is There an Advantage to Using the Periodic System?

The first German textbook to mention the periodic system was written by Victor von Richter.[109] The Russian version of this book had already been published in 1874.[110] Richter, a Baltic-German, worked in a laboratory at the St. Petersburg Technological Institute from 1864 to 1872, and reported on Mendeleev's system in the *Berichte der Deutschen Chemischen Gesellschaft*. From 1872 to 1874, he was a professor at an agricultural institute in Poland. In 1875 he completed his Habilitation at Breslau with a paper on the periodic system and the newly discovered gallium.[111] In the second edition[112] of his textbook, he dedicated a particular chapter to the periodic system (eight pages out of 475). He emphasized that the classification scheme presented was not yet in its final form. Nevertheless, the introduction of the periodic system—an important tool for combining numerous information, which before had no established relationships to one another—in chemical textbooks was no longer pushed aside. Richter believed it was possible to reason from the periodic dependence of the properties of atomic weight that the different elemental atoms are aggregates or a condensation of one and the same primary matter, but not hydrogen.

He also emphasized that the classification scheme presented was not yet in its final form. The atomic weight of some elements was not definitively measured; this is why some changes had to be made and the relationships had to be displayed in another form.[113] This textbook provided an introduction to the periodic system in Russia and Germany. It is unclear if the same was true in Poland.

In 1878 and 1879 inorganic chemist August Michaelis (1847–1916) also mentioned the periodic system in his textbooks,[114] applying the explanation from Meyer's third edition of *Die modernen Theorien der Chemie*.[115] It is important to note that Michaelis was in direct contact with Meyer at Polytechnikum in Karlsruhe.[116]

From 1881 to 1890, textbooks written by Wilhelm Ostwald, Carl Arnold (1853–1929), and the now unknown Otto Hausknecht mention the periodic system.[117] Max Zängerle developed his own so-called natural system, based on "ylem atoms," and published it in his school program.[118] Of the books examined from the years of 1870 to 1890, about 14 percent mention the system. The differences in Brush's data can be explained entirely by the number of books used. An exact number of all textbooks at that time was not available, and thus all statistics are unreliable.

References to the periodic system in textbooks increased considerably from 1891 to 1900; about every second reviewed book mentioned the periodic system. Reviews in pedagogical journals also mentioned the periodic system in one of two ways: either as a guideline for textbooks, or as a result of the new knowledge of elements and substances. Nevertheless, there were some who believed that the use of the periodic system in the classroom did not make sense. As a general guideline, it was used for systematization. But with respect to didactic purposes, the grouping into metals and non-metals was retained, and hydrogen and oxygen were put in front of all other elements.[119] The unusual grouping of elements provided a reason for rejection.[120] Discussions about the methodically ill-suited display of the system can also be found.[121]Approximately 50 percent of the reviewed books from 1901 to 1910 mention and use the periodic system. But the skeptical comments did not stop.[122] Summarizing, it can be stated that an introduction to chemistry using a methodical approach is possible without any classification of the elements. For continued studies, a classification with respect to the properties (physical and chemical ones) works well. Irregularities in the atomic weights, unusual grouping of dissimilar elements, and problems with the display of the system were reasons for its limited use in textbooks.

4. THE PERIODIC SYSTEM AND THE PUBLIC SPHERE

In the second half of the nineteenth century, there were several journals in Germany whose conception was due to the popularization of natural science knowledge. They can be characterized into different categories.[123] The following journals were reviewed with respect to the periodic system:

(a) *Natur*[124] and *Naturwissenschaftliche Wochenschrift*,[125] well-established journals with popular papers from every natural science, which played an important role in the educational discussion;
(b) *Jahrbuch der Erfindungen*,[126] *Naturwissenschaftliche Rundschau*,[127] *and Jahrbuch der Naturwissenschaften*,[128] popular journals for the more educated, which contained scientific papers;
(c) *Himmel und Erde*,[129] a journal that was primarily dedicated to papers on astronomy and earth science.

Few papers in these journals addressed chemical problems. A small number were devoted to the history of the periodic system, the discovery of new elements, and the connections between the elements. Most journals reported on the sessions of the German Natural Scientists and Physicians. Victor Meyer's previously mentioned *Chemische Probleme der Gegenwart* [The Chemical Problems of Today] or Clemens Winkler's *Die Frage nach dem Wesen der chemischen Elemente* [The Question of the Nature of Elements] were reviewed. The unidentified reviewer calls attention to Winkler's assumption "that those chemical substances, which are regarded as substantial indivisible, stem from more simple substances, and that the new formation of elements continues with the gradual cooling-down of the earth." He also wrote that the chemical elements "did not exist from the beginning, that they are products of the transmutation of a primary substance."[130]

The composite nature of the elements was also discussed[131] in connection with Victor Meyer's lecture on the *Probleme der Atomistik* [Problems of Atomism].[132] The small number of theoretical papers regarding the periodic system treats it as a marker for the perception of nature, as a tool in the search for primary matter, and as proof of the mathematical determination of nature.

Actual results concerning the periodic system are discussed, such as the noble gases, radioactivity, and isotopes. It was asked in which manner the new results could be integrated into the system. Around 1890, the discussions turned to non-chemical problems. It should be remembered that in 1887, William Crookes (1832–1919) reported on the origin of the elements at the Royal Institution. This lecture was reviewed.[133] William Preyer then translated it and published his own article,[134] which was also reviewed in the popular journals. Gustav Wendt, who is unknown to today's reader, presented a similar idea.[135] In the background of the renewed discussion on the evolutionary theory, the periodic system became important. There were also trials to apply the descent theory to other (inorganic) fields. If it is valid in biology that all species came from one or more primitive forms, one could hope to find the same for the elements. Ernst Krause (1839–1903, pseudonym Carus Sterne) explicitly described this idea in *Die Entwicklung der chemischen Elemente* [The Evolution of the Chemical Elements]. Krause wrote about the discovery of chemical elements forming natural families like plants and animals, stating that there are periodic relationships in these family groups that can be expressed numerically.[136] This explanation of natural families of elements was used to apply Darwin's theory of evolution to the (inorganic) elements and to discuss its development to a higher level, as well as primary matter. A similar statement can be found in *Die Welträtsel* [The Riddle of the Universe], written by Ernst Haeckel (1834–1919) in 1899. In Chapter 12 he describes the group relationships of elements and the periodic system (in connection with the names of Meyer and Mendeleev). This should be comparable to the different types of animals and plants, so it can be expected that the elements are made up of primary atoms with different numbers and different places. Haeckel reviewed the speculations of Wendt, Preyer, and Crookes about the origin of the elements from primary matter, and does not forget to mention the transmutation of elements.[137] After 1896 the discovery of radioactivity renewed interest in the theory of transmutation.

To summarize, popular journals did not often mention the periodic system. The main interest was not of a chemical nature, but rather the cognition of nature and the connection to the descent theory.

5. CONCLUSIONS

In the second half of the nineteenth century, chemistry became a scientific discipline in Germany. The chemical community started to institutionalize itself. The chemical industry was flourishing to an extent unknown at that time. Organic chemistry was the dominating discipline, and inorganic chemistry was less recognized. Physical chemistry was still in a very early stage of development. Chemical journals, particularly the *Berichte der Deutschen Chemischen Gesellschaft*, reported in a timely manner on the regularities between the atomic weights and the properties of the elements or the so-called natural system of elements discovered by

Mendeleev of Russia, and Meyer of Germany. But the system was not celebrated as a triumphant success. Instead, it was used only to a very small extent as a guideline for research work and did not lead to a systematic search for new elements. Efforts to precisely calculate the atomic weights were directly related to Meyer's system.

Relevant reports from other countries were available due to the establishment of correspondents for the *Berichte* and the fact that foreign chemists could submit their (translated) papers.

German chemists evaluated Meyer's contribution to the periodic system differently. In contrast to Mendeleev, Meyer never recognized his system as a law. Meyer's contributions to the periodic system became increasingly forgotten, despite several papers in which he attempted to explain his purpose for creating the system. Mendeleev's successful predictions of elements were better understood, and overshadowed Meyer's physicochemical ideas.

The periodic system was discovered in the process of writing a textbook. This fact led to the assumption that it was quickly adopted in textbooks for schools and universities. However, this was not confirmed for the case of Germany. The reception process was hindered by general discussions about the best didactic approach for the structure of a textbook and, especially, confusion on the placement of dissimilar elements and irregularities in the atomic weight. The periodic system was presented in popular journals in connection with the origin of the elements, possible transmutation, and the descent theory. It was associated with efforts to explain the evolution of inorganic matter.

Even though Meyer worked and published in Germany and Mendeleev's ideas were published in Germany, the periodic system did not have a prompt and celebrated reception. This finding for Germany is consistent with those for other countries studied in this book.

NOTES

1. Karl Seubert, "Anmerkungen." In *Das natürliche System der chemischen Elemente*, ed. Karl Seubert, Ostwald's Klassiker der exakten Wissenschaften, no. 68, 119–134 (Leipzig: W. Engelmann, 1895), 124.
2. Seubert, ibid., 125.
3. Karl Seubert was born on April 6, 1851, in Karlsruhe. He worked in a technical chemical laboratory in Breslau, as an assistant to Carl Birnbaum (1839–1887) in Karlsruhe, and in the chemical laboratory at Tübingen University from 1878 onward. At Tübingen he was promoted and received his *Habilitation*. In 1885 Seubert became *planmaeßiger a.o. Titularprofessor* at Tübingen, and in 1895 he became professor of Inorganic and Analytical Chemistry in Hannover, where he retired in 1921. He died on January 31, 1942, in Hannover. See Regine Zott, *Briefliche Begegnungen* (Berlin: Verlag für Wissenschafts- und Regionalgeschichte, 2002), 125–126; Armin Wankmüller, "150 Jahre Pharmazie an der Universität Tübingen." In *Physik, physiologische Chemie und Pharmazie an der Universität Tübingen* (Tübingen: Mohr, 1980), ed. Wolf von Engelhardt, 11–129.
4. Hubert Laitko, "Die Disziplin als Strukturprinzip und Entwicklungsform der Wissenschaft—Motive, Verläufe und Wirkungen von Disziplingenese", *Verhandlungen zur Geschichte und Theorie der Biologie*, 8 (2002), 19–55.

5. See Johannes W. van Spronsen, *The Periodic System of Chemical Elements* (Amsterdam (et al.): Elsevier, 1969); Michael D. Gordin, *A Well-Ordered Thing—Dmitrii Mendeleev and the Shadow of the Periodic Table* (New York: Basic Books 2004); Eric R. Scerri, *The Periodic Table: The Story and Its Significance* (New York: Oxford University Press, 2007); Klaus Danzer, *Dmitri I. Mendelejew und Lothar Meyer: die Schöpfer des Periodensystems der chemischen Elemente* (Leipzig: Teubner, 1974); Siegfried Engels and Alois Nowak, *Auf der Spur der Elemente* (Leipzig: Deutscher Verlag für Grundstoffindustrie, 1983).

6. See the chapter on Spain by José Ramon Bertomeu Sánchez and Rosa Muñoz Bello in this book.

7. See, for example, Rudolf Arendt, *Lehrbuch der Anorganischen Chemie nach den neuesten Ansichten der Wissenschaft, auf rein experimenteller Grundlage für höhere Lehranstalten und zum Selbstunterricht* (Leipzig: Voss, 1868); Carl Baenitz, *Lehrbuch der Chemie und Mineralogie: unter besonderer Berücksichtigung der chemischen Technologie in populärer Darstellung* (Bielefeld [et al.]: Velhagen & Klasing, n.d.); Friedrich Rüdorff, *Grundriß der Chemie für den Unterricht an höheren Lehranstalten* (Berlin: Guttentag, 1868); Julius Adolph Stöckhardt, *Die Schule der Chemie* (Braunschweig: Vieweg, 1846); August Wiegand, *Grundriß der Experimentalchemie* (Halberstadt: Dölle, 1842).

8. Arendt, ibid., 65–66.

9. See also Günter Schwanicke, *Aus dem Leben des Chemikers Julius Lothar Meyer*, ed. Heimatverein Varel e. V. *Vareler Heimathefte*, Heft 8, 1995; Karl Seubert, "Nekrolog Lothar Meyer," *Berichte der Deutschen Chemischen Gesellschaft* **28** (1896): 1109–1146; Danzer (note 5).

10. See Lothar Meyer, "Anmerkungen." In Stanislaus Cannizzaro, *Abriß eines Lehrgangs der theoretischen Chemie*, ed. Lothar Meyer (Ostwald´s Klassiker der exakten Wissenschaften 30), 51–61, pp. 58–60.

11. Lothar Meyer, *Die modernen Theorien der Chemie und ihre Bedeutung für die chemische Statik* (Breslau: Maruschke & Berendt, 1864). Five editions of the book were published and it was translated into English and French; the 1880 edition of the book, *Grundzüge der theoretischen Chemie* (Leipzig: Breitkopf & Härtel, 1890), based on *Moderne Theorien*, was translated into English, Russian, and Japanese.

12. Meyer, *Die modernen Theorien der Chemie* (note 11), 14–15.

13. Meyer, ibid., 135.

14. Meyer, ibid., 136–138.

15. Seubert (note 9) 1136–1137 and 1140–1146.

16. See also Scerri (note 5), 92–98.

17. Meyer (note 10), p. 59; Schwanicke (note 9), 24.

18. Seubert (note 1), 6–7.

19. Jeffrey A. Johnson, "GERMANY: Discipline—Industry—Profession. German Chemical Organizations, 1867–1914." In *Creating Networks in Chemistry: The Founding and Early History of Chemical Societies in Europe*, ed. Anita Kildebaek Nielsen and Soňa Štrbáňová (Cambridge: The Royal Society of Chemistry, 2008), 113–138.

20. Viktor von Richter, "Bericht aus St. Petersburg vom 17. Oktober 1869," *Berichte der Deutschen Chemischen Gesellschaft* **2** (1869): 552–554.

21. D. I. Mendelejeff, "Über die Beziehungen der Eigenschaften zu den Atomgewichten der Elemente," *Zeitschrift für Chemie* **12** (1869), 405–406.

22. V. A. Krotikov, "Dve oshibki v pervykh publikatsiyakh o periodicheskom zakone D. I. Mendeleeva [Two mistakes in the early publications on Mendeleev's periodic law]," *Voprosy istorii estestvoznaniya i tekhniki*, **29** (1969): 129–131.

23. See also Karl V. Bening, 1911. *D. I. Mendeleev i Lotar Meier* [D. I. Mendeleev and Lothar Meyer] (Kazan: Centralnaya tipografiya, 1911).

24. Lothar Meyer, "Die Natur der chemischen Elemente als Funktion ihrer Atomgewichte," *Annalen der Chemie und Pharmacie*, VII. Supplementband (1870): 354–364.

25. D. I. Mendelejeff, "Die periodische Gesetzmäßigkeit der chemischen Elemente," *Annalen der Chemie und Pharmacie*, VIII. Supplementband (1871): 133–229.

26. D. I. Mendelejeff, "Zur Frage über das System der Elemente," *Berichte der Deutschen Chemischen Gesellschaft* **4** (1871): 348–352.

27. Lothar Meyer, "Zur Systematik der anorganischen Chemie," *Berichte der Deutschen Chemischen Gesellschaft* **6** (1873): 101–106.

28. Julius Thomsen, "Über die Basicität und Constitution der Überjodsäure," *Berichte der Deutschen Chemischen Gesellschaft* **6** (1873): 2–9.

29. Wurtz propagated a "science française" and overestimated nationalism in the science. For a biography of Wurtz, see Alan Rocke, *Nationalizing science: Adolphe Wurtz and the battle for French Chemistry* (Cambridge, Mass.: MIT Press, 2001); for nationalism and internationalism in chemistry, see Christoph Meinel, "Nationalismus und Internationalismus in der Chemie des 19. Jahrhunderts." In *Perspektiven der Pharmaziegeschichte. Festschrift für Rudolf Schmitz*, ed. Peter Dilg (Graz: Akademische Druck- und Verlagsanstalt, 1983), 225–242.

30. "Protokoll der Vorstandssitzung vom 11. Januar 1880," *Berichte der Deutschen Chemischen Gesellschaft* **13** (1880): 6–8.

31. This is clearly mentioned by the editor, J. Rosenthal, in the foreword of the book. He assumed that Wurtz used Meyer's explanations for his book and changed the content of some paragraphs to emphasize Meyer's contribution. See J. Rosenthal, foreword to *Die atomistische Theorie* by Adolphe Wurtz (Leipzig: Brockhaus, 1879).

32. Lothar Meyer, "Zur Geschichte der periodischen Atomistik," *Berichte der Deutschen Chemischen Gesellschaft* **13** (1880): 259–265, p. 259.

33. This is in contrast to Meyer's statement that he could combine both tables only after the redetermination of the atomic weights of Mo, Nb, Vd, and Tantal at the end of 1869. See Meyer (note 32), 261.

34. This opinion of Meyer is unclear. He himself started to recalculate atomic weights.

35. Meyer (note 32), 261.

36. Quite to the contrary, Meyer wrote in 1870 that Mendeleev arranged the elements in one row, which was subdivided in segments (Meyer (note 24)).

37. Also later on, the double periodicity was attributed to Mendeleev. See Eberhard Hoyer, "Die Doppelperiodizität im Mendelejewschen System der Elemente und ihre Bezeichnung—ein oft übersehener Zusammenhang," *Chemie in der Schule* **24** (1977): 139–141.

38. D. Mendelejeff, "Zur Geschichte des periodischen Gesetzes," *Berichte der Deutschen Chemischen Gesellschaft* **13** (1880): 1796–1804.

39. Lothar Meyer, "Zur Geschichte der periodischen Atomistik," *Berichte der Deutschen Chemischen Gesellschaft* **13** (1880): 2043–2044.

40. Masanori Kaji, "Mendeleev's Discovery of the Periodic Law: The Origin and the Reception," *Foundations of Chemistry* **5** (2003): 189–214, p. 207. See also Masanori Kaji, "Social Background of the Discovery and the Reception of the Periodic Law of the Elements," *Annals of the New York Academy of Sciences*, **988**, 1–5 (2003) and the chapter on Russian case in this book.

41. In 1896 Theodor Poleck thanked Karl Seubert for the issue on Meyer's papers on the periodic system because they show Meyer's priority. *Archive of Berlin-Brandenburgische Akademie der Wissenschaften*, Sammlung Chemikerbriefe Nr. 92.

42. Scerri (note 5), 140–143.
43. See Stephen G. Brush, "Dynamics of Theory Change: The Role of Predictions," *Proceedings of the biennial meeting of the Philosophy of Science Association* (1994), 133–145.
44. Hans-Günther Körber, *Aus dem wissenschaftlichen Briefwechsel Wilhelm Ostwalds*, 2nd part, Briefwechsel mit Svante Arrhenius und Jacobus Hendricus van't Hoff (Berlin: Akademie-Verlag, 1969), 104.
45. Zott (note 3), 44, 46–47. See also the private papers of Wilhelm Ostwald in the archive of the Berlin Brandenburgische Akademie der Wissenschaften, Nr. 1995, letters from Lothar Meyer in 1886 and 1889. The conflict between Meyer and Ostwald in the last fifteen years of Meyer's life is also apparent in a letter from Beilstein to Seubert (archive of Berlin-Brandenburgische Akademie der Wissenschaften, Chemikerbriefe Nr. 7, 1896).
46. Lothar Meyer and Karl Seubert, *Die Atomgewichte der Elemente aus den Originalzahlen neu berechnet* (Leipzig: Breitkopf & Härtel, 1883).
47. Britta Görs, *Chemischer Atomismus* (Berlin: E-R-S-Verlag, 1999).
48. Lothar Meyer, "Über den Vortrag der anorganischen Chemie nach dem natürlichen Systeme der Elemente," *Berichte der Deutschen Chemischen Gesellschaft* **26** (1893): 1230–1250.
49. Meyer, ibid., 1231.
50. Meyer, ibid., 1233.
51. An original plate or model has not yet been found.
52. Meyer (note 48), 1237.
53. Lothar Meyer, ed. *Die Anfänge des natürlichen Systems der Elemente*: Abhandlungen von J. W. Doebereiner (1829) und Max Pettenkofer (1850) (Ostwalds Klassiker der exakten Wissenschaften 66) (Leipzig: Engelmann, 1895) (Reprint 1983. Leipzig: Geest & Portig).
54. His lecture was published as a booklet. Lothar Meyer, *Über naturwissenschaftliche Weltanschauung* (Tübingen: Druck von W. Armbruster & Co., 1895).
55. Michael D. Gordin, "Translating Textbooks: Russian, German, and the Language of Chemistry" *Isis* **103** (1) (2012): 88–98, p. 96–97.
56. Anneliese Greiner and Hermann Klare, eds., *Chemiker über Chemiker Wahlvorschläge zur Aufnahme von Chemikern in die Berliner Akademie 1822–1925 von Eilhard Mitscherlich bis Max Bodenstein* (Berlin: Akademie-Verlag, 1986), 133.
57. Seubert (note 1), 125.
58. Winkler was born in Freiberg in 1838. From 1857 to 1859, he studied at the Mining Academy, where he was influenced by mineralogist Johann August Friedrich Breithaupt (his uncle) and Ferdinand Reich, the co-investigator of indium (his godfather). From 1859 to 1873, he worked in manufactories of colors. In 1864 he was promoted at the University Leipzig. In 1873 he became professor at the Mining Academy. He died in Dresden on October 8, 1908. See Klaus Volke, "Clemens Winkler—zum 100. Todestag, "*Chemie in unserer Zeit* **38** (2004): 360–361; Mike Haustein, *Clemens Winkler* (Frankfurt/Main: Harry Deutsch, 2004).
59. This first known sample of germanium is still stored at the Institute of Chemistry at the Technical University Bergakademie Freiberg.
60. Clemens Winkler, "Germanium, Ge, ein neues, nichtmetallisches Element," *Berichte der Deutschen Chemischen Gesellschaft* **19** (1886): 210–211.
61. The dates for events in Russia are given in the Julian calendar, which lags twelve days behind the Gregorian calendar in the nineteenth century (editor).

62. Mike Haustein, "Die Lücke im Periodensystem—Germanium," *Chemie in unserer Zeit* **45** (2011): 398-405, pp. 402–403.

63. Haustein (note 58), 67–82.

64. Victor Meyer, *Chemische Probleme der Gegenwart* (Heidelberg: Hörning, 1889).

65. V. Meyer, ibid., 27.

66. See the chapter on Russia in this book (editor).

67. Robert Höltje, *Clemens Winkler und das Periodische System der Elemente* (Berlin: VDI-Verlag, 1940), 12.

68. Max Zängerle studied chemistry and natural history in Munich. He was a teacher at a Realschule and also a Realgymnasium in Bamberg, Landau, Lindau, and Munich, and published several textbooks.

69. Max Zängerle, "Über Atomgewichtsregelmässigkeiten," *Berichte der Deutschen Chemischen Gesellschaft* **4** (1871): 571–574.

70. Heinrich Baumhauer studied mathematics and natural science in Bonn. He was a teacher in Frankenberg and Hildesheim, and also at the Agricultural School in Lüdinghausen. Later, he was a professor at the Catholic University in Fribourg. He published papers in the field of geology and mineralogy.

71. Heinrich Baumhauer, "Über das natürliche System der chemischen Elemente," *Berichte der Deutschen Chemischen Gesellschaft* **6** (1873): 652–655.

72. Physicist Friedrich Wächter studied in Zurich, Vienna, and Heidelberg, and worked in several physical and chemical laboratories.

73. Friedrich Wächter, "Beziehungen zwischen den Atomgewichten der Elemente," *Berichte der Deutschen Chemischen Gesellschaft* **11** (1878): 11–16.

74. Hermann Friedrich Wiebe studied at the Technische Hochschule in Berlin, Aachen, Karlsruhe and at the University of Berlin. He worked at the Physikalisch-Technische Reichsanstalt in Berlin.

75. Heinrich Friedrich Wiebe, "Die Ausdehnung der starren Elemente als Function des Atomgewichts," *Berichte der Deutschen Chemischen Gesellschaft* **11** (1878): 610–612. "Starre Elemente" are fixed or solid elements in contrast to gases showing an expansion.

76. Thomas Carnelley, "Mendelejeff's periodisches Gesetz und die magnetischen Eigenschaften der Elemente," *Berichte der Deutschen Chemischen Gesellschaft* **12** (1879): 1958–1961.

77. D. I. Mendelejeff, "Zur Geschichte des periodischen Gesetzes," *Berichte der Deutschen Chemischen Gesellschaft* **13** (1880): 1796–1804, p. 1802.

78. Thomas Carnelley, "Das periodische Gesetz und das Vorkommen der Elemente," *Berichte der Deutschen Chemischen Gesellschaft* 17 (1894): 2287–2291.

79. The paper lists the author's name as J. Traube from the Laboratory of Organic Chemistry at the Technical High School in Berlin. It should be Isidor Traube, docent in physical chemistry.

80. Isidor Traube, "Die Grundlagen eines neuen Systems der Elemente," *Berichte der Deutschen Chemischen Gesellschaft* **27** (1894): 3179–3181.

81. Frank Richter, "Der Physiologe William Thierry Preyer (1841–1897): dem Darwinismus verpflichtet," In *Wegbereiter der modernen Medizin*, ed. Christian Fleck, 169–182, (Jena [et al.]: Bussert & Stadeler, 2004).

82. William Preyer, "Argon und Helium im System der Elemente," *Berichte der Deutschen Chemischen Gesellschaft* **29** (1896): 1040–1041.

83. Clemens Winkler, "Über die Entdeckung neuer Elemente im Verlaufe der letzten fünfundzwanzig Jahre und damit zusammenhängende Fragen," *Berichte der Deutschen Chemischen Gesellschaft* **30** (1897): 6–21.

84. William Ramsay, "Über die neuerdings entdeckten Gase und ihre Beziehung zum periodischen Gesetz," *Berichte der Deutschen Chemischen Gesellschaft* **31** (1898): 3111–3121.

85. Britta Görs discusses a change in the importance of atomic weights due to the periodic system. See (Görs (note 47), 196).

86. Karl Rammelsberg, "Über das Atomgewicht des Urans," *Berichte der Deutschen Chemischen Gesellschaft* **5** (1872): 1003–1006.

87. Karl Anton Henninger, "Correspondenzen, vom 11. Oktober 1875 aus Paris," *Berichte der Deutschen Chemischen Gesellschaft* **8** (1875): 1344–1356, pp. 1355–1356.

88. A. Kuhlberg, "Correspondenzen. Aus St. Petersburg über die Sitzung der russischen chemischen Gesellschaft vom 6./18. November 1875," *Berichte der Deutschen Chemischen Gesellschaft* **8** (1875): 1680–1684, pp. 1680–1681; Karl Anton Henninger, "Correspondenzen, vom 15. Dezember 1875 aus Paris," *Berichte der Deutschen Chemischen Gesellschaft* **9** (1876): 58–70, pp. 60–61.

89. Lothar Meyer, "Über das Atomgewicht des Berylliums," *Berichte der Deutschen Chemischen Gesellschaft* **11** (1878): 576–579.

90. Bohuslav Brauner, "Über das Atomgewicht des Berylliums," *Berichte der Deutschen Chemischen Gesellschaft* **11** (1878): 872–874; Lars Friedrich Nielsen and Otto Pettersson, "Über das Atomgewicht des Berylliums—Erwiderung an Hrn. Lothar Meyer," *Berichte der Deutschen Chemischen Gesellschaft* **11** (1878): 906–910.

91. Lars Friedrich Nielsen, "Über Scandium, ein neues Erdmetall," *Berichte der Deutschen Chemischen Gesellschaft* 12 (1879): 554–557.

92. Lars Friedrich Nielsen, "Über das Atomgewicht und einige charakteristische Verbindungen des Scandiums," *Berichte der Deutschen Chemischen Gesellschaft* 13 (1880): 1439–1450; Lars Friedrich Nielsen and Otto Pettersson, "Über das Atomgewicht und die wesentlichen Eigenschaften des Berylliums," *Berichte der Deutschen Chemischen Gesellschaft* **13** (1880): 1451–1459; Lars Friedrich Nielsen and Otto Pettersson, "Über Molekularwärme und Molekularvolumina der seltenen Erden und deren Sulfate," *Berichte der Deutschen Chemischen Gesellschaft* **13** (1880): 1459–1465.

93. "Protokoll der Vorstandssitzung vom 1. Dezember 1897," *Berichte der Deutschen Chemischen Gesellschaft* **30** (1897): 2955–2956.

94. "Bericht der Commission zur Festsetzung der Atomgewichte," *Berichte der Deutschen Chemischen Gesellschaft* **31** (1898): 2761–2768.

95. "Auszug aus dem Protokoll der Vorstandssitzung vom 28. November 1898," *Berichte der Deutschen Chemischen Gesellschaft* **31** (1898): 2949; "2. Bericht der Commission für die Festsetzung der Atomgewichte," *Berichte der Deutschen Chemischen Gesellschaft* **33** (1900): 1847–1883.

96. Görs (note 47).

97. Stephen G. Brush, "The Reception of Mendeleev's Periodic Law in America and Britain," *Isis* 87 (1996): 595–628.

98. It is very complicated to distinguish between the different types of textbooks. A lot of them were recommended for schools, universities, and self-instruction. Books written by chemists at the university level were recommended for school-level instruction.

99. Brush (note 97), 617.

100. In some cases only reviews of the books were available.

101. This refers only to the territory belonging to the German Empire after 1870.

102. Some textbooks from the period before 1869 were included to understand qualitative changes.

103. "Verzeichnis der gegenwärtig an den preußischen Gymnasien, Progymnasien, Realschulen und höheren Bürgerschulen eingeführten Schulbücher," *Centralblatt für die gesammte Unterrichts-Verwaltung in Preußen* 1 (1880): 76–77.

104. Bernadette Bensaude-Vincent, José Ramón Bertomeu Sánchez, and Antonio Garcia Belmar, "Looking for an Order of Things: Textbooks and Chemical Classifications in Nineteenth-Century France," *Ambix* **49** (2002): 228–251.

105. The distinction between "compound" and "element" as a macroscopic or a microscopic object is a general problem and cannot be discussed in detail.

106. See Rudolf Arendt, *Organisation, Technik und Apparat des Unterrichts in der Chemie an niederen und höheren Lehranstalten* (Leipzig: Voss, 1868); Rudolf Arendt, "Über den naturwissenschaftlichen Unterricht an den höheren und niederen Schulen," *Pädagogische Vorträge und Abhandlungen* **1** (1868): 205–254; Ferdinand Wilbrand, *Leitfaden des methodischen Materials in der Chemie* (Hildesheim: Lux, 1870).

107. Norbert Just, *Geschichte und Wissenschaftsstruktur der Chemiedidaktik* (Mülheim/Ruhr: Westarp, 1989).

108. Rudolf Arendt, *Grundzüge der Chemie und Mineralogie* (Hamburg and Leipzig: Voss, 1894), 239–242.

109. Viktor von Richter, *Kurzes Lehrbuch der Anorganischen Chemie wesentlich für Studirende auf Universitäten und Polytechnischen Schulen sowie zum Selbstunterrichte* (Bonn: Cohen, 1875); 2nd ed. (Bonn: Cohen, 1878).

110. For details on Russian editions, see the chapter on Russia of this book.

111. Gotthold Prausnitz, "Victor von Richter," *Berichte der Deutschen Chemischen Gesellschaft* **24** (1891): 1123–1130.

112. The first edition was not available for us.

113. Richter (note 109) 2nd ed., 265–266.

114. August Michaelis, *Ausführliches Lehrbuch der Chemie* (Braunschweig: Vieweg, 1878); August Michaelis, *Einführung in die allgemeine Chemie und die physikalisch-chemischen Operationen* (Braunschweig: Vieweg, 1879).

115. Michaelis, *Einführung in die allgemeine Chemie* (note 114), 115.

116. Gisela Boeck, "August Michaelis: Erschöpft sich seine wissenschaftliche Leistung in der Michaelis-Arbusow Reaktion?" *Mitteilungen der Fachgruppe Geschichte der Chemie der GDCh* **16** (2002): 20–29.

117. Carl Arnold, *Repetitorium der Chemie mit besonderer Berücksichtigung der für die Medizin wichtigen Verbindungen sowie der "Pharmacopoea Germanica" namentlich zum Gebrauche für Mediziner und Pharmazeuten* (Hamburg and Leipzig: Voss, 1885); Otto Hausknecht, *Lehrbuch der Chemie und chemischen Technologie für Real-Gymnasien, Ober-Real- und Gewerbeschulen* (Hamburg and Leipzig: Voss, 1883); Otto Hausknecht, *Lehrbuch der Chemie und chemischen Technologie für Real-Gymnasien, Ober-Real- und Gewerbeschulen* (Hamburg and Leipzig: Voss, 1883); Wilhelm Ostwald, *Grundriss der allgemeinen Chemie* (Leipzig: Engelmann, 1889); 2nd. ed. (Leipzig: Engelmann, 1890). This edition was explicitly dedicated to Lothar Meyer. This is in contrast to the above-mentioned conflicts between Ostwald and Meyer.

118. Max Zängerle, *Über die Natur der Elemente und die Beziehungen der Atomgewichte derselben zu einander und zu den physikalischen und chemischen Eigenschaften* (München: J. G. Weiss, 1882).

119. Arnold (note 117), 52.

120. In some cases the periodic system was recommended, but with the advice that the instruction was to start with elements that are of greater importance, namely the organogenes. See Friedrich Krafft, *Kurzes Lehrbuch der Chemie—Anorganische Chemie* (Leipzig &Wien: Deuticke, 1891).

121. For example, E. Loew, "Versuch einer graphischen Darstellung für das periodische System der Elemente," *Zeitschrift für physikalische Chemie* **23** (1897): 1–12.

122. For example, Walter Roth, "Review of the textbook *Anorganische Chemie* by Ira Remsen issued by Karl Seubert," *Zeitschrift für physikalischen und chemischen Unterricht* **22** (1909): 61.

123. Andreas Daum, *Wissenschaftspopularisierung im 19. Jahrhundert, Bürgerliche Kultur, naturwissenschaftliche Bildung und die deutsche Öffentlichkeit 1848–1914* (München: Oldenbourg, 2002) 341–342.

124. *Die ᵢᵥₐₜᵤr—Zeitung zur Verbreitung naturwissenschaftlicher Kenntniß und Naturanschauung für Leser aller Stände* [Nature—a newspaper distributing natural scientific knowledge and the perception of nature by readers of all statuses] was part of the "Deutscher Humboldt-Verein" (German Humboldt Association) and was issued from 1852 to 1902. It was founded by Otto Ule (1806–1876). Karl Mueller (1813–1894) joined him as a co-editor in 1856. Hugo Roedel later replaced Ule. It was published in Halle by the publishing house Schwetschke. It addressed a wide variety of subjects: chemistry, anthropology, archaeology, astronomy, biology, botany, ethnology, geography, geology, and so on.

125. The weekly natural science journal *Naturwissenschaftliche Wochenschrift* was founded in 1887. The journal was edited by Henry Potonié and published in Ferdinand Dümmler's publishing house in Berlin. It contained general papers on natural history and natural philosophy, as well as physics and chemistry. It also provided advice for instruction.

126. This was published from 1865 to 1901. Approximately 40 percent of the papers were dedicated to chemistry.

127. *Naturwissenschaftliche Rundschau* [Review of Natural Science] was published by Friedrich Vieweg and Son in Braunschweig starting in 1886. Under the professors who contributed to the journal was also Victor Meyer. Every week, the journal published papers on the progress in all fields of natural science. It was issued until 1912.

128. *Jahrbuch der Naturwissenschaften* [Annual Book of Natural Science] was published in Max Wildermann's publishing house in Freiburg from 1885/86 to 1913/14. The book was concerned with demonstrating "the magnificent progress in the fields of: physics, chemistry and chemical technology, mechanics, astronomy and mathematical geography, zoology and botany, forestry and agriculture, mineralogy, geology and seismology, anthropology and prehistory, health care, medicine and physiology, regional geography and ethnology, trade and industry, traffic and means of transport." It shall be shown that "the most important achievements to the reader who is not specialized or a scholar is an exoteric and inspiring language" (foreword 1885/86, p. VI). Every year, the important results from the previous year in the field of natural science were published.

129. *Himmel und Erde—Illustrierte naturwissenschaftliche Monatsschrift* [Sky and Earth— Illustrated Monthly Natural Science Paper] was edited from 1889 to 1915 by the popular education association "Urania" in the Paetel Brothers publishing house in Berlin. The head of the editorial staff was M. Wilhelm Meyer.

130. "Bericht von der Versammlung der Gesellschaft Deutscher Naturforscher und Ärzten," *Naturwissenschaftliche Wochenschrift* **5** (1890): 414–416, p. 416.

131. "Bericht von der Versammlung der Gesellschaft Deutscher Naturforscher und Ärzten," *Naturwissenschaftliche Wochenschrift* **10** (1895): 625–630.

132. Victor Meyer, *Probleme der Atomistik* (Heidelberg: Winter, 1896).

133. Hugo Roedel, "Die Genesis der Elemente" *Die Natur* **15** (1889): 41–43.

134. William Preyer, "Das genetische System der Elemente," *Naturwissenschaftliche Wochenschrift* **6** (1891): 523–525.

135. Gustav Wendt, *Die Entwicklung der Elemente. Entwurf zu einer biogenetischen Grundlage für Chemie und Physik* (Bonn: In Comm. d. Hirschwald'schen Buchhandlung, 1891).

136. Ernst Krause, "Die Entwicklung der chemischen Elemente," *Himmel und Erde* (1892): 236–243.

137. Ernst Haeckel, *Die Welträtsel—Gemeinverständliche Studien über monistische Philosophie* (Bonn: Strauß, 1899), 227.

PART II

Early Response at the Center of Chemical Research

CHAPTER 4

⌀

British Reception of Periodicity

GORDON WOODS

1. INTRODUCTION

The discovery of periodicity in the properties of the elements and its connection to their atomic weights is one of the most important advances in nineteenth-century chemistry. This chapter will consider the tables of John Newlands (1837–1898) and William Odling (1829–1921), which preceded that of Dmitrii Ivanovich Mendeleev (1834–1907). Mendeleev's table was published in 1869, prior to his being aware of the UK precedents of his tabulation. The major portion of this chapter will extend the ideas advanced by Stephen Brush[1] in *The Reception of Mendeleev's Periodic Law in America and Britain* but will restrict itself to the dissemination of the periodicity concept within the United Kingdom. This will be monitored by recording its appearances in textbooks and examination papers, and in a wider context, by extracting data from *Google Books*.

2. THE PHRASE "PERIODIC TABLE"

2.1 A Peculiarly English Phrase?

The periodic table has a rich history since its inception. It has evolved into many shapes, and indeed dimensions, yet retaining its essential periodic underpinning. In the United Kingdom it is seen as a "table," whereas the French prefer "classification" and the Germans and Russians "system." Mendeleev himself referred to his periodic law in his *Faraday Lecture*[2] and never used the term "table," thus it is ironic that his fame is linked to words that he appears never to have uttered.

2.2 Recognition of the Phrase by Non-Chemists

The arrangement of the elements in rows and columns is seen as a table, but why label it periodic? A related, more familiar word to non-chemists is *periodical*, normally referring to a magazine that appears at regular time intervals.

2.3 The Growth of Periodic Phrases

Google Books is a powerful modern tool for investigating the usage of selected words or phrases over selected time intervals. The writer chose to use its advanced search for books in the English language. This meant that sources other than British, notably North American, are also included but the observed patterns are probably true for British books. The data compare the number of times the terms *periodic table, periodic law, periodic classification,* and *periodic acid* occurred in five-year intervals between 1870 and 1919. *Periodic system* (Mendeleev's original 1869 phrase) had to be ignored as many references, indeed the majority during 1870s, were to meteorology rather than chemistry.[3] The figures for *periodic acid* were initially included for curiosity since the pronunciation is *per-iodic* rather than *peri-odic*; however, there is a much smaller range of values and no consistent trend. Tables 4.1 and Figure 4.1 below display the above data as numbers and graphically.

   Periodic law, the term later used by Mendeleev in 1871, initially grew rapidly, but its growth leveled out by the end of the century. The use of *periodic table* increased steadily, first spreading slowly but eventually becoming greater than that of *periodic law* after Mendeleev's death in 1907, as the English-speaking world preferred its own phrase rather than copying that of Mendeleev and German speakers. *Periodic classification*, always much less popular than *periodic law*, was overtaken by *table* earlier than was *law* and decreased in usage as *periodic table* became the commonest term around 1918. It might be interesting to investigate whether the chronological development from general terms (*law, classification*) to the more restrictive concrete *table* was copied in other languages.

Table 4.1 The use of different periodic terms for five year intervals 1870–1919

Years	1870 -74	1875 -79	1880 -84	1885 -89	1890 -94	1895 -99	1900 -04	1905 -09	1910 -14	1915 -19
Periodic table	0	1	2	44	144	337	703	767	1530	2490
Periodic law	47	420	1280	2130	2590	3340	3060	3970	2720	1970
P. classification	0	1	27	117	130	281	628	550	642	439
Periodic acid	260	264	126	264	171	127	371	269	167	136

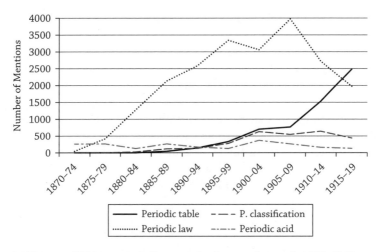

Figure 4.1 The use of different periodic terms for five year intervals 1870–1919

3. BRITISH CONTRIBUTIONS

This section first describes two British charts classifying elements that predate Mendeleev's work. It then follows the gradually increasing uptake by chemists during the next twenty-five years (1870–1894) of the periodic system. Finally, it considers British discoveries that tested the boundaries of this periodic system.

3.1 Before 1869. Two British Classifications of Elements

British involvement in acceptance of periodicity began with the presence of some of their leading chemists at the world's first chemical congress at Karlsruhe (Germany) in 1860. Among those were Britons Odling, George Carey Foster (1835–1919), and Henry Enfield Roscoe (1833–1915), who might have met, among others, the Russian Mendeleev, German Julius Lothar Meyer (1830–1895), and Italian Stanislao Cannizzaro (1826–1910).[4] This mutual meeting of minds would have influenced their future thinking. John Newlands was not invited, being outside the inner circle of British chemists.

The congress grappled with conflicting values for atomic weights of elements, which, with historical hindsight, one realizes stemmed from opposing views on formulae. Whether water is HO or H_2O affects the atomic weight of oxygen and hence of other elements whose atomic weight was determined from the composition of their oxides. The contribution of Cannizzaro, recalling an idea of his compatriot Amedeo Avogadro from half a century earlier, produced a coherent list of atomic weights. This proved crucial for the subsequent recognition of periodicity.

3.2 J. A. R. Newlands

John Alexander Reina Newlands was born in the inner London suburb of Southwark, of a Scottish father and Italian mother who were responsible, respectively, for his second and third names. He was an analytical chemist in the sugar industry.

Newlands had a particular interest in looking for patterns in the atomic weights of different elements, about which he wrote seventeen articles in *Chemical News* over two decades starting in 1863.[5] Although absent from Karlsruhe, Newlands was able to use the atomic weights calculated by Alexander Williamson that resulted from it. Newlands arranged approximately sixty known elements in order of increasing atomic weight and noted a chemical similarity between any element and the seventh element after it. (Counting the elements at both ends, there are thus eight elements, as there are eight notes in a musical octave. Thus, probably unwisely, Newlands called his new idea *The Law of Octaves*.)[6] The elements were in order of increasing atomic weight, with the symbol accompanied by an ordinal number—1, 2, 3 representing first, second, and third. He had thus *unknowingly* used atomic *numbers*, not then a current term! The element order and the repetition of properties are principles of periodicity. His 1866 table is shown below as Table 4.2.[7]

His presentation on March 1, 1866, to the Chemical Society (CS) was disastrous. Valid criticisms included putting two elements in the same space and departing from strict numerical order to produce similar elements on the same horizontal line. Professor Carey Foster from Glasgow ridiculed him by suggesting that Newlands might have listed the elements in alphabetical order! The next day, the publication committee decided not to publish his paper. Later, when Mendeleev's work had shown that Newlands had been almost correct, Odling justified the committee's decision because it did not publish purely theoretical ideas.[8] Odling, a noted chemical theorist, acted correctly by not publishing his own paper through the CS but in *The Quarterly Journal of Science*.

Table 4.2

No		No		No.		No.		No.		No.		No.		No.	
H	1	F	8	Cl	15	Co&Ni	22	Br	29	Pd	36	I	42	Pt&Ir	50
Li	2	Na	9	K	16	Cu	23	Rb	30	Ag	37	Cs	44	Os	51
G	3	Mg	10	Ca	17	Zn	24	Sr	31	Cd	38	Ba&V	45	Hg	52
Bo	4	Al	11	Cr	19	Y	25	Ce&La	33	U	40	Ta	46	Tl	53
C	5	Si	12	Ti	18	In	26	Di&Mo	34	Sn	39	W	47	Pb	54
N	6	P	13	Mn	20	As	27	Zr	32	Sb	41	Nb	48	Bi	55
O	7	S	14	Fe	21	Se	28	Ro&Ru	35	Te	43	Au	49	Th	56

3.3 W. Odling

Although born in Southwark like Newlands, Odling's chemistry background was very different. A prominent figure in the chemistry establishment, he was secretary of the CS (1856–1869) before becoming vice president and president. He then held various offices until 1891 in the Institute of Chemistry. He became professor of Chemistry at Oxford in 1872, retiring in 1911, aged eighty-two![9]

Prior to Karlsruhe, Odling had recognized the stepwise decrease in valency for sets of three related elements or triads headed by carbon, nitrogen, oxygen, and fluorine linked to the increase in atomic weights. In 1864, benefiting from more accurate atomic weights post Karlsruhe, he produced the table (Fig. 4.2)[10] below, which is remarkably similar to Mendeleev's 1869 table.[11]

His horizontal rows, particularly for groups 14–17, evidence periodicity, but there are more errors than in Mendeleev's later tables. He saw more clearly than Mendeleev the need to separate all the transition metals, not merely the precious metals, from the main group elements. However, his frequent use of the symbol " suggests too many undiscovered elements and he made no attempt at predicting their properties. The article looked at numerous relationships between atomic weights of similar elements, noting that a difference of about sixteen often occurs. He further noted an atomic weight difference in range 84.5-97 occurred on some of first and third elements of Döbereiner triads (sets of three similar elements) and suggested that other pairs of similar elements with that range difference might have an undiscovered middle one. In effect, this modest prediction was extending Döbereiner's patterns. The article ends prophetically that although some such similarities may be coincidental, they are too numerous and definite not to depend upon some hitherto unrecognized general law.

No evidence was found of Odling contesting Mendeleev's claim for primacy of discovery, and indeed he appears to have published relatively little after 1875, although he contributed several entries to Watts,[12] Dictionary of Chemistry. He wrote about Atomic Weights,[13] "Cannizzaro's general propositions are in the present state of knowledge, too great to admit of their adoption," so some values are equivalent weights. However, a telling footnote says "This view has now been modified by the writer, see Metals, Atomic weights and Classification of," [14] referring to Mendeleev's 1871 article. He pointed out the two-dimensional nature of the table showing different trends both across and down.

Thus, not only had Odling not mentioned his own table, but he had second, and positive, thoughts about the values of Mendeleev's work. Two Britons had therefore produced charts containing some of the criteria that would be more fully shown by Mendeleev. However, Newlands was not a member of the inner circle of chemists unlike Odling but he was a general chemist whose main interest was not periodicity, perhaps because he was busy with CS administration.

3.4 1866–1875, A Quiet Decade in Britain

In Europe, while some chemists were thinking of property patterns, Mendeleev and Lothar Meyer published papers on periodicity that were not initially recognized as

				Ro 104	Pt 197
				Ru 104	Ir 197
				Pd 106·5	Os 199
...... H 1	„	„		Ag 108	Au 196·5
„	„	Zn 65		Cd 112	Hg 200 ......
...... L 7	„	„		„	Tl 203
G 9	„	„		„	Pb 207 ......
... B 11	Al 27·5	„		U 120	„
C 12	Si 28	„		Sn 118 ... . „	
... N 14	P 31	As 75		Sb 122	Bi 210
O 16	S 32	Se 79·5		Te 129 ... . „	
..... F 19	Cl 35·5	Br 80		I 127	„
..... Na 23	K 39	Rb 85		Cs 133	„
Mg 24	Ca 40	Sr 87·5		Ba 137 ... . „	
	Ti 50	Zr 89·5		Ta 138	Th 231·5
	„	Ce 92		„	
	Cr 52·5	Mo 96		V 137	
	Mn 55			W 184	
	Fe 56				
	Co 59				
	Ni 59				
	Cu 63·5				

Figure 4.2 W. Odling's table in 1864 from his paper "On the Proportional Numbers of the Elements," *Quarterly Journal of Science*, 1 (1864): 642–648, p. 643.

important, until in 1874 de Boisbaudran spectroscopically identified a new element from the Pyrenees, which he patriotically named gallium. Mendeleev recognized gallium as the element he had predicted to fit below aluminum, hence his name *eka*-aluminum.

De Boisbaudran, working in Paris, published his results in *Comptes Rendus*, followed by an English translation[15] "If further research confirms the identity of the properties just assigned to *eka*-aluminum with that of gallium." de Boisbaudran had concluded, "it will be an instructive example" of a predicted element. He also stated that he did not know of Mendeleev's description of the properties of the hypothetical metal.

Initial support of the significance of this discovery came from Pattison Muir (1848–1931)[16] in early 1876, ten years after Newlands's humiliation. "It ought to be remarked that, previous to the discovery of the periodic law, it was not possible to predict the existence or foretell the properties of undiscovered elements" also pointing. He also stressed the significance of the similarities between gallium and *eka*-aluminum.[17]

3.5 1876–1886, News Spreads of Discovery of Predicted Elements

Until 1875 in Britain, neither the earlier writings of Newlands and Odling nor later ones of Mendeleev had attracted much attention. The discovery of a predicted element changed this, although initially it met with skepticism.

One can visualize this new information as water at the tip of a cone, trickling down and spreading out to a wider but less qualified audience. Brief articles in journals can be quickly produced, whereas a textbook takes longer and is costlier to produce. Gradually, professional chemists came to accept new ideas, which they conveyed to university students, until, finally, school pupils and the general public were also made aware.

Pattison Muir had a much longer article on Chemical Classifications in 1877, which included information about Mendeleev's law and thoughts as to where it would lead. "Mendeleev's law presents us with a definite variable, viz. atomic weight, and it attempts to represent the chemical and physical properties of both elements and compounds as functions of this variable. A further more careful study of the exact properties of groups of compounds out of elements will doubtless enable us to make a nearer approximation to the nature of the function in question. Supposing that the periodic law be clearly established we shall be met with questions like: Are there periods within periods? Is the divergence from the law itself periodic?" [18]

3.6 Earliest Book References

A separate later section considers more of the books examined, but first, the two earliest books that mentioned periodicity will be reviewed in greater detail.

The earliest book found, not written by someone personally involved in projecting periodicity, was the twelfth edition of Fownes's *Manual of (Elementary) Chemistry*,[19] which was arranged in the order then favored: non-metals, laws of chemical combination, atomic theory, and then metals. An early Table of Elementary Bodies listed gallium but gave no atomic weight. The most important section[20] starts, "A very remarkable relation has been shown to exist between the quantivalence of the elements and the order of their atomic weights. Arranging the elements in vertical columns . . . we find that, with certain exceptions belonging to the iron and platinum groups, they all arrange themselves such that the first horizontal line is occupied by monad elements, the second by dyad elements.[21] Hydrogen stands alone, there being no element between it and the monad metal lithium. This relation . . . called the 'periodic law' was first pointed out by Newlands in 1864, then developed by Odling and Mendelejeff." The section ends with: "The blank spaces in the preceding table indicate places of elements which probably exist but have not been discovered.

An anticipation of this has actually been realised. . . . The discovery of gallium, with atomic weight 68, has verified this prediction" by Mendeleev. By contrast, the eleventh edition, published four years earlier, neither listed gallium in the atomic weights table nor mentioned Mendeleev. This implies that gallium's discovery was crucial to this early recognition of the periodic law.

Miller's textbook[22] had repeated revisions. Its sixth edition mentioned the recent discovery of two elements, gallium[23] and davyium.[24] An important innovation was the chapter headings for the metals Group I, then Group II, Group III; the periodic system had become *the* method to *classify* metals. A later section[25] is headed "Relations between the Atomic Weights: Periodic Law of Newlands and Mendelejeff." Note the chronological order and use of both names. It continued, "If the elements are arranged in order of their atomic weights . . . it will be found that, with certain gaps, the relation between their properties, their quadrivalence, and their atomic weights take the form of a periodic function." (Note the early (first?) use of *periodic function* by someone not working in the periodic law field.) Thus, two long-running textbooks, with original authors now dead, mentioned the periodic law soon after recognition of gallium as *eka*-aluminum. Revision was by chemists working for chemistry societies who would have easy access to British and overseas journals. However, a later section will show that this did not mean all, or even most, new editions included this new development.

We will return to consideration of publications, but will first examine the impact of *events* on the acceptance of the periodic law. A scientific principle may be considered true when accepted by most people able to make a reasoned judgment. A second predicted element, scandium, was discovered shortly after (1879) by the Swede Lars Nilson. Although a second discovery is a lesser landmark, it was valuable corroborative evidence since the similarities between gallium and *eka*-aluminum could merely have been coincidental.

3.7 Primacy Rewarded and Shared(?)

It is thus significant that on November 30, 1882, Britain's premier scientific society, the Royal Society, awarded the Davy Medal jointly to Mendeleev and Lothar Meyer, not wishing to indicate sole primacy of discovery of the periodic law. The president (mathematician William Spottiswoode) spoke of how the labors of Mendeleev and Lothar Meyer had extended our knowledge of the relations between the atomic weights of the elements and their respective chemical and physical properties, and atomic volume curves (Fig. 4.3).[26]

Further evidence of Mendeleev's reputation was his invitation to the 1887 British Association in Manchester where he met other European chemists. It is significant that in the photograph below Mendeleev is in the front row between Lothar Meyer and Roscoe with German and Manchester professors behind.

Newlands continued to press his claim as the originator of periodicity and in 1884 republished his past papers, adding carefully selected comments of others.[27] Among them were Mendeleev's "It is indeed possible that Newlands had prior to me enunciated something similar to the Periodic Law" and Odling's "Mr. Newlands was the first chemist to arrange the elements in such a seriation that the new ones

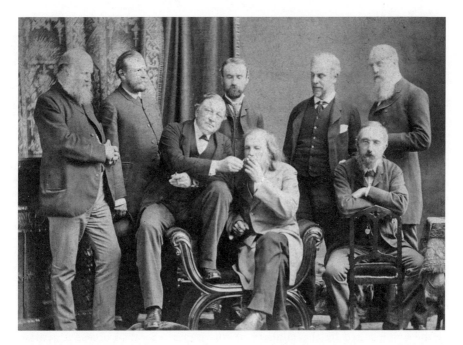

Figure 4.3 Mendeleev with a cigar sat between Henry Roscoe (left) and Carl Schorlemmer (right) with J. P. Joule and Lothar Meyer on the left of the back row.September 1887, British Association for the Advancement of Science meeting at Manchester.Courtesy of the University of Manchester Library.

might be predicted to exist where certain gaps are observed in the seriation of atomic weights." Note, however, that Newlands had made no predictions regarding the properties of elements yet to be discovered.

Clear support for periodicity came from Dundee Professor Thomas Carnelley (1852–1890), speaking at the British Association (BA) Aberdeen meeting. He started with, "The truth of the Periodic Law, as enunciated by Newlands, Mendelejeff and Lothar Meyer is now generally allowed by most chemists. Nevertheless but little has been done towards attaining a reasonable explanation of the Law."[28]

Professor George Stokes, in his address as the 1887 Royal Society president, after referring to the 1882 joint award, continued that we must not forget our own countrymen, and thus Mr. J. A. R. Newlands was to receive the Davy Medal "for his discovery of the Periodic Law of the Chemical Elements. Though in the somewhat less complete form in which the law was enunciated by him, it did not attract the attention of chemists, still in so far as the work of the two foreign chemists above mentioned was anticipated, the primacy belongs to Mr. Newlands." [29] For Newlands it was a share in primacy recognition within a diplomatically balanced citation.

Two years later, Mendeleev's reputation led to invitations to lecture at both the Royal Institution (RI) and the Chemistry Society (CS). He addressed the RI on *An Attempt to Apply to Chemistry One of the Principles of Newton's Natural Philosophy.* This lecture to scientists of all kinds had very few remarks about the periodic law, which was the subject of the second lecture to the CS.

Four days later, Mendeleev was due to deliver the prestigious *Faraday Lecture*[30] on *The Periodic Law of the Chemical Elements* to the CS Fellows but hurried back home on hearing of the illness of his son, leaving CS Secretary Professor Henry Armstrong (1848–1937)[31] to deliver his prepared text, summarized below.

His 1869 *Zeitschrift* paper had been helped by extra data being found in the intervening twenty years: The discovery of the spectroscope to identify elements, more accurate atomic weights, and more information about the properties of rarer elements. Many workers looked for numerical relations between the atomic weights of analogous elements, which helped him recognize atomic weight as a method of listing elements. He then listed the eight points in his 1869 paper. Detail was given of how some atomic weights were amended because of their properties (Y, Be, Th, and Pt). Mendeleev still believed that the periodic *law* allowed no exceptions to listing of elements in increasing atomic weight order, hence tellurium's atomic weight *must* be < 126.5 and Brauner's experiments had confirmed this.[32] The final section dealt with repeated similar trends in formulae and properties of elements and their compounds; this second dimension to the chart of elements was Mendeleev's special, and repeated, feature. Others had compared trends in families, but the periodic law ultimately had produced trends *both* across rows and down columns.

In Russia Mendeleev had resigned from his position at St. Petersburg University as he felt insulted by the minister of education's refusal to accept a petition Mendeleev had presented on behalf of the students. Gordin[33] stresses it was indeed for the perceived insult rather than the rejection of the students' claims. Nevertheless, honors continued to be bestowed on him as the first formulator of periodicity. England's two oldest universities each gave him honorary degrees in June 1894. His reputation was ably summarized by the Cambridge citation, which concluded: "It is indeed a great thing to have discovered among so many elements a relationship occurring at fixed intervals, as if periodically, and from observation of the known to have also forecast the unknown. By the genius of this man, also previously unknown elements have been foretold by his singular mental insight, and subsequently discovered in nature itself. These elements named after famous nations gallium, scandium and germanium have rendered his name more illustrious, and in so far as it relates to him, the reputation of the Russians. I present to you, Professor of chemical science D. I. Mendeleef." [34]

The following year, Newlands passed away. A century later, the Royal Society of Chemistry placed a blue commemorative plaque on his home at 19 West Square, London SE11. Maybe the words are a belated apology for the CS response to his 1866 *Law of Octaves* lecture.

4. 1894–1920, THREE BRITISH DISCOVERIES ACCOMMODATED

4.1 1894–1897 Inert Gases Found

Two future Nobel laureates, Scottish chemist William Ramsay (1852–1916) and English physicist John Rayleigh (1842–1919), by elegant experiments discovered a gas forming approximately 1 percent of air with a *molecular* weight of about 40. This was colorless, odorless, and even less reactive than nitrogen. Ramsay named it argon (Greek *argos*, inactive), believing it was an element, as he could neither decompose it nor make it from other elements.

- Consider the partial 1894 periodic table below, with atomic weights beneath. Ramsay resorted to specific heat measurements to determine the atomicity of the gas since this would give the *atomic* weight needed to place the gas in the periodic table. It appeared to be monatomic, with an atomic weight of about 40.

O	F	Na	Mg	Al
16	19	23	24	27
S	Cl	K	Ca	Sc
32	35.5	39	40	45

Mendeleev could see no space for such an element, and suggested the gas was N_3, analogous to O_3.[35] However, coupled with the identification of terrestrial helium, Ramsay boldly used the periodicity principle to predict *another complete column* in the table, hence the prediction supported, not opposed, the table.[36] Further support came from Ramsay's 1897 presidential address to the BA chemistry section, "An Undiscovered Gas," which described how he had used the periodic table to seek an element between helium and argon. It was likely to be chemically inert and monatomic, with an atomic weight of about 20.[37] Neon was just that!

Ramsay's work was assisted by Morris Travers (1872–1961), a youthful future professor. Travers's account of their work includes the following statement: "In 1894 the Periodic Law was generally accepted, however 30 years had elapsed since Newlands had first formed the Law of Octaves, and 25 years since Mendeleev had confirmed Newlands' suggestion and had elaborated on their system of classification of the elements, its importance was by no means fully recognized. Ramsay was probably the first teacher in the country to base his course of lectures upon the periodic classification of the elements, which he regarded as the most important generalization relating to chemistry."[38]

4.2 Atomic Weight/Atomic Number Reversals

Table 4.3 lists atomic weights for the three pairs of elements, from which the second element has the lower atomic weight. These pairs differ from Mendeleev's law, which listed elements in order of increasing atomic weight. Mendeleev's 1871 table placed Ni after Co and I after Te because of their valency and other properties with the atomic weights "corrected" (e.g., Te = 126) so the periodic *law* was not broken. The discovery of argon further challenged the *law* but it appears that most chemists were becoming content to produce a table with some anomalous, possibly incorrect,

Table 4.3						
Elements in accepted table order	Ar	K	Co	Ni	Te	I
Atomic weights	39.9	39.1	58.9	58.7	127.6	126.9

atomic weights in order to place the elements in the chemically correct group. However, Mendeleev in 1905 still listed argon's atomic weight as 38.

4.3 1913 Radioactivity Explained, Isotopes Identified

Nature had thrown a puzzle at scientists, namely radioactivity. A fundamental tenet of chemistry is that one element cannot be changed into another. Hence, at the start of the twentieth century, investigators could not understand what happened with alpha and beta decay. Frederick Soddy (1877–1956) realized that by showing properties different from the starting material, the decay products had been changed to elements two columns to the left (α) or one to the right (β).[39] In a short second part,[40] Soddy related this to the periodic law and wrote that radioelements had atoms of different mass, for which he coined the word *isotopes* (Greek for same place, in the periodic table, of course). Extension of this to stable elements helped account for atomic weight/atomic number reversals, but clearer support soon followed.

4.4 1915 Atomic Number Identified

Even greater support for Mendeleev followed shortly from another Briton. He was Henry Gwyn Jeffreys Moseley (1887–1915), born into a distinguished scientific family,[41] a scholar of Eton College, then Oxford University. Moseley investigated the wavelength, hence the frequency, of X-rays from approximately sixty metallic elements. He found a *linear relationship* between the square root of the frequency and the position, i.e., ordinal number, of the element in the periodic table. This number is the atomic number, which he also stated as the number of positive units of electricity in the atomic nucleus. Thus, he could state that there must be ninety-two elements up to uranium, the heaviest known atom, hence identifying exactly which spaces remained to be filled, justifying some gaps Mendeleev had left for predicted elements. The recognition of an ordinal number (i.e. atomic number) also explained the three atomic weight inversions, since it is atomic *number* rather than atomic *weight* which governs the order of elements.[42] Consequently, since once again the periodic system was able to *accommodate* a new idea, as simply by replacing atomic weight with atomic number, the periodicity principle was maintained. Physicist Maurice de Broglie wrote, "La loi de Moseley justifie la classification de Mendeléeff: elle justifie mêmes les coups de pouce que l'on avait été obligé de donner à cette classification [Moseley's contribution justifies Mendeleev's classification; it even accounts for modifications needed to add to the classification.]."[43] (Moseley, surely a future FRS, later volunteered for army service and sadly, was killed by a Turkish sniper in the Dardanelles.)

Several British chemists displayed their belief in periodicity by publishing alternative representations of a periodic table, notably between 1882and 1898, as periodicity became generally accepted.[44]

Thus, more than fifty years since 1869, two early British periodic charts were surpassed by that of Mendeleev, whose periodic law began to be accepted to a greater degree after some elements he predicted were discovered. Further discoveries, including those by Ramsay and Moseley, were accommodated by modifications of the periodic law.

5. BOOKS FOR CHEMISTS

5.1 Histories of Chemistry

More detail was expected in books focusing on history of chemistry, such as Pattison Muir's[45], published in the year Mendeleev died. The periodic law's initial publication and increasing acceptance were both within the working life of the author, who described Mendeleev's 1871 paper on "The periodic regularity of the chemical elements as one of the most important contributions ever made in the advancement of accurate knowledge of natural phenomena."[46] The writer used the *Faraday Lecture* as source of Mendeleev's eight *Zeitschrift* points, after which he described short and long periods, correction of atomic weights, and consideration of properties of unknown elements and their compounds.

Hilditch [47] identified three advantages and five disadvantages of the periodic system. He saw the designation *law* to the periodic *system* as an overstatement, suggesting that the real law may appear from the emerging constitution of the atom. Lowry[48] devoted one-tenth of his book to the classification of the elements, including Mendeleev's correction of atomic weights and accurate predictions in a broad survey ending with Moseley's very recent work.

5.2 Texts for University and Beyond

Texts for university students and practicing chemists were mostly written by chemistry professors, often by two authors each responsible for their specialism in organic and inorganic, as Manchester professors Schorlemmer and Roscoe, respectively. Some books ran to numerous editions with revisions, sometimes continuing after the death of the original author, as with Miller, Fownes, and Bloxam. A few books will be discussed in more detail to illustrate general points as representative of many texts.

Bloxam[49] (1890) mentions gallium in a half page of small print (less than about periodic acid!) and even his son's 1910 revision describes gallium, germanium, and periodicity in 6/850 pages. This is tepid approval in a long-running text, but still a greater fraction than Roscoe's[50] 5/1,800 pages, which mentions the discovery of gallium but not Mendeleev.

Some support was, however, given by Frankland and Japp.[51] Considerable detail is provided of the features of Mendeleev's periodic system, correction of atomic weights, discovery of gallium, and the possibility that scandium is *eka*-boron. After summarizing the classification of the elements, they wrote, "Newlands' system therefore in all essential points is identical to that which Mendeleef published in 1869, except that Newlands failed to recognize the existence of the transitional elements"—Mendeleev's eighth group. The fact of Mendeleev's system being more perfect in details has led some to make Mendeleev discoverer of the periodic law. This chapter ends with: "It is quite inconceivable that the remarkable relation of the periodic law should be the work of chance. No explanation of the periodic law has been offered. At present it is an empirical law established by careful experiment and comparison." The text supported periodic law, but favored Newlands for primacy.

Two books relegate periodicity and radioactivity to the final chapters, where they can be economically introduced without altering the rest of the book. Kipping and Perkin[52] laud the accurate predictions about gallium, yet partly because the periodic system is based on valency and some elements have variable valency, the system is rated only to help advanced students. They use the term *families*, not *groups*, to avoid confusion with groups in qualitative analysis, which was more familiar to them. Mellor[53] prefers his own layout for the table, seeing no reason to prefer any one in particular. After a clear comparison table of *eka*-silicon and germanium, he concludes that the dramatic verification of the predictions is less positive proof of the periodic law than some suppose. He gives considerable detail but even in 1914 he is a skeptic.

Pattison Muir's[54] book is outstanding, written for the questioning student. He regrets Mendeleev's book was not published in Western European languages, though subsequently Mendeleev's *Osnovy khimii* [The Principles of Chemistry] appeared in English, French, and German editions.[55] In Pattison Muir's text parallel columns are used to compare recently discovered gallium and scandium with their *eka-* counterparts, and the discovery of *eka*-silicon is expected . . . which soon was the case! In its second edition he added, "The discovery and study of germanium by Winkler entirely confirmed Mendelejeff's predictions. *Eka*-silicon and germanium are the same element."[56] This is a powerful praise from a Cambridge don, who the following year became joint editor of the second edition of *Watts, Dictionary of Chemistry*.

Ramsay's[57] preface makes clear that his book contrasts with previous inorganic texts, as no systematic textbook had been written in English based upon the periodic arrangement of the elements. He suggests that this neglect stems from (a) ancient division between metals and non-metals, (b) excess importance being given to differences between acids and bases, obscuring the fact that they are both hydroxides, and (c) commercial considerations.

An immediate difference is the chapter arrangement. A brief introduction is followed by chapters on the *elements themselves*, but starting with *metals in group order*. Then follow chapters on *compounds of non-metals* starting with modern groups 17 and 16. The huge Group 16 section (300/630 pages) deals systematically with all oxy and hydroxyl compounds. The penultimate chapter draws together threads, survey-ing trends in numerical properties of elements *across* periods, then dealing with Mendeleev's use of periodicity for prediction of new elements and atomic weight corrections. Recent manufacturing processes are relegated to a final section. This leading professor acknowledged the central role of the periodic system, writing, "the following table is named the periodic table." [58] Note his use of a large, bold font.

5.3 When Did Books Devote More to Periodic Ideas?

More than thirty books for academic chemists were analyzed by decades to assess the percentage of pages devoted to periodicity. Books were given a point score (0, 1, 2, or 3) according to whether the percentage was 0, <1, 1–5, or >5, respectively.

This coarse analysis suggests that by the 1890s, coverage of periodicity in books

Books with Decade	No points	Some points	Total points	Average points/book
71–80	4	3	3	0.4
81–90	2	5	8	1.1
91–00	0	6	10	1.**7**
01–10	0	2	3–4	1.5–2.0
11–20	0	10	16	1.6

had reached a plateau. Two contemporary British professors' opinions of when periodic ideas were mostly accepted were stated earlier in this chapter as 1886 Carnelley and 1894 Travers. The appendices in Brush's article were restricted to between 1871 and 1890. They indicate whether prediction was made of new ele-ments. Using five-year periods and deleting, probably incompletely, identified US sources, the fraction of books with prediction increased 0.0, 0.05, 0.25, and 0.8— that is, 80 percent of books published between 1886 and 1890 had predictions. Analysis of the corresponding data for journals gave an earlier date, probably because journal articles are shorter but also because many of their writers worked in periodicity and thus wanted to publish.

5.4 A New Role?

Some names (Armstrong, Pattison Muir, Thompson, and Watts) have been men-tioned more than once. All but Pattison Muir were Royal Society Fellows and most were not professors but rather had editorial or secretarial positions with journals or chemical societies. They, and perhaps Groves,[59] represent the emergence of chem-ists who focus on writing rather than research.

6. FOR GENERAL READERS

6.1 Books

Very few books for the generalist were found. "Periodic law" was first met in Bernays's short book,[60] from an unexpected publisher. The book is without chemical equations, but its penultimate paragraph mentions Newlands, Mendeleev, and the periodic law.

Tilden's book of chemical history[61] resulted from a lecture course given for working men at the Royal School of Mines to mark Queen Victoria's Diamond Jubilee. Chapter 4 of the book summarized the work of Döbereiner, Odling, Newlands, Mendeleev, and Lothar Meyer.

6.2 *Encyclopaedia Britannica*

Since 1770 this multivolume publication has provided scholarly information, written by specialists, particularly for academic generalists. The ninth edition was published over twenty-four years, so that its supplement was also the tenth edition. The *Chemistry, Inorganic* entry in the ninth edition was written by Henry Armstrong.[62] It had a considerable amount of detail, including discussion of similar elements grouped together. The final page, headed *Periodic Relation of the Elements*, includes "Hence the whole of the elements may be arranged in a number of groups, each group consisting of members of the same natural family following each other in the same order." It continues with the formulae of equivalent compounds and states that the periodic character is especially evident with physical properties. The final paragraph reads, "The establishment of the Periodic Law may truly be said to mark a line in chemical science, and we anticipate that its application and extension will be fraught with most interesting consequences." This long article, published in the same year as de Boisbaudran's article about gallium's discovery, may have had the periodicity section added at the end. Armstrong certainly recognized its importance, as his tenth edition entry started by quoting that final paragraph.[63]

After praising Mendeleev's work for its classification of elements, prediction of elements to be discovered, and amendment of atomic weights, Armstrong suggested there may be a special reason for the Te-I atomic weight inversion. However, he then proposed an alternative table to that of Mendeleev, with a space for every integral atomic weight so that there are many spaces unfilled by elements. The inert gases were considered diatomic, hence halving their atomic weights. The general reader could learn about periodicity, alas with mistakes, as the alternative table was rejected by professional chemists.

6.3 Newspapers

There was very little the non-scientist could casually learn in the press about science until nearly 1900. *The Times Digital Archive*'s first mention of "periodic table" was in the Special Correspondent's two-hundred-word report of the BA1897 annual meeting presidential address. During the next two decades there were five articles of similar length, mostly from comparable meetings. Overall, sparse coverage was given, even in this leading daily newspaper.

Ramsay's BA presidential address (August 31, 1911) at Portsmouth ranged widely over the current science scene. The page-long report, considerably larger than the sum of the above six articles, included the statement: "With the help of this arrangement Mendeléef predicted the existence of unknown elements under the names of *eka*-boron, *eka*-aluminum and *eka*-silicon[,] since named scandium, gallium and germanium by their discovers Cleve, Lecoq de Boisbaudran and Winkler." It had taken forty years for the generalist to be able to read about periodicity. In the 1920s science coverage increased, with nine more articles mentioning the periodic table by, a new player, the *Science Correspondent*.

Ramsay also wrote in an American journal,[64] which is of special interest as it mentioned *periodic table* seven times as a way of systematizing properties of elements. Ramsay pointed out that the air gases, from their atomic weights, fit between Group VII (including hydrogen) and Group I. Furthermore, he saw that because of their lack of electrical polarity and chemical reactivity, they form a connecting link between the two, thus bridging the gap between the electropositive and electronegative elements. The reason for this perceptive comment followed with knowledge of electronic structures. Overall, there was little found from examined sources for the general reader until 1920. Of five British Nobel laureates in Chemistry until the 1920s (William Ramsay (1904), Ernest Rutherford (1908), Frederick Soddy (1921), Francis William Aston (1922), and Arthur Harden (1929)), Ramsay wrote most for the public.

7. SCHOOLS AND UNIVERSITIES

7.1 The British Education Pattern

The periodic law is a subset of chemistry, itself a subset of science education. First, the British education system will be briefly explained, since it is unreasonable to expect all readers to be aware of its special features and its changes over the years.[65]

Approximately 10 percent of school pupils are privately educated at fee-paying schools, confusingly often called public schools, whose teachers in the late nineteenth century came almost entirely from Oxford and Cambridge (Oxbridge), two ancient universities still today globally significant. Between 1870 and 1920 many "red-brick" universities were established in large cities like Manchester and Leeds. Elementary education, for ages five through eleven, focused on the three Rs, Reading, wRiting and a'Rithmetic, while compulsory education continued to age fourteen and, optionally, beyond. Science was seen as less important than classics (Latin and Greek), particularly at private schools and Oxbridge, which fostered their mutual interdependence.

7.2 Different Schools and Different Examination Boards

Edward Frankland (1825–99) introduced summer schools for chemistry teachers at the Royal School of Science in London in 1869.[66] A 120 teachers attended this course, designed to instruct in practical skills, as described by Frankland's assistant George Chaloner.[67] An 1886 advertisement in *Chemical News* referred to a "Summer Course of Lectures and Laboratory teaching of Chemistry (or an introduction for it), its Technical Applications by Professor Armstrong FRS. The course will extend for 2 weeks, 10–5 daily, Saturdays excepted, starting Monday July 5th. Fee £2." It is difficult to know how much, if any, theory was covered in the lectures, except from two later references.[68] Ramsay urged chemistry teachers to try out Lothar Meyer's lecture method based on the periodic arrangement. A Woolwich Artillery College teacher added that it was an excellent plan for non-beginners, and it was attempted for teachers attending the summer schools as early as 1879.

Science was at that time seen as less important than classics, English, and mathematics. The London (state) Schools Board introduced chemistry examinations in approximately 1883. There were 198 chemistry candidates, compared with 3,113 taking algebra. Chemistry and physics were finally put on a par with classics, English, and foreign language in 1904, as a result of the 1902 Education Act.

Clifton College (CC) in Bristol stands out as a forerunner in science education, despite having won less Oxbridge scholarships in science than in classics, though not to the extent of other elite private schools. The caliber of the CC staff is clear in that by 1910 seven former science masters, including William Tilden, had left to become heads of university departments. Between 1867 and 1869 three laboratories were built and in 1869, the year of Mendeleev's *Zeitschrift* paper, CC Science Society held its first meeting.[69] The photographs below (Fig.4.4) show one of these laboratories with a blackboard *displaying a periodic table* top left which part has been greatly enlarged in the second image. Three features of the table are of special historical interest. (i)The table is headed periodic *Law* not table. (ii) It is attributed to both Newlands and Mendeléeff, showing an English view of the table's initial recognition. (iii) The absence of the noble gases. The latter two points suggest a date between 1886 and 1894.

7.3 Schoolbooks

Shenstone's book had a chart[70] showing the periodic system of the classification of the elements and also referred to the significance of discovery of predicted elements and raised the possibility of inclusion of air gases.

A more historical view was provided by Fisher,[71] whose book was aimed at the Oxford Local certificate and university medical school entry. He mentioned periodicity involving Newlands, then went on to say, "The elements Gallium 70 and Germanium 72 were unknown when the table was first drawn up, but their probable existence was predicted by the distinguished Russian chemist Mendelejeff, and he also foretold their general characters and atomic weights with what afterwards

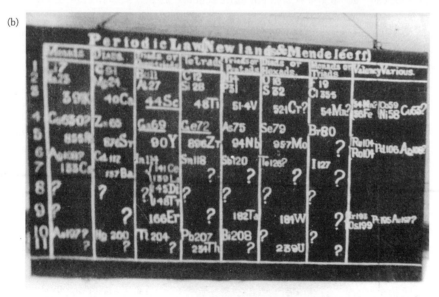

Figure 4.4 Clifton College laboratory c. 1890, with top left a periodic table and top right a valency list (see the close-up photo of the table). Courtesy of Clifton College archives.

proved a near approach to reality." He expected that other gaps would be filled with other new elements.

The only book found written by state-school teachers that mentioned periodicity was by Burnell & Dicks.[72] It discussed different methods of classifying elements (metal/non-metal, valency) and then advised the reader to use the periodic classification as a guide to learning. The topic was backed by questions like "In what way does valency vary periodically with atomic weight?" This was a book ahead of its time.

At the edge of the investigated period, the prolific author Holmyard[73] (also CC) detailed Mendeleev's eight points about periodicity, originally from *Zeitschrift*. However, the book did not merely mention the periodic law but arranged its chapters sequentially by groups, starting with 0 (the inert gases). Nevertheless, most books published in the next twenty years still ignored periodicity, as it was not in the School Certificate examination for pupils age sixteen.

7.4 Curriculum Concerns

Considerable concern about school science education at both elementary and secondary levels is apparent from the discussions at annual BA meetings and reports from the Association of Public Schools Science Masters (1907), as well as the Board of Education about the equivalent state schools (1911), which centered on broad curriculum issues like the relative time allocation for different subjects.

In 1919 the government appointed a committee under physics Nobel laureate Joseph J. Thomson to consider the earlier reports and to give general principles for future courses.[74] It concluded that the course for twelve- to sixteen-year-olds should be physics and chemistry, including some study of plant and animal life and directed to objects and experiences from life. The metric system should also be adopted. For the minority of children still at school, aged sixteen to eighteen, specialist subjects should be covered during 50 to 65 percent of the time, though scientists should nevertheless continue with some literary study and vice versa; indeed, advanced scientists should have a reading knowledge of French or German.

7.5 2010 Syllabus

Compare the above with the AQA chemistry examination syllabus for today's sixteen-year-old pupils.

- to explain how attempts to classify elements in a systematic way, including those of Newlands and Mendeleev, have led through the growth of chemical knowledge to the modern periodic table.
- to explain why scientists regarded a periodic table of the elements first as a curiosity, then as a useful tool and finally as an important summary of the structure of atoms.

- Newlands, and then Mendeleev, attempted to classify the elements by arranging them in order of their atomic weights. The list can be arranged in a table so that elements with similar properties are in columns, known as Groups. The table is called a periodic table because similar properties occur at regular intervals.
- The early periodic tables were incomplete and some elements were placed in inappropriate Groups if the strict order of atomic weights was followed. Mendeleev overcame some of the problems by leaving gaps for elements that he thought had not been discovered.

The syllabus indicates a great acceptance of the periodic table in considerable detail!

7.6 Examinations

Brush gauged acceptance mainly by studying books and periodicals. Course details can be discerned by reading the syllabus, *if* available. An alternative approach studies past examination papers, which show in more detail the information expected. This focus can be different in emphasis from the textbook. Mark schemes showing how marks were awarded rarely exist. There will be questions demanding direct recall but also explanation, comparison, or construction of an argument. The practical problems of this investigation are the general unavailability of questions from more than a century ago and the narrow field being researched.

7.6.1 School-University Interface

No evidence was found for use of periodicity up to age sixteen, hence investigation must use the next examination hurdle that the pupils scale to reach university. They apply and expand syllabuses, thus offering evidence of how periodicity is regarded.

7.6.2 Oxford Entrance

Future students, across all subjects, sat an examination which included a 'general paper' with including an essay having a pithy title like "Moderation is good." At Oxbridge universities, entry examination papers were set by the component colleges rather than by subject departments. In 1899 only four Oxford colleges (Magdalen, Balliol, Christ Church, and Trinity[75]) had their own basic laboratories and set the entrance scholarship papers, including the following questions selected for their periodic table context. The data below shows how the number of such colleges, out of approximately twenty-five, increased gradually.

1899	1910	1917	1918	1922
4	5	7	8	10

1895. "Mendeleef predicted that an element now known as gallium would be discovered by spectral analysis and atomic weight about 70. Give an account of the reasoning involved." This question requires factual chemical knowledge and an ability to explain its relevance. The examiner accepts the periodic law and focuses on a key discovery to this acceptance.

1910. "Discuss the position of (a) copper (b) manganese (c) argon in the periodic classification of the elements."

1923. "Give an account of the chemistry of either arsenic or silicon with respect to their position in the periodic table." For both these questions the candidate needs knowledge to process based upon an appreciation of trends both across a period and down a group.

About every other year from 1900 onward there was a question about the work of one or two eighteenth- or nineteenth-century chemists such as Cavendish, Lavoisier, Dalton, Davy, and Faraday. Other historical questions included "Describe, and explain, the principle of the spectroscope. How has this led to the discovery of new elements and to a knowledge of the sun and stars?," which links astronomy with the discovery of the periodic system. Another question, from 1918, was, "In what respect has the discovery of radioactivity modified our conception of the chemical atom?" There were, however, only a few historical/philosophical questions, with most questions being more factual, such as, "How does ammonia react with chlorine, mercuric chloride, potassium, ethyl iodide, ethyl oxalate?"[76]

7.6.3 Entrance to London Colleges

Entrance papers were set by the various colleges in London, which later formed London University. Various authors, often examiners, wrote booklets that quoted the actual question followed by specimen answers. Among the few questions related to classification were: 1888: "The molecules of most elementary substances contain 2 atoms. Name the exceptions"; 1889: "Group the following elements according to their valency or atomicity." The question then listed some elements. It did not mention periodicity but appears to gives a meaning to atomicity different from today's usage.

More closely related to periodicity was this 1890 question: "What is the 'specific heat' of an element? How is it determined, and what value is it to chemical classification?" The specimen answer used data for thallium to help place it in the alkali metals rather than the lead group! Specific heat was used with Dulong and Petit's Law to help determine atomic weights, which link to periodicity. With hindsight, one realizes that thallium is a hard-to-place element, as neither given answers is correct. Intriguingly, the examination was given from 7 to 10pm, so candidates were encouraged to arrive early to secure seats near the gaslights needed later in the May evenings![77]

Overall, apart from Oxbridge, little evidence has been found of periodicity being examined for university entrance despite the encouragement given by Ramsay and others to teachers. Was it considered to be relevant only to the brightest pupils?

7.6.4 For University Students

Questions set for coursework at Cambridge in c.1884 included: "Describe the elements arsenic, antimony and bismuth in respect of their behaviour as metals and the behaviour of their compounds as metallic compounds" and "Give a sketch of the chief states of the oxides of gallium and indium, particularly insofar as they bear upon the question of the place of each of the metals in classification." The use of "place" and "classification" links with the periodic law, though it is not specifically mentioned. The influence of Pattison Muir, then at Cambridge, an early supporter of periodicity whose research focused on bismuth, is also possible.[78]

Six years later, Elliott Steel's substantial book[79] of a thousand questions set by fifteen different examining authorities had none about periodicity, but he devised three given below.

1. "Describe some family of the chemical elements showing the gradation of the chemical and physical properties."
2. "On what grounds did Mendeleev foretell the existence of scandium (*eka*-boron)? Explain the principles according to which he described its properties before the body had been actually obtained." (*Note his use of body rather than element.*)[80]
3. (*A more advanced question.*) "Who predicted the existence of gallium and what did he call it? Did the properties he assigned to it agree with the properties it was found to possess?"

Although the book was published after the discovery of germanium, the lack of past examination questions implies examiners had not yet seen periodicity as important, so the author needed to redress this absence.

7.6.5 1898 Cambridge Tripos Part I. (at the End of Year 1) Included the Questions Below.[81]

1. "From ordinary air how would you prepare pure specimens of each of its pure constituents?" (*Question was perhaps set as a link to Ramsay's recent work on Argon.*)

2. "Chromium, tungsten, uranium and molybdenum are classified together in Mendeleef's arrangement of the elements; discuss the various reasons for and against this grouping. What is the position of manganese with respect to the other elements?" This question is of interest because it dealt with four elements considered to be in the same group until c.1950, when Th-Pa-U were understood to be actinides and not transition metals.

A compulsory question from the 1906 London Intermediate Science Pass degree gave two extracts, one in German, one in French. "Translate into good English and illustrate its statements by taking 3 groups from Mendeleeff's table as examples. Nous venons de considérer le nouveau système de poids atomique comme fournissant des éléments nouveaux à la classification des corps simples." Competence in a foreign language was required even for a pass degree! Today, British scientists have an advantage that English has become the global science language.

Periodicity was now accepted as a classification principle to be applied in occasional questions by university students, sometimes considered in an historical context. It would await the elucidation of atomic structure before it was seen as more important. Today it is met by most school pupils age sixteen.

8. CONCLUSIONS

Three elements Mendeleev had predicted were discovered by European chemists. Two Britons, Newlands and Odling, are among six originators of the periodic system identified by both Scerri and van Spronsen.[82] In the second half of the research period, 1895–1920, several very significant discoveries were made by other British scientists including Ramsay, Thomson, Soddy, and Moseley, mostly future Nobel Prize winners.

Brush[83] concluded that by the late 1880s most textbooks published in the United States and Great Britain discussed the periodic law to some extent, which is similar to my analysis of the percentage of books devoted to periodicity and the views of Carnelley and Travers about general acceptance of the periodic law. These are comparable conclusions but using somewhat different criteria.

Examinations are a further way to gauge acceptance of periodicity as to how and when it appears. Until 1920 it featured only occasionally in some university examinations and certainly not as a central theme of inorganic chemistry.

ACKNOWLEDGMENTS

I am grateful for the patience shown by librarians at Cambridge, Edinburgh, London, Manchester, Birmingham, and Oxford Universities, as well as the Royal Society, and RSC. I also thank reviewers for their considered criticisms freely given.

BRITISH ACRONYMS

BA   British Association (for the Advancement of Science), now BSA (British Science Association)

FRS   Fellow of the Royal Society

RSC   Royal Society of Chemistry, formed after The Chemical Society (CS) joined with the Institute of Chemistry

NOTES

1. Stephen G. Brush, "The Reception of Mendeleev's Periodic Law in America and Britain," *Isis* **87** (1996): 595–628.
2. Dmitrii Mendeléeff, "*The Periodic Law of the Chemical Elements* (Faraday Lecture of Mendeleev),"*Journal of the Chemical Society*, **55** (1889): 634–656.
3. The data analyzed only refers to those books on their database, so the absolute values will be greater but the size sequence is likely to be correct.

4. Clara De Milt, "The Congress as Karlsruhe," *Journal of Chemical Education*, **28** (1951): 421–425 lists congress attendees.

5. For further details, see Eric R. Scerri, *The Periodic Table, Its Story and Its Significance*, (New York: Oxford University Press, 2007): 72–82 and Jan van Spronsen, *The Periodic System of Chemical Elements: A History of the First Hundred Years* (Amsterdam: Elsevier, 1969), 102–112.

6. John A. R. Newlands, "On the Law of Octaves," *Chemical News* **12** (1865): 83.

7. Newlands, "[Extract from report of meeting of the Chemical Society, March 1st, 1866]" *Chemical News* **13** (1866): 113.

8. Newlands, "[Extract from report of meeting of the Chemical Society, June 19, 1873]," *Chemical News* **27** (1873): 318.

9. For details of Odling's work, see Scerri, *The Periodic Table* (note 5), 82–92 and van Spronsen, *The Periodic System,* (note 5), 112–116.

10. William Odling, "On the Proportional Numbers of the Elements," *Quarterly Journal of Science*, **1**, (1864): 642–648.

11. Dmitrii Mendelejeff, *"Ueber die Beziehungen der Eigenschaften zu den Atomgewichten der Elemente. Zeitschrift für Chemie"* **12** (1869): 405.

12. Henry Watts (1815–1884), specialized in chemistry writing. He was editor of the *Chemical Society Journal* and later the Society's librarian. Much British written material would be seen by him and appear in successive editions of the eponymous dictionary.

13. William Odling, "Atomic Weights," *Watts, Dictionary of Chemistry*: 2nd ed. vol. I., (London: Longmans Green, 1879–82): 452–473.

14. Odling, Ibid.. 2nd ed., vol. III,: 957–976. "Metals, Atomic Weights and Classification of" was referring to *Annalen der Chemie und Pharmacie,* (8, supp. 1871): 133–229, readable in English in Jensen, *Mendeleev on the Periodic Law, Selected Writings, 1869–1905* (New York: Dover, 2002): 38–109. Odling's elements table had dashes ready for Ga, Sc, and Ge.

15. Paul-Emile Lecoq de Boisbaudran, "Chemical and Spectroscopic Character of the New Metal Gallium, Discovered in the Blende of the Mine of Pierrefitte in the Valley of Argeles in the Pyrenees," *Chemical News* **32** (1875): 159; idem, "On certain properties of Gallium," *Chemical News* **32** (1875): 294.

16. In 1876 Matthew Moncrieff Pattison Muir was at Owens College, Manchester, but in 1877 he moved to Cambridge, where he taught chemistry and was in charge of the Gonville and Caius college laboratory. His continued support for Mendeleev's periodic classification will become apparent later.

17. Pattison Muir, "Remarks on the Discovery of Gallium," *Philosophical Magazine* supp. **1** (1876): 542.

18. Pattison Muir, "Chemical Classifications," *Philosophical Magazine* **4** (1877): 265–273.

19. George Fownes, *Manual of Chemistry, theoretical and practical* (London: Churchill, 1877) 12th ed. vol. 1. Fownes (1815–1849) was professor of Practical Chemistry at University College, London, but died young. His book continued under various editors (e.g., Henry Bence Jones and Augustus Hofmann). Fownes's former assistant Henry Watts revised editions 10–13.

20. Ibid., 264–268. *Relation between Atomic Weight and Quantivalence.* Quantivalent, quadrivalence were used for valency.

21. Monad, dyad . . . of group valency one, two . . .

22. William A. Miller, *Elements of Chemistry. Pt 2. Inorganic Chemistry.* (London: Longmans Green, Reader & Dyer 1878) 6th ed. Revised by Charles Groves. Groves was secretary of the Institute of Chemistry. He worked with Herbert McLeod as co-editor of Part 1.

23. Ibid., 550–552.
24. Ibid., 908.
25. Ibid., 974–977.
26. *Proceedings of the Royal Society*, **34** (1883): 329.
27. John Newlands, *The Discovery of the Periodic Law and on Relations between Atomic Weights* (London: Spon, 1884). He quoted Mendeleev from *Chemical News*, **43** (1881): 15 and Odling from *Pharm. Journal*, (August 25, 1877): 144.
28. Thomas Carnelley, *Nature* **31** (1885): 539 and *Chemical News*, over several weeks starting **53** (1886): 157.
29. *Proceedings of the Royal Society* **43** (1887): 195.
30. Dimitrii Mendeleev, (note 2).
31. Armstrong mentioned Periodic Law in *Encyclopaedia Britannica*, 9th ed., in its Chemistry entry.
32. Later, from further measurements, Brauner changed his opinion.
33. Michael D. Gordin, *A Well-Ordered Thing, Dmitrii Mendeleev and the Shadow of the Periodic Table* (New York: Basic Books, 2004), 199–203.
34. Translation by a classics colleague from Latin in *Cambridge University Reporter*, (July 4, 1894). All orations at Oxford and Cambridge about honorands continue today to be in Latin.
35. Dmitrii Mendeleev, "Professor Mendeléeff on argon," *Nature*, **51** (1895): 543. Also in Jensen, (note 14): 189.
36. A photo hangs in Mendeleev's study of four supporters of the Periodic Law, namely the three discoverers of predicted elements and his colleague Bohuslav Brauner. I once read that Ramsay was added as a fifth supporter.
37. William Ramsay, "An Undiscovered Gas," *Nature* **56** (1897): 378–382.
38. Morris Travers, *Discovery of the Rare Gases* (London: Arnold, 1928), 78.
39. Frederick Soddy, *Chemistry of the Radio-elements, Pt I* (London: Longmans Green, 1911).
40. Soddy, *Chemistry of the Radio-elements … and the Periodic Law, Pt II*. (London: Longmans Green, 1914), 1–9.
41. His father and both grandfathers were all Royal Society Fellows, probably uniquely so.
42. Henry Moseley, "The High Frequency Spectra of the Elements Part I and Part II," *Philosophical Magazine* **26** (1913): 1024ff, and **27** (1914): 703ff.
43. Maurice de Broglie, *Scientia* **27** (1920): 105. See also J. L. Heilbron, *H G J Moseley* (Berkeley: University of California Press, 1974): Flyleaf.
44. Edward G. Mazurs, *Graphic Representations of the Periodic System during One Hundred Years*, 2nd ed. (Alabama 34586: University of Alabama Press, 1974). Illustrations include H. Bassett, (1892): 38, T. Bayley, (1884): 84, T. Carnelley, (1886): 59, W. Crookes, (1898): 51, and J. Reynolds, (1886): 52.
45. Pattison Muir, *History of Chemical Theories and Laws* (London: Chapman & Hall, 1907): 353–378.
46. Pattison Muir was referring to *Annalen der Chemie und Pharmacie* **8** Supp. (1871): 133–229, readable in English in Jensen (note 14): 38–109.
47. Thomas P. Hilditch, *A Concise History of Chemistry* (London: Methuen, 1911): 38–41.
48. Thomas M. Lowry, *Historical Introduction to Chemistry* (London: Macmillan, 1916).
49. Charles L. Bloxam, *Inorganic and Organic Chemistry with Experiments* (London: Churchill): 5th ed. 1890, 10th ed., 1910.
50. Henry E. Roscoe and Carl Schorlemmer, *A Treatise of Chemistry, vol. 2* (London: Macmillan, 1888).

51. Edward Frankland & Francis R. Japp, *Inorganic Chemistry*. (London: Churchill, 1884): 70–80. Both taught at London Royal School of Mines.

52. Frederic S. Kipping and William H. Perkin Jr., *Inorganic Chemistry* (London: Chambers, 1911), 713–729. These professors were brothers-in-law.

53. Joseph W. Mellor, *Modern Inorganic Chemistry* (London: Longmans Green, 1914), 807–815. An early book by a prolific author.

54. Pattison Muir, *Principles of Chemistry* (Cambridge: Cambridge University Press, 1884).

55. See the chapter on Russia of this book (editors).

56. Ibid. 2nd ed. (1889), 232–245.

57. William Ramsay, *A System of Inorganic Chemistry* (London: Churchill, 1891).

58. Ibid., 21.

59. On Groves, see note 22.

60. Albert J. Bernays, *Chemistry* (London: Society for the Propagation of Christian Knowledge, 1888).

61. William A. Tilden, *Short History of the Progress of Scientific Chemistry* (London: Longmans Green, 1899), 271–274. He also wrote Mendeleev's obituary, *Proceedings of the Royal Society* **84** (1911): xvii–xx.

62. *Encyclopaedia Britannica*, 9th ed., 467–544 (1876). NB. 9th ed. (1875–89) 24 vols. 10th ed. (1902-1903) 11 vols, 11th ed. (1910–11) 28 vols.

63. *Encyclopaedia Britannica*, 10th ed., vol. 26, 708ff. Note that the volume numbering continues from 9th ed.

64. William Ramsay, *Popular Science* (Oct. 1901): 581

65. Suggested further reading is David A. Layton, *Science for the People: The Origins of the School Curriculum in England* (London: Allen & Unwin, 1973) and Edgar W. Jenkins, *From Armstrong to Nuffield; studies in twentieth-century science education in England and Wales* (London: Murray 1972), particularly chapters 1 and 2. As implied by the titles, the Scottish education system is (somewhat) different. Armstrong is best known for his retirement career as an educationalist advocating the heuristic method.

66. See Colin A. Russell, *Edward Frankland, Chemistry, Controversy and Conspiracy in Victorian England* (Cambridge, New York, etc.: Cambridge University Press, 1996).

67. Edward Frankland, *How to Teach Chemistry*, Ed. G. Chaloner (London: Churchill, 1875).

68. William Ramsay, *Chemical News* **68** (1893): 173 and W. R. Hodgkinson, Ibid., 184.

69. See http://www.cliftoncollegeuk.com/ocs/history/scienceatclifton/, accessed November 16, 2013.

70. William A. Shenstone, *Elementary Chemistry for Schools and Colleges*, 5th ed., (London: Arnold, 1905), 150–154.

71. Walter W. Fisher, *Elementary Classbook of Chemistry*, 5th ed. (Oxford: Clarendon, 1905), 162.

72. Sidney W. Burnell & Arthur J. Dicks, *Chemistry* (London: Ralph, Holland, 1911), 247–254.

73. Eric J. Holmyard, *Inorganic Chemistry for Schools and College* (London: Arnold, 1922).

74. *Natural Science in Education.* 1918. HMSO (His Majesty's Stationery Office).

75. H. G. J. Moseley was a physics student at Trinity between 1906 and 1910, one of only seven physics students in his intake. He was disappointed to obtain a second class degree, but there were no firsts.

76. Various Oxford college entrance papers, 1891–1924.

77. W. Jerome Harrison. *Advanced Inorganic Chemistry* (London: Blackie, 1896).

78. By permission of the Master and Fellows of St. John's College, Cambridge.

79. Robert Elliott Steel, *Inorganic Chemistry* (London: Methuen, 1890).
80. This writer's brief comments on some questions appear thus (*italics*).
81. As note 78.
82. Scerri (note 9), chs 3–4 and van Spronsen (note 9), ch 5.
83. Brush (note 1), 617.

CHAPTER 5

⚬⋀⚬

Mendeleev's Periodic Classification and Law in French Chemistry Textbooks

BERNADETTE BENSAUDE VINCENT AND
ANTONIO GARCÍA BELMAR

The most striking feature of the diffusion of Mendeleev's system in France is that his great achievement prompted no real debates, no controversy among French academic chemists. It is not that his work was totally ignored. Rather, it was integrated as a *non-event* in the daily work focused on the discovery and characterization of chemical elements thanks to new techniques (spectroscopy, crystallization, and so on). In science journals Mendeleev's system attracted attention only insofar as it could lead to the discovery of new chemical elements.[1]

After briefly mentioning when and how Mendeleev's ideas were presented in French primary, secondary, and higher education chemistry textbooks and mentioned in official programs, we will try to understand the reasons for preferring alternative criteria for classification in chemistry textbooks. In addition to the explicit arguments advanced by those who mentioned Mendeleev's proposals, we will attempt to interpret the silence that most textbook authors kept. In a third section, we will symmetrically focus on the small group of chemists who promoted Mendeleev's periodic classification and try to disentangle their motivations and modes of appropriation. We will then conclude that, far from being a form of resistance to Mendeleev's specific system, the overall skepticism expressed in French chemistry textbooks was the expression of an enduring *statu quo* resulting from a long debate over the best chemical classification in educational milieus.

1. NO BIG DEAL

In September 1879 the Department of Haute-Marne organized at the Hôtel de la Préfecture de Chaumont an *exposition scolaire* aimed at exhibiting the innovative activities developed by teachers and students of local primary school institutions. It was intended to contribute to the reform of primary education following the trauma caused by the defeat in the war against Prussia. As one of its organizers claimed, "it is primary instruction, and its patriotic direction, which made the strength of our enemies. It should make ours."[2] It was in this context of educational reform and post-war tensions that we found the first reference to didactic use of Mendeleev's periodic system in France. A local newspaper, *L'Union de la Haute-Marne,* published the comments of an outraged anonymous visitor on a "tableau de chimie", signed by a teacher named Lesourd who arranged the chemical elements in alphabetical order and was awarded an "honorable mention."*"Que je vous plains! Mes pauvre senfants!"*claimed the anonymous visitor, who suggested the alternative following arrangement:

> If only the bodies were grouped in families, the synoptic table would possibly have a *raison d'être*. Consider, for example, Mendéleeff's classification, which allows to find a priori the thirty bodies considered as simple still to be discovered, and demonstrate through this table that the general properties are a function of atomic weights, then it is all right; the table will have its *raison d'être*.[3]

This first reference found in a local newspaper, away from Paris and as part of discussions on the reform of primary education, comes as a striking contrast to the conclusions drawn from our inquiry on textbooks. A survey of about one hundred chemistry textbooks published by seventy-nine French authors between 1870 and 1920 clearly suggests that most authors considered that both the periodic classification and the periodic law on which it was based were of a great scientific and philosophical interest but of no utility for didactic purposes.

Our survey shows that only seventeen authors decided to include references to Mendeleev's system in at least one of their chemistry textbooks. Mendeleev's views were exposed in eleven out of the nineteen university and high technical education treatises and textbooks, thirteen out of the sixty-five secondary schools chemistry textbooks, and one out of the nine textbooks written for future primary school teachers attending courses at the *écoles normales*. There was zero reference in the twelve primary school manuals consulted.

These figures suggest a number of remarkable features. The presence of Mendeleev's system increased according to the level of education: it is totally absent in primary schools texts, occasionally present in secondary education textbooks, and more frequent in higher education textbooks. To interpret such differences it is important to remember that primary and secondary education textbooks had to follow official national programs, which was not the case for higher education textbooks. While high school programs recommended explaining the reasons for the

distinction between metals and non-metals and for whatever system of classification adopted, they did not mention the periodic system until 1893, in connection
with prescription for using the atomic notation.[4] From 1893 onward, all textbooks,
whether they be new or reprinted, made at least brief mention of Mendeleev's
system.[5]

As higher education textbooks were not guided by a national curriculum,
they could be expected to more intimately mirror the personal choices of their
authors and more broadly the choices of the academic community. Let us take
a look at the responses to Mendeleev's announcement in academic journals. In
fact, Mendeleev's system was rarely discussed in academic journals. A survey of
thirteen French journals by LudmillaNekoval in 1994 listed 150 references to
Mendeleev's periodic system between 1874 and 1920.[6] Among them 10 percent
were abstracts or translations of papers published in foreign journals by foreign chemists, including Mendeleev himself. In addition, most of the references
occurred in popular or semi-popular journals rather than in academic journals
such as the *Comptes rendus de l'Académie des sciences* or the *Annales de chimie*.[7]
This imbalance is the most striking evidence that Mendeleev's system was a
non-event.

Similarly, chemistry textbooks highlighted the predictive character of
Mendeleev's system and presented it as an approximate law providing a useful
tool for the determination of atomic weights and a guide for future research.
Among others, Édouard Grimaux praised the periodic system for its capacity of
"predicting the existence of not yet isolated elements and pointing to how many
elements were still to be discovered."[8] This predictive power was also emphasized by Gabriel Chesneau (1859–1937), a professor at the National School of
Mines, who considered that Mendeleev's classification provided significant
services to chemistry because it prompted innumerable efforts to get more
precise determinations of atomic weights or to fill up the gaps in the series of
elements.[9]

The modest attention paid to Mendeleev's system does not proceed from any
indifference with regard to taxonomic issues. Quite the contrary. Classification
was a major issue discussed in French chemistry textbooks. Since the late eighteenth and early nineteenth centuries French authors of textbooks had been
comparing the various existing options to introduce order in the multitude of
chemical substances so as to organize the contents of their books.[10] An entire
chapter was usually dedicated to classificatory systems and the few references to
Mendeleev occurred in this context. Whereas academic papers seldom discussed
issues related to classification, textbook authors pondered the advantages and
drawbacks of various systems and desperately sought to justify their choice.
While this contrast confirms the standard view that the quest for a classification of chemical elements was primarily a didactic concern, it also requires that
Mendeleev's system be evaluated in the context of this long-standing debate
conducted by textbook authors about the respective merits of natural and artificial classifications(Figure 5.1).

2. A Long-Standing Controversy

The debate was framed in the early nineteenth century, as an alternative between the so-called "natural classifications" (based on all the characters of the substances to be classified) and "artificial classifications" based on one single character. A natural classification in chemistry would take into account the most numerous and most essential analogies, while an artificial classification would select or prioritize one single property and order the substances according to its variations. André-Marie Ampère's attempt to introduce a classification modeled after botanical and zoological classifications in chemistry was the standard reference for natural classifications.[11] Although this attempt raised a lot of criticisms, it was unanimously considered as the most satisfactory in its inspiration.[12] Nevertheless, Louis Jacques Thenard, author of a big treatise of chemistry, made the alternative choice of a classification retaining Jons Jacob Berzelius's major divide between non-metals and metals and ordering the simple substances in the latter category according to one single criteria: their affinity for oxygen.[13] As Thenard's treatise became an exemplar for all textbook writers in the 1820s and 1830s, his artificial classification was extremely influential and continuously revised and improved over the numerous reprints of Thenard's treatise.

In the 1840s Jean-Baptiste Dumas made an attempt to develop a natural classification of non-metals but never extended his attempt to the group of metals. He subsequently maintained Thenard's classification based on their oxygen affinity, which proved to be extremely useful for didactic purposes, although it was generally considered as "artificial," "arbitrary," and "irrational." Thus, a hybrid artificial/natural system came to prevail in the mid-nineteenth century, gradually and continuously updated according to new criteria such as "atomicity" or isomorphism. Despite being adopted by a majority of textbooks, however, these hybrid classifications never ceased to be described as imperfect and provisional. In the mid-century, renewed attempts at natural classifications improving on Ampère's seminal essay were published. And even those authors who adopted artificial classifications kept claiming that natural classifications were the ideal goal toward which all efforts should be directed.

In this context, Mendeleev's periodic classification being based on the unique criterion of increasing atomic weights could be perceived as one more attempt to establish an artificial classification, although it embraced all known simple bodies and predicted new elements (Figure 5.1). It was seen as a disguised artificial classification and prompted the usual reproaches addressed by the partisans of natural classifications to artificial ones: that it left "in the shadow certain similarities in favour of others rightly or wrongly considered as predominant."[14] Moreover, although the periodic system confirmed the "natural" character of well-established families of elements, it was not suitable for a "methodical exposition of facts," since it prompted illegitimate rapprochements. As Edmond Willm pointed out in 1888, "if it were adopted, it would lead to move closer elements which are too far by features of all kinds."

$$H = 1.$$

1	2	3	4	3	2	1			
Li 7,01	Gl 9,08	Bo 10,9	C 11,07	Az 14,01	O 15,06	Fl 19,06			
Na 22,00	Mg 23,01	Al 27,01	Si 23	P 30,00	S 31,08	Cl 35,37			
K 39,03	Ca 39,91	So 43,0?	Ti 50,23	V 51,1	Cr 52,45	Mn 51,8	Fe 55,88	Ni 58,0	Co 58,0
Cu 63,18	Zn 64,88	Ga 69,9	Ge 72,4	As 74,9	Se 78,87	Br 79,76			
Rb 85,2	Sr 87,3	Y 89,6	Zr 90,4	Nb 93,7	Mo 95,9	•	Ru 103,5	Rh 101,1	Pd 100,2
Ag 107,66	Cd 111,7	In 113,1	Sn 117,35	Sb 119,6	Te 127,7	I 126,51			
Cs 132,7	Ba 136,86	La 138,5	Ce 141,2	Di 145	•	•			
•	•	Yb 172,6	•	Ta 182	W 183,6	•	Os 195	Ir 192,5	Pt 191,3
Au 196,2	Hg 199,8	Tl 203,7	Pb 206,39	Bi 207,5	•	•			
•	•	•	Th 231,96	•	U 239,3	•			

Figure 5.1 In the appendix to his secondary school chemistry textbook, Paul Lugol included a "table called of the periods" as an interesting attempt to "overcome the somewhat artificial nature of the division of simple bodies into metalloids and metals" (Lugol, Paul, *Cours élémentaire de chimie: à l'usage des élèves de l'enseignement secondaire classique et des candidats au baccalauréat, ouvrage rédigé conformément au dernier programme officiel.* . ., Paris, Belin frères, 1898, p. 467).

Since Mendeleev's system was primarily discussed in high-education textbooks, the distinction between textbook and academic literature will be somewhat blurred in the following pages. Some authors were ready to excuse Mendeleev, maintaining that such difficulties were due to the limited knowledge acquired about the physical and chemical properties of a large number of elements as well as to the uncertainty of some values. Even the most positive supporters of Mendeleev's system insisted that the periodic law was not a natural law, but rather was simply "the index of a more general law that still escapes us," as Grimaux put it. Rodolphe Engel (1850–1916), a chemistry professor at Montpellier Science Faculty and author of secondary school textbooks, stressed the approximate character of the periodic law in order to excuse the defects of Mendeleev's table: "The periodic grouping of elements is the index of more general and more precise relations that still escape us, because of insufficient knowledge about the properties of elements."[15] In 1902 a textbook for *baccalauréat* students by P. Revoy claimed that the deficiencies observed in the various attempts at a classification taking into account all the properties of elements, including "celle du chimiste russe Mendeléef (1870)," were due to our imperfect knowledge of the properties of elements.[16] Thus, the general distrust for Mendeleev's system was motivated not by a rejection of the periodic law but rather by the impression that it was premature. It expressed an overall skepticism based on a deep conviction of the radical imperfection and limitation of chemical knowledge.

Additional critic voices, however, regarded the deficiencies of Mendeleev's classification as the proof that he had chosen the wrong principle of classification. According to Paul Schützenberger, a professor of mineral chemistry at the Collège de France, the classificatory principle used by Mendeleev —the increasing atomic weight values – was hardly applicable to all elements: "Elements with slightly different atomic weights were extremely dissimilar from a chemical point of view, or, on the contrary, could offer marked similarities."[17] For Schützenberger, such defects clearly indicated that "nature does not proceed through demarcated families." Still, Schützenberger conceded that Mendeleev's periodic table presented a "highly elevated philosophical value" because it points to a theory of matter. For Schützenberger the "periodic law" that Mendeleev's classification disclosed provided a "correlation between the chemical characters of elements, their atomic valence and the value of their atomic weight," which constituted a "link of solidarity among the different simple bodies and provides in this way a powerful support to the idea of the unity of matter."[18]

Here is a major irony of the reception of Mendeleev's periodic system. Since the periodic law rests on arithmetic relations between atomic weight values, many chemists considered the periodic system as inspired by a numeric vision of the universe. Armand Gautier, for instance, claimed: "According to Mendeleef, one can derive the major properties of each substance from the consideration of its atomic weight. Isn't it that Pythagoras had a vague intuition of the truth, when he claimed that numbers are the principle of everything?"[19] The emphasis on arithmetic relations induced associations between the periodic table and William Prout's hypothesis about a primary matter, alleged source of all chemical elements. Despite Mendeleev's strong and constant opposition to this hypothesis, many French chemists claimed that his work was an attempt to demonstrate the unity of matter. And strangely enough whether they advocated Prout's hypothesis or not, they ended up condemning Mendeleev's system. For instance, Marcellin Berthelot ventured a few remarks on Mendeleev's system in a historical volume on *Les Origines de l'alchimie* (1885).[20] After presenting the theoretical views of medieval alchemists on the composition of elements, he turned to nineteenth-century attempts at classifying chemical elements. His view of the periodic system was thus framed by the alchemical quest for the transmutation of elements. Mendeleev's system is presented as a "bold attempt, close to a chimera, to construct numeric series encompassing all simple substances to be discovered in the future." Far from presenting Mendeleev's table as a true "system," Berthelot insisted that it was no more than a collection of periodic series resulting from artificial and ingenuous arrangements based on arithmetic data rather than on experimental facts. Berthelot assumed that Mendeleev's parallel series resulted from "the absolute values of atomic weights and not from their periodic differences," so that the chemical analogies were included in his table "as a necessary side-effect."[21] The result was an artificial classification, a regular and approximate scheme that could be convenient (*commode*) but not really heuristic. Berthelot thus minimized the significance and novelty of Mendeleev's system:

> The relations between atomic weights or volumes and physical or chemical properties have been established a long time ago in chemistry and long before any distribution of elements in parallel series. Without absolutely excluding such conceptions, we should not overestimate the scientific value of such elastic frameworks; we should refrain from crediting them with past or future discoveries, to which they do not lead in reality in a precise and necessary manner.[22]

Finally, Berthelot's condemnation concerned not just the periodic law but all attempts at a natural classification, from Ampère to Dumas, Newlands, and Lothar Meyer. Berthelot's verdict embraced all classifications:

> We must conclude that apart from the ancient natural families of elements which have been recognized since long, we have got but artificial assemblies. The system of periodic series, just as the system of the multiples of hydrogen, so far did not provide any certain and definite rule for discovering either the simple substances recently discovered, or those that we do not know yet.[23]

Berthelot's skepticism about all tentative classifications was shared by his pupil Alfred Ditte. On the occasion of the centennial 1900 world exhibition in Paris, he published a broad panorama of one century of chemical classifications, which included the traditional division between metals and non-metals, classifications based on atomicity and a lengthy discussion of Mendeleev's system.[24]He argued that all successive attempts to achieve a natural classification failed because "both simple and complex bodies are just forms of a sole ponderable matter, these forms being characterized by their atomic or molecular weight and by the nature of their particular movement."[25] Ditte conceded that Mendeleev discovered "interesting relations between properties and atomic weights" and "felt confident enough to state a general law." But he listed all possible arguments against the generality of the periodic law, concluding on a skeptical tone: none of the attempts at a rational classification of simple substances has been successful, none of the principles upon which such classifications were based is satisfactory. Just as Schützenberger in 1880 was in favor of "multiple considerations, many often impossible to reconcile in a simple way," twenty years later Alfred Ditte was content to form "groups more or less natural, but incompletely defined and badly delimited."[26]

Despite a number of nuances and divergences, French chemists reached a consensual conclusion: Mendeleev's chemical system has significant limitations that made its use in education premature, if not impossible. As Maurice Hanriot stated in 1883, "although M. Mendelejeff's classification table is very illuminating in many respects, it can still be regarded as a first step towards a true classification." Mendeleev's achievement is not rejected for its intrinsic deficiencies. It is the state of the art in chemistry that made it impossible to establish a systematic classification like the one proposed by Mendeleev.

The pragmatic compromise that had prevailed since the 1840s was not displaced by Mendeleev's periodic system. Hybrid natural/artificial classifications were still largely used in French chemistry textbooks up to the first decades of the twentieth century, while the ideal natural classification remained in a far distant future (Figure 5.2). While this general attitude may partially explain why most chemistry authors obviated Mendeleev's ideas in their textbooks, it nevertheless raises a question: How is it that a few authors paid attention to an imperfect and unfeasible classificatory system?

3. MODES OF APPROPRIATION

Turning our attention to the small but significant group of authors who adopted Mendeleev's views, we will focus on two leading figures: Adolphe Wurtz, a professor at the Paris Medicine Faculty, and Édouard Grimaux, a professor at the École polytechnique. Why did they care for Mendeleev's classification and how did they appropriate it to their own projects?

It is tempting to think that the supporters of Mendeleev were less skeptical than their colleagues and ready to claim that the periodic system was closer to the perfection of natural classifications than the conventional hybrid system. Yet judging from Edmond Willm's four-volume chemistry treatise for medical students, the attention paid to Mendeleev's system did not proceed from a positive choice meant to emancipate chemistry from compromises and hybrid solutions. Willm maintained the old division between metalloids (organized according to a revised version of Dumas's natural classification) and metals still arranged according to the criterion of affinity elected by Thenard. Nevertheless, in both categories Willm rearranged the elements according to their atomicity so that his overall classification "does not deviate too much from that of Mr. Mendeleef." In trying to include atomicity, which he considered as still poorly defined, Willm claimed that he was able to take into account a maximum of analogies.

It is also tempting to assume that the large publicity given to the discovery of elements predicted by Mendeleev changed the attitude of French chemists and convinced them that Mendeleev was right in presenting the periodic law as a law of nature, an equivalent of Newton's law in the realm of chemical individuals. Again, this is a wrong inference. The confirmation of Mendeleev's prediction of eka-aluminium, eka-boron, and eka-silicon did not convince French chemistry authors of the validity and universality of Mendeleev's law. It was clear in France that Lecoq de Boisbaudran, the discoverer of Gallium in 1874, was not aware of Mendeleev's prediction of eka-aluminium. He himself had predicted the existence of a metal on the basis of spectral data and claimed that having heard about Mendeleev's prediction would have misled him since he would have worked on ammonia precipitates instead of ammonia solutions.[27]

The predictive power of Mendeleev's table was not the motivation of its French champion supporters. Adolphe Wurtz (1817–84) was among the first to spread the periodic system in France and Mendeleev acknowledged that he greatly contributed to its popularization.[28] Wurtz, co-organizer of the Karlsruhe Conference with August von Kekulé (1829–96) in 1860,[29] cared for Mendeleev's periodic classification because it was a vehicle for spreading the atomic weights system recommended at the Karlsruhe Conference.

Wurtz gave a detailed exposé of the periodic system in *La théorie atomique* [The Atomic Theory].[30] In the prologue of a posthumous edition of *La Théorie atomique*, Charles Friedel (1832–99) introduced it as "the clearest exposition of the atomic theory by the end of the nineteenth century."[31] In fact this volume has been extremely influential for spreading the system of atomic weights and the notion of atomicity.[32] Wurtz argued that the atomic theory and the periodic system were mutually reinforcing: on the one hand, Mendeleev's discovery would have been impossible without the system of atomic weights adopted at the Karlsruhe Conference (as Mendeleev himself argued); on the other hand, "the discoveries of the eminent Russian chemist" provided "a solid argument in favor of the new system of atomic weights."[33] Wurtz was less impressed by Mendeleev's predictions than by his power of synthesizing. The periodic system transformed an "immense collection of facts" into "a science able to classify and coordinate them." He consequently emphasized the distance between Mendeleev's system and earlier taxonomic attempts by Dumas, in particular. For Wurtz, the periodic classification distinguished itself because first, "it embraces for the first time, all chemical elements," and second, it includes "all physical and chemical properties." As he emphasized the dependence of many properties on the atomic weights of elements, Wurtz undoubtedly was the most inclined to consider it as a "natural classification." Yet the natural classification embracing all analogies remained an ideal, a goal out of reach.[34] Wurtz did not ignore the exceptions and deficiencies of the periodic law that he presented as just the approximation of a natural law:

> In a word, if it is true in general to claim that the properties of bodies endure periodic modifications according to increasing atomic weights, the law of these modifications escapes us and presumably this law is not simple; for, on the one hand, one observes that this increase is not regular, as the differences between the atomic weights of neighbor elements vary between relatively wide limits, without allowing us to discover regularities in these variations; on the other hand, one should acknowledge that the degradation of properties, or the more or less important

variation between the properties of neighbor elements, do not seem to depend on the value of the differences between their atomic weights. These are difficulties.[35]

In brief, Mendeleev did not bring about the final solution to the long-standing quest for a natural classification. And apparently he did not even bring about an optimal didactic tool, since Wurtz did not try to reorganize his *Leçons élémentaires de chimie moderne* [Elementary Lessons of Modern Chemistry], published in the same year (1879) according to Mendeleev's classification. He continued to use the same old classification that he employed in his previous lectures, without even mentioning *en passant* the periodic system.[36]

While Wurtz only used the periodic classification in the service of the atomic theory, his former students and disciples were the first to utilize it for chemistry teaching. Édouard Grimaux (1835–1900), Armand Gautier (1837–1920), Edmond Willm (1833–1910) and Maurice Hanriot (1854–1933), who supported Wurtz's atomic views, made all possible efforts in the 1880s to introduce it in the French educational system, both in high schools and universities. Even more than Wurtz, Grimaux emphasized the relation of mutual reinforcement between the atomic theory and the periodic classification.

> Mendeleev's classification was only possible through the use of atomic weights and could not be deduced from equivalents system. This pioneering classificatory system is a new evidence of the interests and the resources that contemporary atomic weights provided. [37]

Thus, the periodic system was introduced as part and parcel of a French battle for the atomic theory. When the atomic weight notation and system supported by Wurtz became compulsory in the official curricula for secondary education, a new generation of textbook authors clearly announced the adoption of the atomic theory in their titles. However, only a few of them included the chemical properties observed by a "Russian chemist" in the package of the atomic theory. And even in their secondary school and university textbooks, the old hybrid natural/artificial classification proved extremely resilient.

4. CONCLUSION: SILENCE, RESISTANCE, DELAY?

How are we to interpret such ways of dealing with Mendeleev's system? The strong link between the introduction of Mendeleev's periodic classification in French education and the battle for the introduction of atomism could lead us to the conclusion that the delayed adoption of Mendeleev's ideas in France was due to the long-standing opposition to atomism in this country. Such was the interpretation suggested by the chemist Georges Urbain (1872–1938) in an article published in 1934 for the centenary of Mendeleev's birth.[38] The overwhelming silence that most French chemistry authors kept about Mendeleev's ideas would thus be the expression of the rejection of modern theories under the influence of a prevailing positivism.

However, this would be a misleading interpretation based on a presentist view of the periodic system. For there was no evidence in the nineteenth century that the periodic system provided a useful didactic tool. Here points a paradox: While it is established that didactic purposes fostered the intensive search for a classification of chemical substances throughout the nineteenth century, it is important to emphasize that the periodic system did not come as "THE" solution to the problems of chemistry teachers. Mendeleev himself, who had formulated the periodic law in the process of writing a textbook, never reorganized his *Principles of Chemistry* in the subsequent editions according to the periodic classification.[39] In the case of French chemists, the periodic system did neither rival nor overthrow the established status quo adopted by textbook authors in the course of a long debate over the best classification. Despite its generally recognized artificial and arbitrary character, the distinction between metals and metalloids adopted in all French chemistry textbooks from the early decades of the nineteenth century became a self-evident organizing criterion. The sequence Metalloids-Metals-Organic compounds had the great advantage of dividing the course of chemistry in three almost equal sections. It facilitated the distribution of the contents of chemistry courses and was established in the official programs for all levels of the educational system. It also fits very well with the publishers' commercial strategies, since general chemistry textbooks could be divided into several independent volumes, making their purchase more affordable for students.

The consensus about a hybrid system combining Dumas's classification for metalloids with Thenard's classification for metals provided such a robust and stable basis that textbook authors could dispense with any justification of this choice in their introductory chapters. This attitude of inertia was encouraged by the assumption that a discussion about classificatory issues was irrelevant and even impossible without having previously studied the physical and chemical properties of simple and compound bodies. Chapters on classification consequently lost their position as introductory chapters in elementary textbooks and official programs. The issue of classification thus became a subject reserved for advanced students. The limited didactic value and the apparent complexity of Mendeleev's periodic classification was an additional reason not to include it—or to mention it in appendixes or additional paragraphs written in low characters that readers could easily overlook.

To sum up, most French chemistry textbook authors considered Mendeleev's periodic classification as a defective and unfeasible solution for a problem already solved in a more pragmatic and satisfactory way. More than a symptom of a strong anti-atomist resistance to the periodic classification, the lack of interest in Mendeleev's achievement could be the expression of the increasing autonomy of the educational sphere inhabited by a crowd of teachers and textbook writers who were mainly concerned with didactic efficiency. For them, the benefits of Mendeleev's system did not compensate for the loss of the "practical" division between metals and non-metals and nobody was ready to break with this long-standing tradition in chemistry teaching (Figure 5.2).

NOMENCLATURE

12. Corps simples. — Le tableau ci-joint contient les noms des principaux corps simples, les symboles par lesquels on les désigne dans l'écriture abrégée et leurs équivalents en poids. Dans la première partie contenant les métalloïdes on a ajouté l'équivalent en volume.

MÉTALLOIDES

Nom	Symbole	Équivalent en poids	Équivalent en volume	Nom	Symbole	Équivalent en poids	Équivalent en volumes
Oxygène	O	8	1	Azote	Az	14	2
Soufre	S	16	1	Phosphore	Ph	31	1
Sélénium	Se	39,5	1	Arsenic	As	75	1
Tellure	Te	64,5	1	Carbone	C	6	1?
Fluor	Fl			Bore	Bo	11	
Chlore	Cl	35,5	2	Silicium	Si	14	
Brôme	Br	80	2	Hydrogène	H	1	2
Iode	I	127	2				

MÉTAUX

Nom	Symbole	Éq.	Nom	Symbole	Éq.	Nom	Symbole	Éq.
Potassium	K	39	Fer	Fr	28	Cuivre	Cu	31,75
Sodium	Na	23	Nickel	Ni		Plomb	Pb	103,5
Lithium	Li		Cobalt	Co		Bismuth	Bi	208
Calcium	Ca	20	Chrome	Cr				
Strontium	Sr		Zinc	Zn	33	Mercure	Hg	100
Baryum	Ba		Cadmium	Cd		Palladium	Pd	
			Uranium	Ur		Argent	Ag	108
Magnésium	Mg	12				Platine	Pt	97,5
Manganèse	Mn	27,5	Etain	Sn	59	Iridium	Ir	96,5
Aluminium	Al	13,55	Antimoine	Sb	120	Or	Au	98,2

Figure 5.2 In 1888 Leduc kept grouping metalloids in Dumas's four "natural families" and metals in Thenard's six artificial sections (Leduc, Anatole, *Cours élémentaire de chimie: rédigé conformément au programme de 1885 pour la classe de rhétorique et le baccalauréat ès lettres* (2e édition), Vve E. Belin et fils, Paris, 1888, p. 9).

APPENDIX: LIST OF THE FRENCH CHEMISTRY TEXTBOOKS WITH REFERENCES TO MENDELEEV PERIODIC CLASSIFICATION

Basin, J. *Leçons de chimie, à l'usage des élèves de première (sciences)*, Paris, Nony, 1893.

Basin, J. *Leçons de chimie (chimie générale, chimie organique, analyse chimique): à l'usage des élèves de première (sciences) et des candidats au Baccalauéat de l'enseignement secondarie moderne (2em série),4em édition, par ... professeur agrégé au Lycée de Lille*, Paris, 1901.

Cadot, A. Bourgarel, Paul, *Leçons de chimie à l'usage des élèves des classes préparatoires aux écoles du gouvernement (programme de la commission interministérielle approuvé par arrêté du 26 juillet 1904)*, Paris, Garnier frères, 1908.

Chesneau, G.Lois générales de la chimie: introduction du cours de chimie générale professéà
l'École nationale des mines par, . . . Ingénieur en chef des mines, Paris, Libraire polytechnique C. Béranger, éditeur, 1899.

Copaux, Hippolyte Perpérot, H., Introduction a la chimie générale. Lois fondamentales de
l'atomisme et de l'affinité exposées a des chimistes débutants, par . . ., professeur à l'école
de physique et de chimie industrielles de la ville de Paris, Paris, Gauthier-Villars, 1919.

Engel, Rodolphe, Traitéélémentaire de chimie. Métaux. Chimie Organique et manipulations
d'analyse, à l'usage des candidats au Baccalauréat es sciences et au baccalauréat moderne, au certifiât d'études physiques, chimiques et naturelles a l'école centrale des arts
et manufactures et aux écoles du gouvernement, Paris,—J.-B. Baillière et fils, 1896.

Gautier, Armand, Cours de chimie, Paris, F. Savy, 1887–1892.

Gautier, Armand, Cours de chimie minérale, organique etbiologique, Paris, Masson,
1895–1897.

Gautier, Henri, Charpy, Georges, Leçons de chimie, à l'usage des élèves de mathématiques
spéciales (Quatrième éd., entièrement refondue, conforme au programme du 27 juillet
1904)/par . . . Paris, Gauthier-Villars et fils, 1905.

Grimaux, Édouard, Introduction à l'étude de la chimie, théories et notations chimiques: premières leçons du cours professéà l'École polytechnique,Dunod, Paris, 1883.

Haller, Albin; Muller, Paul Thiébaud Traitéélémentaire de chimie: à l'usage des candidats au
certificat d'aptitude des sciences physiques, chimiques et naturelles, et des candidats aux
baccalauréats scientifiques, Paris, G. Carré, 1896.

Joannis, Alexandre, Cours élémentaire de chimie: professéà la Faculté des sciences de Paris
pour les candidats au certificat d'études physiques, chimiques et naturelles (P. C. N.).
Paris, Baudry, 1897.

Joly, Alexandre, Cours élémentaire de chimie (notation atomique). Métaux. Chimie Organique,
premier fascicule,Paris, Hachette, 1894.

Joly, Alexandre, Éléments de chimie (notation atomique), rédigés conformément aux pro-
grammes officiels à l'usage des candidats aux Baccalauréats classiques, 4em édition,
par . . . Professeur adjoint la Faculté des sciences de Paris,Maitre de conférences a la
Ecole Normale supérieur, Paris, Hachette, 1895.

Joly, A.; Lespieau, R., Cours élémentaire de chimie . . . Métaux. Chimie organique. 4e édition
entièrement refondue . . ., Paris, Hachette, 1902.

Joly, A., Éléments de chimie, notation atomique, rédigés . . . pour la classe de philosophie et les candidats aux baccalauréats classiques (2e partie), par . . . 3e édition, Paris, Hachette, 1902.

Lugol, Paul, Cours élémentaire de chimie: à l'usage des élèves de l'enseignement secondaire classique et des candidats au baccalauréat, ouvrage rédigé conformément au dernier programme officiel . . . (2e édition, revue et corrigée), Paris, Belin frères, 1898.

Lugol, P. Compléments de chimie, à l'usage des classes de l'enseignement secondaire, par, . . .
Ouvrage rédigé conformément au programme officiel de 1902 . . . Second cycle, classe de
mathématiques A et B, Paris, Belin frères, 1904.

Malette, J., Éléments de physique et chimie, à l'usage du conducteur des ponts et chaussées et
des candidats à cet emploi, Paris, O. Doin, 1909.

Naquet, Alfred;Hanriot, Maurice, Principes de chimie fondée sur les théories modernes, par . . . professeurs agrégés a la Faculté de Médecine de Paris, París, F. Savy,
1883–1885.

Poiré, Paul, Tanquerey, A., Leçons de chimie, ouvrage rédigé conformément aux pro-
grammes . . . du 4 août 1905, 2 vol. (Bibliotheque des Ecoles normales) Paris, C.
Delagrave, 1906.

Pozzi-Escot, Marius Emmanuel, Traite élémentaire de physico-chimie: ou, Lois générales et
théories nouvelles des actions chimiques, à l'usage des chimistes, des biologistes et des
élèves des grandes écoles, Paris, C. Béranger, 1905.

Revoy, P. *Notions de chimie générale à l'usage des candidats aux baccalauréats d'ordre scientifique et aux écoles du gouvernement*, Paris, Nony, 1902.

Schützenberger, Paul, *Traité de chimie générale: comprenant les principales applications de la chimie aux sciences biologiques et aux arts industriels, par . . . professeur au Collège de France, 7 vol.*, Paris, Hachette, 1880–1894.

Schützenberger, Paul, *Leçons de chimie générale, professées au Collège de France pendant l'année 1895-96, par . . .; publiées par les soins de O. Boudouard, 1 vol. (VII-586 p.): fig.; in-8 . . .*, Paris, O. Doin, 1898.

Troost, Louis, *Traitéélémentaire de chimie (11e éd. rev. et corr.)*, Paris, Masson, 1895.

Troost, Louis, *Précis de chimie (34e édition)*, Paris, G. Masson, 1902.

Willm, Edmond; Hanriot, Maurice, *Traité de chimie minérale et organique*, 4 vol, Paris, G. Masson, 1888–1889.

Wurtz, Adolphe, *La théorie atomique*, Paris, G. Baillière, 1879.

NOTES

1. This prevalent concern can be inferred from the picks in the number of references that are always connected to the discovery of new elements: between 1876 and 1878: gallium and scandium; in the 1890s and 1900s: inert gases, rare earths, and radioactive elements.

2. Duponnois, *Exposition scolaire de la Haute-Marne, de 1879* [School Exposition of the Department of Haute-Marne], . . . Chaumont, impr. de Cavaniol, 1881, 1.

3. *L'Union de la Haute-Marne, journal politique, littéraire, administratif et d'annonces, paraissant les lundi, jeudi et samedi*, 1, octobre, 1879, 3.

4. Bruno Belhoste, *Les Sciences dans l'enseignement secondaire français* [Sciences in the Secondary Education in France], t. 1: *1789–1914*, Paris: INRP-Economica, 1995.

5. For example, the best-seller *Traité élémentaire de chimie* [Elementary Textbook of Chemistry] *by* Louis Troost, introduced Mendeleev in its eleventh edition, Paris, Masson, 1895.

6. Ludmilla Nekoval-Chikhaoui, *Diffusion de la classification périodique de Mendeleïev en France entre 1869 et 1934* [The Diffusion of Mendeleev's Periodic Classification in France between 1869 and 1934], Thèse de l'Université Paris Sud, 1994.

7. Among the 150 references, 21 occurrences in the *CRAS*, 17 in the *Bulletin de la Société chimique de Paris*, 16 in *Le radium*, 8 in the *Journal de physique*, 3 in *Revue générale de chimie pure et appliquée*, 1 in *Annales de chimie et de physique*.

8. Édouard Grimaux, *Introduction à l'étude de la chimie, théories et notations chimiques: premières leçons du cours professéà l'École polytechnique* [An Introduction to the Study of Chemistry, Theories and Chemical Notations: First Lessons of the Courses, taught in the ÉcolePolytechnique], Paris, Dunod, 1883, p. 117.

9. Gabriel Chesneau, *Lois générales de la chimie: introduction du cours de chimie générale professéà l'École nationale des mines* [General Laws of Chemistry: An Introduction to the Courses of General Chemistry, taught in the École National des Mines], Paris, Libraire polytechnique C. Béranger, éditeur, 1899, p. 57.

10. See A. Garcia Belmar, J.R. Bertomeu Sanchez, B. Bensaude Vincent, "The Power of Didactic Writings: French Chemistry Textbooks of the Nineteenth Century" (en coll avec) in David Kaiser (ed.) *Pedagogy and the Practice of Science: Historical and Contemporary Perspectives*, Cambridge, Mass., MIT Press, 2005, p. 219–251.

11. A-M. Ampère, "Essai d'une classification naturelle pour les corps simples,"*Annales de Chimie* (I) 1816: 120–125.

12. See José-Ramon Bertomeu, Antonio Garcia-Belmar, B. Bensaude-Vincent, "Looking for an order of things. Textbooks and chemical classifications in Nineteenth-Century France,"*Ambix*, 49, nov. 2002, 227–250.

13. Thénard, J. *Traité de chimieélémentaire, théorique et pratique* [A Textbook of Elementary Chemistry, Theoretical and Practical] (6th ed. 1834–36), Paris, Crochard, 1813–1816.

14. Paul Schützenberger, *Traité de chimie générale* [A Textbook of General Chemistry], Paris, Hachette, 1880-1894, vol. I, p. 324.

15. Rodolphe Engel, *Traitéélémentaire de chimie. Métaux. Chimie Organique et manipulations d'analyse, par . . .À l'usage des candidats au Baccalauréat ès sciences et au baccalauréat moderne, au certificat d'études physiques, chimiques et naturelles a l'école centrale des arts et manufactures et aux écoles du gouvernement* [An Elementary Textbook of Chemistry. Metals. Organic Chemistry and Manipulations of Analysis by . . . For the Use of Candidates of Baccalaureates of Science and of the Modern Baccalaureates, of Certificates for the Studies of Physics, Chemistry and Natural History in the ÉcoleCentrale des Arts et Manufactures and in the Écoles du Gouvernement], Paris, - J.-B. Baillière et fils (Paris), 1896, 48.

16. Revoy, P. *Notions de chimie générale à l'usage des candidats aux baccalauréats d'ordre scientifique et aux écoles du gouvernement*, Paris, Nony, 1902, 166.

17. Paul Schützenberger, *Traité de chimie générale: comprenant les principales applications de la chimie aux sciences biologiques et aux arts industriels* [A Textbook of General Chemistry, Including Major Applications of Chemistry to Biology and Industrial Technology], Paris, Hachette, 1880–1894, I, 324.

18. Ibid.

19. Armand Gautier, *Cours de chimie* [Courses of Chemistry]. Paris, F. Savy, 1887–1892, 43

20. Berthelot, *Les origines de l'alchimie* [The Origins of Alchemy], Paris, Steinhel, 1885, p. 302–313.

21. Ibid., p. 309.

22. Ibid., p. 311–312.

23. Ibid., p. 313.

24. Alfred Ditte, *Revue des cours scientifiques de la France et de l'étranger*, 46, N°20, 15 nov. 1890, 209–219. Ditte was a professor of chemistry at the Sorbonne and author of a *Traitéélémentaire de chimie fondée sur les principes de la thermochimie* [A Textbook of Chemistry, founded on the Principles of Thermochemistry] (Dunod, 1884) Ditte advocated a mechanical view of individual elements as he concluded that each substance was a step in a continuum on ponderable matter and that a natural classification should be based on the quantity of movement and mass.

25. Alfred Ditte, *Introduction à l'étude des métaux: Leçons professées à la Faculté des sciences* [An Introduction to the Studies of Metals: Lessons Taught in the Faculty of Sciences] (Paris, Société d'éditions scientifique, 1902), 473.

26. Ibid., 471.

27. Lecoq de Boisbaudran, "Sur quelquespropriétés du gallium, présenté par M. Wurtz," *ComptesRendues des Séances de l'Académie des Sciences* (Séance du lundi 6 décembre 1875), 1875, 2e semestre, T.81, n° 22, 1100–1105 (p.1104). Lecoq was a wine trader in Cognac who conducted research on spectral analysis. He published the volume *Les spectreslumineux* [The Emission Spectra], Paris, Gauthier-Villars, 1874.

28. *ArkhivMendeleeva: Avtobiograficheskiematerialy* [Mendeleev's Archive: autobiographical materials]. Vol. 1. (Leningrad: Leningrad State University, 1951), 53, quoted by Ludmilla Nekoval Nekoval (note 6), 78.Wurtz's book, first published in 1879, has been widely circulated until 1911 in France (ten editions) as well as in the United Kingdom (eight editions of the English translation to which should be added several American editions). It has been translated in German (1879), Spanish (1879), and Russian (1889).

29. See Mary Jo Nye, *The Question of the Atom: From the Karlsruhe Congress to the Solvay Conference, 1860–1911* (Los Angeles: Tomash, 1983). B. Bensaude-Vincent,

"Karlsruhe, septembre 1860: l'atome en congrès," *Relations internationales, "Les Congrès scientifiques internationaux,"* 62 (1990):149–169.

30. Adolphe Wurtz, *La théorie atomique* (Paris: F. Alcan, 1879), 112–126.
31. Adolphe Wurtz, *La théorie atomique, précédée d'une introduction sur la vie et les travaux de l'auteur par Ch. Friedel*, 4th edition (París: F. Alcan, 1886), lii.
32. See Alan J. Rocke, *Nationalizing Science. AdolpheWurtz and the Battle for French Chemistry*, (Cambridge, Mass. & London: The MIT Press, 2001), 301–331.
33. Wurtz (note 30), 126.
34. Ibid.
35. Ibid. p. 117.
36. Adolphe Wurtz, *Leçons élémentaires de chimie moderne*, 4e édition revue et augmentée (Paris: G. Masson, 1879). The first edition was published in 1867–68 (Paris: V. Masson et fils).
37. Grimaux (note 8), 119.
38. G. Urbain, "Comment les idées de Mendéléef ont été accueillies en France [How were Mendeleev's ideas accepted in France?],"*Revue scientifique*, 657 (1934), 657.
39. Authors on Russian in this book have different opinions on the structure of later editions of Mendeleev's *Principles of Chemistry*. See the chapter on Russia (editor).

PART III

Response in the Central
European Periphery

CHAPTER 6

cℵo

Nationalism and the Process of Reception and Appropriation of the Periodic System in Europe and the Czech Lands

SOŇA ŠTRBÁŇOVÁ

1. INTRODUCTION

The 1870s marked the onset of an exceptionally fruitful and dynamic period in the development of chemistry in the Czech Lands. University education and research in chemistry was taking place at several universities and technical universities, where the structure of the main chemical subjects developed gradually into organic, inorganic, analytical, physical, fermentation, and medical chemistry, just to mention the main specialties. At the same time, the process of the Czech National Revival led to the cultural, linguistic, social, and political emancipation of the modern Czech nation and stepwise almost entirely separated the linguistically Czech and German scientific communities in all their representations, including university education.[1] In Prague, the divided German and Czech Polytechnics (and later Technical Universities) existed since 1869, whereas the Charles-Ferdinand University split into its Czech and German counterparts only in the years 1882 and 1883. The chemical community was organized in several professional associations that also reflected the ethnic division of the scientific scene. The Society of Czech Chemists,[2] founded in 1866, had almost exclusively Czech membership, while a specialized German chemical association has never been created in the Czech Lands.[3]

This study deals with two closely intertwined themes: the reception of the periodic system in the Czech Lands and in Europe and the crucial role of the Czech chemist Bohuslav Brauner in this process. I am going to demonstrate a specific set of conditions that shaped the process of appropriation of this new scientific idea by not only scholarly argumentation, but also particular circumstances, in this case Slavic nationalism and Russophilia in the Czech society at the turn of the nineteenth century. The course of dissemination and reception of the periodic system also showed linkage to the linguistic emancipation of the Czech nation as reflected in the controversy over the Czech chemical terminology, where the periodic system served as argument to one party of the dispute. Thus, the process of acceptance of the periodic system can serve as a case study shedding more light on the process of nationalization of scientific knowledge and the dichotomy between nationalism and cosmopolitanism in the Czech science at the turn of the nineteenth century.

2. THE PERSONALITY OF BOHUSLAV BRAUNER

The personality of Bohuslav Brauner (1855–1935)[4] embodies the multiethnic and multicultural environment of the Habsburg Monarchy. He was predestined by his outstanding chemical education, parentage, social status, and prominent position of his family in the Czech society, to assume significant position among the Czech and European chemists. Some moments in his biography are a key to understanding his role in the acceptance of the periodic system.[5]

Noteworthy is Brauner's ancestry,[6] which had endowed Brauner with all embracing capabilities. His family background, especially from the maternal side, apparently influenced his professional orientation. Brauner's mother Augusta, née Neumann, was a sophisticated and educated woman. Her father Karl August Neumann (1771–1866) was the first professor of chemistry at the Prague Polytechnic, and her great-uncle Caspar Neumann (1683–1737), one of the great figures of the European science, was professor of chemistry in Berlin and friend of G. E. Stahl. B. Brauner's father, František August Brauner (1810–1880), was a lawyer who became, after 1848, one of the most influential Czech politicians. All of his children, not only the chemist Bohuslav, became visible personalities. B. Brauner's sister Zdenka (1858–1934) was an internationally recognized modern painter and the other sister Anna (1856–1930) married the French writer Élémir Bourges (1852–1925), one of the founders of the Académie Goncourt; his brother Vladimír (1853–1924) was a notable lawyer.

Several teachers influenced Brauner during his Prague university studies. The lectures of the Czech chemistry professor Vojtěch Šafařík (1829–1902), son of the well-known Slavist Pavel Josef Šafařík (1795–1861), pushed him to enroll in the Czech Technical University in 1873, where he pursued research into inorganic chemistry under the guidance of František Štolba (1839–1910), one of the first chemists who lectured in Czech. To obtain the doctorate, Brauner simultaneously signed up at the Prague Charles-Ferdinand University, where he studied under the German organic chemists Adolf Lieben (1836–1914) and Eduard Linnemann (1841–1886). However, Brauner himself considered his genuine mentor the physicist Ernst Mach (1838–1916), with whom he carried out some investigations on fluorescence. "E. Mach awakened . . . my old love for physics' sister science . . . Mach kept his sincere sympathy for me during long years and advised me always in a friendly manner" recalled Brauner in 1907.[7]

It was customary in Brauner's generation that young Czech chemists studied at least for some time abroad, especially in Germany, France, and England. Brauner spent the years 1878 and 1879 in Robert Bunsen's (1811–1899) laboratory in Heidelberg, where he expected to learn modern inorganic chemistry not accessible in the Czech Lands. In 1880 Brauner returned to Prague to obtain his doctorate at the Prague University. He soon left again to work with Sir Henry E. Roscoe (1833–1915) at the Owens College in Manchester. In 1882 Brauner came back to Prague to defend his habilitation thesis in 1883 at the newly established Czech University. Nevertheless, in spite of his *dozent* degree,[8] he had to wait for his appointment of associate professor until 1890 and for his full professorship until 1897. Since 1904 Brauner headed the newly established Department of General, Inorganic and Analytical Chemistry[9] of the Czech University in Prague until his retirement in 1925. Several foremost chemists avowed themselves to be pupils of Brauner, including the Nobel Prize winner Jaroslav Heyrovský (1890–1967), but in reality, only a few of them followed in his footsteps. His honors reflect his merits. Brauner was elected among others honorary member of the Czech, London, American, French, Polish and Russian chemical societies as well as member of the Czech Academy of Sciences and Arts and several foreign academies, was decorated with high Austrian, Russian, and Yugoslav orders, and became Chevalier de la Légion d'Honneur.

3. BRAUNER'S ENGAGEMENT FOR THE ACCEPTANCE OF THE PERIODIC SYSTEM

Chemical research attracted Brauner just in the period when the attention of European chemists started to turn toward Mendeleev's periodic system. Their interest was triggered when the French chemist Paul Émile Lecoq de Boisbaudran (1802–1912) in 1875 discovered gallium, whose existence and properties were predicted by Mendeleev in his periodic table. According to Brauner's testimony[10] of 1907, he as a young student of chemistry learned about the exciting finding in a Czech daily newspaper article in 1876 and afterward from the letters of Karel Otakar Čech, (1842–1895), a Czech chemist who worked in St. Petersburg.[11] Another impulse of Brauner's interest in the systemization of elements was acquaintance with Lothar Meyer's treatise *Die modernen Theorien der Chemie* in 1877.[12] On the other hand, Brauner in his draft letter to Mendeleev of February 1881 states that his first information about the periodic system stemmed from Lothar Meyer's treatise in 1877: "I arrived to the study of the periodic law of which I had not known anything before only at the beginning of 1877. My attention was first called by the 'Modernen Theorien der Chemie' and then I went on reading your original paper."[13] The main impetus behind his decision to devote his research to the periodic system Brauner describes as follows: "Finally I succeeded in the spring of 1877 in borrowing from the library the well-known eighth "supplement" of "Liebig's Annalen,"[14] where a fundamental description of the periodic system was presented. Reading of this article made an enormous impression on me, which can hardly be described; it opened the view of a new world and new fields of chemistry hitherto absolutely unknown to me. It was as if the scales fell from my eyes and I suddenly saw clearly the vast series of wonderful problems of general chemistry . . . When reading Mendělěěv's wonderful communication . . . I recognized soon the direction which I had to follow in my work. I fixed my Life's aim already that moment: it was the experimental research of the solution of problems connected with Mendělěěv's system and the most important of all seemed the solution of the following problem: What is the position on the so-called rare elements and especially those of the rare earths in Mendělěěv's system?"[15]

According to this personal testimony, Brauner, unlike many of the contemporary chemists, became without hesitation a convinced supporter of the periodic system and in 1877 started to publish and lecture about it. In his first informative article, which he wrote still as a student in 1877 for the journal of the Czech Chemical Society *Listy chemické* [Chemical Letters], he aimed to bring up to date the Czech chemists about the periodic system.[16] The paper was based on Brauner's two lectures on the periodic system read at the meetings of the Society of Czech Chemists on June 21 and 28, 1877, apparently the first detailed information on this topic offered to the Czech chemical community. Brauner's first scientific article in a foreign journal on Mendeleev's periodic system appeared a year later in the German *Berichte*[17] as his contribution to the discussion on the atomic weight of beryllium.

In 1878 Brauner finally decided to pursue the problem matter of periodicity, but he lacked specialized training in inorganic chemistry. By his own words he preferred Heidelberg to Uppsala, "where the periodic system had been denied by Nilson[18] and my defense of it was disregarded."[19] However, Brauner, who became Bunsen's (Robert Wilhelm Bunsen, 1811–1899) student, was apparently disappointed also by Bunsen's negative attitude toward the periodic system: "We heard nothing on Mendělěěv's system from Bunsen in his Heidelberg lectures of 1878–79. When I spoke to him about the elements of rare earths . . . and pointed out how well they confirmed the atomic weights proposed by Mendělěěv, he answered 'Leave these conjectures alone.' And there is no doubt that Bunsen knew and studied Mendělěěv's ideas—of course only privately."[20] Another source mentions that when Brauner brought up that the atomic weights of the rare earth elements correspond to Mendeleev's regularities, Bunsen snubbed him by saying that "such regularities can also be found in the exchange rates."[21]

In spite of his criticism of Bunsen, Brauner made good use of his stay in Heidelberg, where he learned about methods that he made use of in the upcoming years, like gas analysis, spectral analysis, and preparation of pure rare earths. Nevertheless, he still was looking for a place where he could satisfy his curiosity in the periodic system. Such location became the laboratory of Sir Henry E. Roscoe (1833–1915), Bunsen's pupil and former collaborator, founder of the Manchester School of Chemistry at Owens College, the precursor of the University of Manchester. Brauner, who spent the years 1880 through 1882 there, was encouraged by Roscoe to carry out his own independent experimental research, since then almost solely dedicated to questions associated with the periodic system.

In Manchester, Brauner investigated the properties of some elements, especially the famed trio—cerium, lanthanum, and didymium (then still believed to be an element). Jointly with his colleague John I. Watts, he corrected the value of the atomic weight of beryllium to 9 and confirmed its place in Group II. Besides experimental work, he became immersed in the rich social and sport life of Manchester and also used the opportunity to acquaint the local chemists with Mendeleev's achievements. His lectures on the works of Mendeleev in the Owens College Chemical Society were part of his enduring effort to disseminate the idea of the periodic system. Among the contemporaries—fellow colleagues—he "indoctrinated" were, especially, Sydney Young (1857–1937), later professor of chemistry at Dublin University, Arthur Smithells (1860–1939), later professor at the University of Leeds, Harold B. Dixon (1852–1930), who succeeded Roscoe at Owens College, the already mentioned J. I. Watts, and the Japanese Toyokitchi Takamatsu (1852–1937), who became professor at the Imperial University Tokyo.[22] We may assume that in this way the early notions of the periodic system were passed along not only within the British Islands but also to Japan.

Brauner's entire scientific work "consisted largely in the exemplification and perfection . . . of Mendeleev's periodic law."[23] Among other things, he proved, along with J. I. Watts, that the correct atomic weight of beryllium was 9 and in this fashion confirmed its position at the head of Group II, as Mendeleev had claimed. Brauner was first to prepare a salt of quadrivalent cerium, revise its atomic weight, and place it in Group IV of the periodic table. The elements of rare earths that Brauner started to explore while in Manchester attracted him for the rest of his life. Their thorough investigation led eventually to his probably most significant achievement—solution of the enigma of their placement in the periodic table. In 1902 Brauner concluded that the rare earths form a separate group of closely related elements that occupy a single place in the periodic table starting with lanthanum (atomic number 57) and ending with hafnium (atomic number 72).[24] This modification of the table became the most frequently used initial model of the contemporary tables (Figure 6.1.).

We may get better insight into Brauner's role in the dissemination of the periodic system in the light of his personal friendship with Mendeleev. Mendeleev's extensive obituary essay written by Brauner in 1907 deserves special attention in this respect. In the obituary, Brauner summarizes his relation to Mendeleev and the periodic system, and recapitulates not only Mendeleev's but also his own scientific life.[25] Another important source illuminating the relationship of the two scientists is the book of the Russian historians B. Kedrov and T. Cheptsova.[26] We learn from these and other biographical documents[27] that the first contact of Brauner and Mendeleev took place in January 1881, when Brauner sent to the Russian chemist his paper, published jointly with his English colleague John I. Watts, where they referred to Mendeleev's *Osnovy khimii* [The Principles of Chemistry]. A letter dated January 17, 1881, where Brauner "expressed [his] regret that [Mendeleev's] excellent treatise was quite unknown in Western countries," accompanied the reprint.[28] Mendeleev answered with a long letter and sent Brauner his photograph. This was the beginning of their correspondence, cooperation, and friendship, which lasted

Periodická soustava prvků dle Mendělejeva v úpravě prof. Braunera.

Řada:	Skupina: 0 — R	I — R_2O	II — R_2O_3	III — R_2O_3	IV — RH_4 R_2O_4	V — RH_3 R_2O_5	VI — RH_2 R_2O_6	VII — RH R_2O_7	VIII — R_2O_8	R_2O_6	R_2O_4	R_2O_3	Hraniční formy slučovací
1.		H 1,008											
2.	He 4,0	Li 7,03	Be 9,1	B 11,0	C 12,00	N 14,01	O 16,00	F 19,0					1ní malá perioda
3.	Ne 20	23,05 Na	24,36 Mg	27,1 Al	28,4 Si	31,0 P	32,06 S	35,45 Cl					2há malá perioda
4.	Ar 39,9	K 39,15	Ca 40,1	Sc 44,1	Ti 48,1	V 51,2	Cr 52,1	Mn 55,0	Fe 55,9	Co 59,0	Ni 58,7	Cu 63,6	1ní velká perioda
5.		63,6 Cu	65,4 Zn	70 Ga	72,5 Ge	75,0 As	79,2 Se	79,96 Br					2há velká perioda
6.	Kr 81,8	Rb 85,5	Sr 87,6	Y 89,0	Zr 90,6	Nb 94	Mo 96,0	Ma	Ru 101,7	Rh 103,0	Pd 106,5	Ag 107,93	2há velká perioda
7.		107,93 Ag	112,4 Cd	115 In	119,0 Sn	120,2 Sb	127,6 Te	126,97 J					3tí velká perioda
8.	Xe 128	Cs 132,9	Ba 137,4	Lu 138,9	Ce 140,25								3tí velká perioda
9.													
10.						Ta 181	W 184	Re	Os 191	Ir 193,0	Pt 194,8	Au 197,2	4tá velká perioda
11.		*197,2 Au	200,0 Hg	204,1 Tl	206,9 Pb	208,0 Bi	Pb	At					
12.		Rd 225			Th 232,5		U 238,5						

Prvky vzácných zemin { Pr 140,5 | Nd 143,6 | Sm 150,3 | Eu 152 | Gd 156 | Tb 159 | Dy 162,5 | Er 166 | Tu 171 | Yb 173,0 etc.

Figure 6.1 Brauner's modification of the Periodic Table of Elements published in 1909 in the textbook of inorganic chemistry for the Czech Technical University. Note that the rare earths are incomplete. Reproduced from Preis and Votoček, *Anorganická chemie* (note 92), 379.

until Mendeleev's death in 1907. The two scientists met in person three times. The first time was in St. Petersburg in autumn of 1883, at the occasion of the Congress of the Russian Natural Scientists in Odessa, where Brauner reported about the atomic weight of tellurium.[29] Mendeleev did not participate in this meeting, but after the congress, Brauner traveled to St. Petersburg to see Mendeleev.

The next encounter of the two chemists took place in Prague in March 1900 (Figure 6.2.) and for the last time in winter of c.1901,[30] again in St. Petersburg during the Congress of the Russian Natural Scientists where Mendeleev invited Brauner and went to hear Brauner's lecture on the position of rare earths in the periodic system.[31] Fedorovich asserts[32] that the Russian chemical community reacted coldly and Mendeleev himself skeptically to Brauner's proposal that the rare-earth elements be placed in a distinctive "interperiodic" group. Mendeleev "stubbornly held the view that the rare earths must be placed in the individual groups . . . nevertheless he commented on Brauner's proposals as follows: 'As

Figure 6.2 Joint photograph of Mendeleev and Brauner made during Mendeleev's visit in Prague in 1900. Reproduced from Brauner, "D.I. Mendělěěv" (note 5), 241.

I do not have a possibility to deny such conclusion, I guess that we should cautiously leave this question open.'" In spite of this misunderstanding, Brauner's paper was published in Russian and its amended version appeared as a separate chapter in the seventh (1903) and eighth (1906) edition of Mendeleev's *Principles of Chemistry*.[33] For Brauner, however, his last encounter with Mendeleev also had a symbolic value, as Mendeleev entrusted him with an honorable task captured in his recollection: "Parting, Dimitrii Ivanovich said: 'I never more shall see you again—I leave you my periodic system—look after it!' In fact I had seen him the last time then."[34]

In order to understand the motivations of Brauner's early adoption of the periodic system and his enthusiasm in promoting it all over Europe, we must take into consideration Mendeleev's Russian nationality and the prevalent Russophilia in the Czech society of that period, amplified by strong anti-German feelings, which made the acceptance of the periodic system also a matter of national sentiments. We can sense this stimulus already in one of Brauner's early letters written at the beginning of February 1881, where Brauner mentioned the Russophilia of his father: "I already had the opportunity to meet often Russians at the house of my father. My father was a true friend of the Russian nation and cultivated in the hearts of his children the Slavic sentiment. He made us feel proud being members of the big Slavic family. I had often the opportunity to read the letters written by father's Russian friends, which father translated and explained to us children (Figure 6.3.)."[35]

It seems that in the subsequent years Brauner even shared to a certain extent the rising militant anti-German nationalism of the Czech society, in spite of his mixed heritage and broad international contacts. This stance of Brauner can be documented by the following startlingly strong chauvinistic statement in Mendeleev's obituary essay, which he wrote in 1907: "We Czechs are often censured that as cultivators of science we only feed of the leftovers from the table of the great German science . . . Although I justly respect German science, and especially the achievements of Germans in chemistry, I will always be proud that I never had to sponge from the table of the German science, while the guiding star of my life was the Slavic part of science established by Dimitrii Ivanovich Mendeleev."[36] In those times, Brauner obviously considered his cooperation with Mendeleev not only scientific but also a patriotic task associated with a deflection from German science to Slavic science. It is clear that during the years to come, especially in the more tolerant atmosphere of democratic Czechoslovakia after 1918, Brauner tempered his nationalism as this and similar proclamations are markedly toned down in the 1930 version of Brauner's essay, which skips the previous statement and instead reads: "I am indeed proud that the guiding ideas in my life have been in the branch of chemical science, which was founded for us and for many following generations by that great Russian and Slavonic genius, Dimitrij Ivanovič Mendělěěv."[37] Jaroslav Heyrovský, who modified and translated Brauner's essay into English at the occasion of his seventy-fifth birthday, deliberately removed this and other exaggerated nationalistic statements, apparently with the consent of the author. In the words of Heyrovský: "Some parts of the

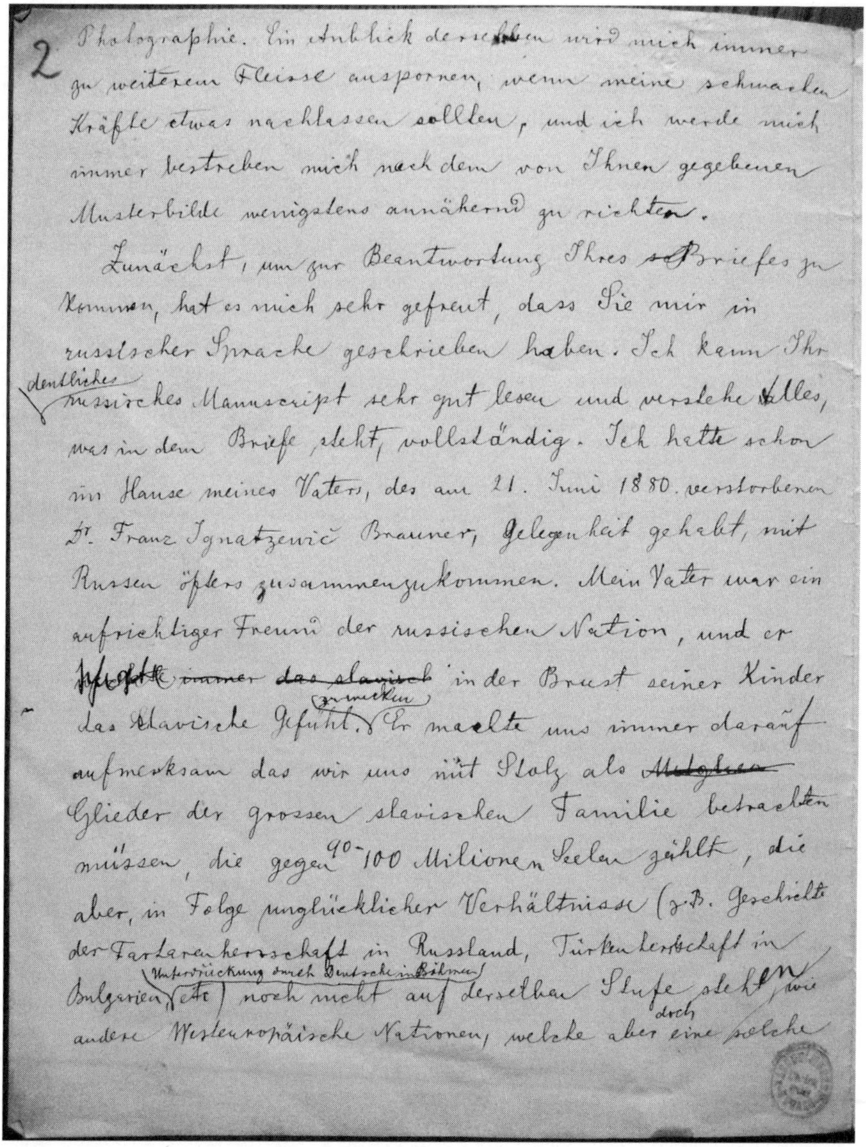

2 Photographie. Ein Anblick derselben wird mich immer
zu weiterem Fleisse anspornen, wenn meine schwachen
Kräfte etwas nachlassen sollten, und ich werde mich
immer bestreben mich nach dem von Ihnen gegebenen
Musterbilde wenigstens annähernd zu richten.
Zunächst, um zur Beantwortung Ihres Briefes zu
kommen, hat es mich sehr gefreut, dass Sie mir in
russischer Sprache geschrieben haben. Ich kann Ihr
deutliches russisches Manuscript sehr gut lesen und verstehe Alles,
was in dem Briefe steht, vollständig. Ich hatte schon
im Hause meines Vaters, des am 21. Juni 1880 verstorbenen
Dr. Franz Ignatzewič Brauner, Gelegenheit gehabt, mit
Russen öfters zusammenzukommen. Mein Vater war ein
aufrichtiger Freund der russischen Nation, und er
in der Brust seiner Kinder
das slavische Gefühl zu erwecken. Er machte uns immer darauf
aufmerksam das wir uns mit Stolz als
Glieder der grossen slavischen Familie betrachten
müssen, die gegen 90–100 Milionen Seelen zählt, die
aber, in Folge unglücklicher Verhältnisse (z.B. Geschichte
der Tatarenherrschaft in Russland, Türkenherrschaft in
Bulgarien, Unterdrückung durch Deutsche in Böhmen etc) noch nicht auf derselben Stufe stehen, wie
andere Westeuropäische Nationen, welche aber doch eine solche

Figure 6.3 Brauner's draft letter to Mendeleev of February 1881; the page where Brauner speaks about the Russophilia of his family. PNP, Brauner Bohuslav, Personal Collection, Correspondence.

essay, bearing on **topics of local interest** or on the political situation in 1907, have been omitted."[38]

This way or the other, Brauner took the "last will" of his master seriously, as documented by his extensive bibliography[39] and efforts to make known and evolve the periodic system. He claimed that thirty-two of his papers were related to the periodic system,[40] in which he paid attention especially to the following aspects: estimation and revision of atomic weights of elements with a focus on rare earth metals; search for new elements predicted by the periodic table; modifications of the periodic table, particularly placing elements and groups of elements into correct positions in the periodic table; reforming the standards of atomic weights; and publicizing the periodic system—articles in journals, chapters in treatises, lectures, and courses. His complex investigations made Brauner an internationally recognized specialist in research on some elements, especially the elements of rare earths, their atomic weights, and their position in the periodic system. Most of his key papers were published in English, German, and Russian journals, for instance in the *Journal of the Chemical Society, Zeitschrift für anorganische Chemie, Berichte de Deutschen chemischen Gesellschaft, Monatshefte für Chemie, Zhurnal Russkogo Fiz.-Khim. Obshchestva* [Journal of Russian Physico-Chemical Society], and others. Brauner also published in the Czech periodicals, but not so frequently. Aside from journal articles, he also wrote specialized chapters for some voluminous treatises. As already mentioned, Mendeleev engaged him in writing the chapters on rare earth metals for the seventh and eighth edition of his *Osnovy khimii* [The Principles of Chemistry] (1906). In Abbeg's extensive *Handbuch der anorganischen Chemie* (1905–13), Brauner was the author of critical chapters on atomic weights of sixty elements. Brauner stimulated the international recognition of the periodic system in his lectures at the traditional congresses of the German natural scientists and physicians: in 1881 in Salzburg,[41] 1889 in Heidelberg, and 1899 in Munich, and through his authority in international chemical organizations. He not only made Mendeleev's work better known all around Europe, but also became, thanks to his reading knowledge of the Russian language,[42] a welcome mediator of contacts between the Russian- and English-speaking chemists. In 1881 the *Journal of the Chemical Society* asked him to report regularly about the Russian chemical publications, a task that Brauner fulfilled for years.[43]

Since 1888 Brauner attempted to put through, jointly with the American chemist F. P. Venable (1856–1934) and other chemists, the atomic weight of oxygen (16) as standard for calculation of relative atomic weights of elements;[44] this proposal was only accepted in 1900 at the 4th International Congress of Applied Chemistry in Paris. In the years between 1921 and 1930 Brauner was member of the International Committee on Chemical Elements and president of its subcommission on atomic weights.

Gerald Druce, one of Brauner's biographers, reminds us that "Brauner's work was not always acclaimed so highly. Some of the critics of the periodic law directed their attacks mainly upon Brauner and his researches." Gregoire Wyrouboff[45] criticized both Brauner and Mendeleev, saying that Brauner "made a speciality of the art, causing reluctant elements to enter into the classification of Mendeléef" and concluding that the periodic system was "a very interesting and highly ingenious table of analogies and dissimilarities of the . . . elements," but that it must be accepted or rejected as a whole because of "certain defects" as the "laws of nature admit no exception."[46]

4. CONTROVERSY OVER THE CZECH CHEMICAL NOMENCLATURE AND THE RECEPTION OF THE PERIODIC SYSTEM IN THE CZECH LANDS

As already mentioned, the Czech chemical community was informed for the first time about the periodic system and Mendeleev's ideas from Brauner's article published in 1877 in the Czech journal *Listy Chemické*.[47] Although this journal has been the main platform of the Czech chemists until today, and read by most of them, in the years to follow Brauner informed the Czech public about his research mostly elsewhere, namely in the prestigious periodicals of the Czech Academy of Sciences, Letters and Arts, the *Rozpravy České akademie* [Debates of the Czech Academy], and the *Bulletin International České Akademie*, which published articles in French and English.[48] In *Listy chemické* only appeared Brauner's summarizing reviews informing the Czech chemical readership about the main results of his research and the general advancement of the periodic system.[49] Thus, Brauner, unlike most Czech chemists, was only an occasional contributor to this national chemical journal, which implies that Brauner with his specific engagement in the periodic system was conscious of his solitary position on the Czech chemical scene and realized that the majority of its readers might not understand the details of his research. Furthermore, Brauner did not participate actively in the doings of the Society of the Czech Chemists and never held any office there. Brauner himself seems somewhat resentful not only for his belated professorship, but also for the lack of interest in his favored topic—inorganic chemistry and the rare earths—among the Czech colleagues. He testified regarding this indifference when talking about the rare earths as the "least popular part of the unpopular and merely tolerated inorganic chemistry."[50] Another source of his discontent became, apparently, the controversy over the reform of the Czech chemical nomenclature, which deserves attention because of its interconnection with the promotion of the periodic system in the Czech Lands.

It is necessary to emphasize beforehand that the introduction of national languages into all domains of life, including science, became important means of national emancipation and formation of national identities in the nineteenth century, especially for the small nations living within the multinational Habsburg Monarchy.[51] Thus, the long process of creation of the modern Czech scientific terminology,[52] lasting until the 1920s, had not only practical but also significant political motives. The firm foundations of the Czech chemical nomenclature were laid in the middle of the nineteenth century, mainly thanks to Brauner's tutor and father-in-law Vojtěch Šafařík, who assembled and modified the existing Czech chemical entries and created a modern system of developing chemical formulas in 1853.[53] Šafařík's chemical nomenclature was used only with minor changes until the end of the nineteenth century, or in some cases survived even in the second decade of the twentieth century. No wonder that already at the end of the nineteenth century, such stagnant terminology was becoming obsolete, especially compared with the international nomenclature, which flexibly reacted to the rapid progress of the discipline. Since the beginning of the twentieth century, debates started within the

Chemical Society emphasizing the necessity of a fundamental reform prompted at the initiative of young chemist Alexander Batěk[54] (1874–1944), Brauner's pupil and one of his few potential successors. Batěk faced a promising career. He defended his doctoral dissertation[55] "Revision of the Cerium Atomic Weight"[56] in 1899 under Brauner's mentorship. This work won him financial support of the Czech Academy, which enabled him to stay at Ramsay's laboratory in London on Brauner's recommendation. Unexpectedly, after his study trip, Batěk left the university for unclear reasons and became a high school teacher, never again pursuing an academic career.[57] In 1900 Batěk published, in the Czech chemical journal, suggestions for correction of the Czech chemical nomenclature,[58] but his ideas did not evoke any interest. In his memoirs, Batěk mentions that he was already aware during his studies of the inconsistencies in the nomenclature, especially incorrect chemical formulas of inorganic compounds that did not reflect the valences of the elements: "I could not understand how people as great as Brauner and other creators of science can tolerate such mess and accept it."[59] Eventually, his new appeal in 1908 at the 4th Congress of the Czech Naturalists and Physicians in Prague was answered by the Czech Chemical Society, which created a Commission for Nomenclature,[60] but only in 1914 was the new nomenclature based on Batěk's proposals enacted by the Society, thanks to backing from Emil Votoček (1872–1950),[61] organic chemist with a great international reputation. In contrast, Brauner, who himself was member of the Commission, became the most ardent opponent of the reform and prevented its official adoption for another four years. Only in 1918 did the decree of the Austrian Ministry of Culture and Education officially introduce at schools of all levels, including universities,[62] the new Czech chemical nomenclature of inorganic compounds, which has been valid with certain modifications until today. Brauner was the only Czech chemist who never accepted it and used the old one (modified by himself where necessary) until the end of his life.[63]

Brauner's persistent refusal to recognize the new Czech chemical nomenclature was related to the new developments in physics and physical chemistry influencing the regularities in the periodic system of elements. In the session of the Czech Chemical Society on October 21, 1916, he gave a lecture entitled "On the Evolvement of the Periodic System,"[64] where he pointed to the new discoveries related to the atomic nucleus, especially those of Rutherford, Soddy, Bohr, Richards, Moseley,[65] and others, which fundamentally affected the periodic system. He focused in particular on the recent research into isotopes and radioactive elements with serious consequences "for theoretical chemistry, especially for the advancement of the periodic system" and noted that "If these modern views upon the atomic nucleus and its charge are correct, then it means that the properties of elements are periodic functions not of the atomic weights, as has been believed, but those of **the positive charge of the nucleus**, or, given that this notion is only hypothetic, the **serial number** in the periodic system."[66] Although Brauner did not mention his acute controversy over the nomenclature reform, which was stirring the Czech chemical community for several years, his lecture clearly offered arguments against the reform. We may find similar reasoning, this time openly directed against the new nomenclature, in Brauner's only textbook, a manual of qualitative analysis published jointly with his assistant Jindřich Křepelka (1890–1964)[67] in 1919. An extensive final chapter, which Brauner called "Foreword II," is entirely devoted to criticism of the reformed Czech chemical terminology along with his own suggestions.[68] Brauner again objects that the reform does not respect the new advances in chemistry and in the periodic system and proposes a wide-ranging modification of the nomenclature that takes into account the positive and negative valences of the elements and recent findings on atomic structure.[69] He also complains that his criticism of the nomenclature and his own proposals had not met with understanding in the Czech scientific circles: "In a special session of the Czech Academy devoted to discussion on the new (corrected) nomenclature, I expressed some of my opinions, which, however, met with strong rejection."[70]

Although Brauner had been immersed in this longtime conflict with the representatives of the Czech chemical community, we know at least two Czech opponents of the periodic system. The first of them was the Nestor of the Czech chemists, Vojtěch Šafařík. Brauner recalled the chemistry courses of Vojtěch Šafařík at the Czech Technical University between 1873 and 1875, where Šafařík did not talk at all about the periodic system. Although definitely aware of Mendeleev's papers and his textbook *Principles of Chemistry*, Šafařík only mentioned Mendeleev's sensitive thermometers and preparation of absolutely pure alcohol.[71] Brauner's first report of 1877 on the periodic system was followed in 1878 by Šafařík's textbook for the Czech Technical University,[72] where Šafařík referred to the periodic system and showed the periodic table for the first time in the Czech monographic literature. The periodic table in the book was entitled "Survey of the Periodic System of Elements According to Mendeleev Supplemented by Recent Research."[73]

In his commentary[74] Šafařík identified the periodic system as a "natural system which was called by Mendeleev periodic system, and compared it with Linné's system of plants, which, however, "cannot be applied in chemistry because we don't know many elements and don't know well enough the features of those that are known." In his skepticism, Šafařík questioned the significance of the periodic system and explained why he had not used it as the basis of his textbook: "Here we have the foundations of a natural system of elements called Mendeleev's periodic system; although it unveils new surprising relations and perspectives, it is still full of gaps and doubts, still in its beginnings, and therefore it is not suitable to become a foundation for a textbook." [75] We have other indications, as well, that Šafařík was not a supporter of the periodic system, and adhered to his negative opinion during his whole life. Obviously, in his textbook written for the students of the Czech University in 1884,[76] Šafařík entirely omits the periodic system and does not mention Mendeleev at all. The same applies to Šafařík's university lectures. While Šafařík's early criticism sounds quite reasonable, his later deprecatory position can also be explained by subjective motivation: his lasting illness, gradual loss of interest in chemistry, and, hence, disinterest in following the new advances in the dynamically developing field.[77] Other clues suggest that his views might have been affected by certain personal tensions between him and Brauner, which persisted in the 1880s when Šafařík became professor of chemistry at the Czech University and most likely stood in the way of Brauner's appointment for professor.[78]

Another isolated voice that opposed the periodic system a few years later belonged to Jaroslav Formánek (1864–1936), one of the worldwide pioneers of modern spectroscopy.[79] Formánek, who was between the years 1912 and 1930 head of the chemistry department of the Forestry and Agriculture Section of the Prague Czech Technical University, published a textbook of inorganic chemistry in 1921,[80] where he points to the shortcomings of the periodic system. In the chapter called "On the Natural System of Elements,"[81] he refers to recent scientific discoveries that made atomic weights inappropriate as basic constants for classification of elements and should be replaced by other values, namely by ordinal numbers, which reflect the charge of the nucleus. For this reason, Formánek replaces the notion of the "periodic system of elements" by "natural system of elements" and in his textbook, instead of the Mendeleev table, uses the table of Otto Hönigschmid[82] (Figure 6.4), with the ordinal numbers as an alternative to atomic weights.[83]

The table is entitled "The System of Elements by Ordinal Numbers" and the comment to the table says: "If we mark in Mendeleev's system in its contemporary (Brauner) arrangement each element by its ordinary number . . ., it turns out that the ordinary number stands exactly for the *charge of the nucleus*. From the attached table of the ordinal system arranged by O. Hönigschmied [*sic*] we learn how the elements follow one after another." Aside from Šafařík, representing the older generation, and the younger Formánek, Brauner did not meet serious opposition in his long-lasting home campaign for the periodic system in the Czech Lands. The reaction was rather polite indifference as also testifies the earlier-mentioned lecture on the periodic system read at the meeting of the Czech Chemical Society by Brauner in October 1916. At the end of his speech, Brauner reminded the auditory about his first paper on the rare earths of 1881 and brought up the fact that he, as the first chemist in Austria-Hungary, employed as guideline for his research the periodic system. At the same time, he also expressed his disappointment that the Czech chemists showed only lukewarm interest in the periodic system and underestimated its importance. Finally, he challenged his colleagues to "show further applicability of the periodic law in their experimental work, namely through research into rare elements, which are still neglected in university lectures. Only then will wake up the interest in the periodic law among the uninterested chemists, only then will the great role of the periodic system in chemistry of the future be undoubtedly appreciated."[84]

The seemingly indifferent approach to Brauner's research and to the periodic system and its theoretical foundations was obviously caused by the fact that inorganic and physical chemistry still belonged at the beginning of the twentieth century among the marginal specialties in the Czech Lands, while organic and analytical chemistry were considered more attractive—the first one because of its dynamic worldwide development and both because of their versatile applicability in practical fields like chemical, sugar, and fermentation industries, ironmongery, and medicine, which were employing most Czech chemists. Nevertheless, in spite of Brauner's apparent frustration, most evidence suggests that the Czech chemical community accepted the periodic system quite smoothly. As indication may

Místo t. zv. »atomové váhy« zavádí se pro určitý prvek *číslo řadové*.
Označíme-li totiž v Mendělejevově soustavě dle dnešního jejího
uspořádání *(Braunerova)* každý prvek řadovým číslem, tedy vodík
H = 1, helium He = 2, lithium Li = 3, atd. až po uran U = 92,
ukáže se, že řadové číslo značí přesně *náboj jádra,* jenž byl by
tedy u helia = 2, u lithia = 3 atd.

Z připojené tabulky řadové soustavy uspořádané *O. Hönig-
schmiedem* seznáme jak prvky za sebou dle řadových čísel následují.

SOUSTAVA PRVKŮ DLE ŘADOVÝCH ČÍSEL.

0	VIII			I		II		III		IV		V		VI		VII	
				a	b	a	b	a	b	a	b	a	b	a	b	a	b
				H 1													
He 2				Li 3		Be 4		B 5		C 6		N 7		O 8		F 9	
Ne 10				Na 11		Mg 12		Al 13		Si 14		P 15		S 16		Cl 17	
Ar 18				K 19		Ca 20		Sc 21		Ti 22		V 23		Cr 24		Mn 25	
	Fe 26	Co 27	Ni 28	Cu 29		Zn 30		Ga 31		Ge 32		As 33		Se 34		Br 35	
Kr 35				Rb 37		Sr 38		Y 39		Zr 40		Nb 41		Mo 42		— 43	
	Ru 44	Rh 45	Pd 46	Ag 47		Cd 48		In 49		Sn 50		Sb 51		Te 52		J 53	
X 54				Cs 55		Ba 56		La 57	Ce 58	Pr 59	Nd 60	— 61		Sm 62		Eu 63	
Gd 64	Tb 65	Dy 66	Ho 67	Er 68	Tu I 69	Ad¹⁾ 70		Cp²⁾ 71		Tu II 72		Ta 73		W 74		— 75	
	Os 76	Ir 77	Pt 78	Au 79		Hg 80		Tl 81		Pb 82		Bi 83		Po 84		— 85	
Em 86				— 87		Ra 88		Ac 89		Th 90		Bv³⁾ 91		U 92			

¹) Aldebaranium (Lutecium). — ²) Cassiopeum (Celcium). — ³) Brevium (Uran X₂)

Figure 6.4 The periodic table arranged by O. Hönigschmid in Formánek's Textbook of Chemistry published in 1921. Reproduced from Formánek, (note 79), 281.

serve the comparatively early occurrence of the periodic table and presentation of Mendeleev's periodic system in the textbooks written by the Czech chemistry professors.[85]

5. THE PERIODIC SYSTEM IN THE CZECH CHEMISTRY TEXTBOOKS

The first appearance of the periodic table in the Czech chemical literature in Šafařík's textbook of 1878[86] was already mentioned, as well as Šafařík's criticism and later his disregard of the periodic system. Such negative approach was, however, an exception. Since the 1880s chapters on the periodic system became an integral part of a number of university textbooks. Bohuslav Raýman (1852–1910), Šafařík's pupil and Brauner's close friend, one of the most influential Czech chemists of his time, [87] who introduced modern organic chemistry into the curricula and also saw after implementation of physical chemistry at the Prague Czech universities, published the first Czech textbook on theoretical chemistry in 1884.[88] In the chapter "Atomic Weights and the Properties of Elements,"[89] the book contains a detailed explanation of the periodic system and the periodic table taken from Mendeleev's *Principles of Chemistry*. For Raýman, "the Mendeleev Law and the classification of elements which is based on it, [is] the foundation of unified conception of physical properties of elements and base of conception of affinity."[90] Raýman's textbook for the first time in Czech literature analyzed in such detail the theoretical background and consequences of the periodic system and acquainted the university students with this revolutionary theory.

For further dissemination of the periodic system among the students in the years to come deserves credit Raýman's pupil, the organic chemist Emil Votoček, teacher of several generations of Czech chemists and author of basic voluminous Czech textbooks on inorganic and organic chemistry. Votoček published in 1902, jointly with the analytic chemist of older generation Karel Preis (1846–1916), the first specialized Czech textbook of inorganic chemistry,[91] which from then on appeared in several revised and amended editions until 1954, with Votoček as the single author since 1922. The first edition of 1902 contained a detailed chapter "On the Periodic System of Elements,"[92] dealing also with the prehistory of the periodic system. The second edition of 1909[93] presented (probably for the first time) the periodic table modified by Brauner[94] with the rare earths as a separate group, although still incomplete. In 1922 Votoček modernized and extended the textbook[95], adding an up-to-date critical chapter on the periodic system,[96] with the newest version of Brauner's modification of the periodic table showing the international atomic weights and serial numbers of the elements[97]

Introduction of physical chemistry both at the Czech Technical University and the Czech University made it necessary to issue a modern textbook of this subject. However, no Czech author was able to fulfill this task at that time. Therefore, at the instigation of Raýman, E. Votoček undertook the mission of translating the advanced physical chemistry published in 1897 by the professor at *Université libre* in Brussels, Albert Reychler (1854–1938).[98] The Czech version of 1902[99] contains the interrelated chapters "Classification of Elements After Mendeleev"[100] and "Classification of Elements After L. Mayer"[101] and a simple modification of the periodic table.[102] Reychler's book also indicates the acceptance of Mendeleev's system in Belgium. Among the other numerous textbooks that have referred to the periodic system deserves attention the laboratory manual published in 1912 by Jaroslav Milbauer (1880–1959), the first professor of specialized inorganic technology at the Czech Technical University in Prague, who considers Mendeleev's periodic system an instrument of analytical chemistry.[103] The periodic table[104] is shown in his book in Brauner's modification. Not only the students of chemistry, but also the Czech medical students became acquainted with the periodic system in the first volume of the widely used *Medical Chemistry* of Jan Horbaczewski (1854–1942)[105]. This textbook contains a separate chapter, "Periodic System of Elements," featuring the periodic table,[106] which, however, does not yet demonstrate the special position of the rare earth proposed by Brauner. After 1920, most Czech university chemistry textbooks contained some explanation of the periodic system with the periodic table in Brauner's modification, where the elements are ordered by the serial numbers (Figure 6.5.).

Exploration of the usage of the periodic system in the Czech secondary school textbooks in the monitored period turned out to be a quite complicated task given the fact that the complex Austro-Hungarian educational system encompassed several types of secondary schools, which used authorized textbooks adapted to specific curricula differentiated by grades. Luckily enough, the National Pedagogic Library of J. A. Comenius[107] keeps a representative (although incomplete) collection of secondary school textbooks with a printed catalogue,[108] which allowed for the undertaking of a thorough survey. It disclosed that references to the periodic system and/or the periodic table only occurred in textbooks for the so-called *Realschule* and *Realgymnasia*.[109] The oldest reference to the periodic system was found in Matzner's *Inorganic Chemistry* of 1903,[110] which contains a simplified version of the periodic table and a short explanation of Mendeleev's periodic law.[111] Many of the textbooks issued after 1908 treated the periodic system in more detail, like the revised edition of Matzner's textbook of 1908,[112] which contains a comprehensive explanation of the periodic law[113] and a more elaborate periodic table,[114] however without displaying the special position of the rare earths. Worth mentioning are some aspects of the *Textbook of Inorganic Chemistry* by F. Mašek and J. Němeček issued by the Czech Chemical Society in 1910,[115] where the chapter on the periodic system includes not only its detailed and qualified explanation but also a short history of its discovery.[116] A special enclosure displays the complete Brauner modification of the periodic table, and in addition, a table with the merging types of oxides and Lothar Meyer's chart that plotted atomic volumes against atomic weights of elements. All these attributes made the textbook a reliable source of knowledge focused on the youngest generation of students. Similar principles were preserved also in the subsequent textbook of these two authors designed for the *Realschule* 1918[117] and accredited by the Chemical Society. Most of the other textbooks used in *Realgymnasia* and *Realschule* referred to the periodic system as well, which indicates

Mezinárodní atomové váhy na rok 1925. **Periodická soustava Mendělějevova.** *Prof. Dr. Bohuslav Brauner.*

Grupa:	I	II	III	IV	V	VI	VII	VIII				
	R_2O	R_2O_2	R_2O_3	R_2O_4	R_2O_5	R_2O_6	R_2O_7	R_2O_8	R_2O_6	R_2O_4	R_2O_3	
				RH_4	RH_3	RH_2	RH			(R)		Nejvyšší normální kyslíčníky / Nejvyšší těkavé sloučeniny vodík.
Řada:	$\frac{I}{R}$	$\frac{II}{R}$	$\frac{III}{R}$	$\frac{IV}{R}$	$\frac{V}{R}$	$\frac{VI}{R}$	$\frac{VII}{R}$		VIII $\frac{0}{R}$ (a R)	$\frac{0}{R}$		Positivní / Negativní
1.	H 1·008 1								He 4·00 2			Rudimentární perioda
2.	Li 6·940 3	Be 9·02 4	B 10·82 5	C 12·000 6	N 14·008 7	O 16·000 8	F 19·00 9	*)	Ne 20·2 10			První jednoduchá perioda
3.	Na 22·997 11	Mg 24·32 12	Al 26·97 13	Si 28·06 14	P 31·027 15	S 32·064 16	Cl 35·457 17		A 39·91 18			Druhá jednoduchá perioda
4.	K 39·096 19	Ca 40·07 20	Sc 45·10 21	Ti 48·1 22	V 50·96 23	Cr 52·01 24	Mn 54·93 25	Fe 55·84 26	Co 58·94 27	Ni 58·69 28	Cu 63·57 29	První dvojnásobná perioda
5.	63·57 Cu 29	65·38 Zn 30	69·72 Ga 31	72·60 Ge 32	74·96 As 33	79·2 Se 34	79·916 Br 35		Kr 82·9 36			
6.	Rb 85·47 37	Sr 87·63 38	Y 88·9 39	Zr 91 40	Nb 93·1 41	Mo 96·0 42	43	Ru 101·7 44	Rh 102·91 45	Pd 106·7 46	Ag 107·880 47	Druhá dvojnásobná perioda
7.	107·880 Ag 47	112·41 Cd 48	114·8 In 49	118·70 Sn 50	121·77 Sb 51	127·5 Te 52	126·932 J 53		Xe 130·2 54			
8.	Cs 132·81 55	Ba 137·37 56	La 138·90 57	Ce 140·25 58	Pr 140·92 59	Nd 144·27 60	61	**)				
9.	Sm 150·43 62	Eu 152·0 63	Gd 157·26 64	Tb 159·2 65	Dy 162·52 66	Ho 163·4 67	Er 167·7 68					Čtyrnásobná perioda
10.	Tu 169·4 69	Yb 173·6 70	Lu 175·0 71	Ct (Hf)? 72	Ta 181·5 73	W 184·0 74	75	Os 190·8 76	Ir 193·1 77	Pt 195·23 78	Au 197·2 79	
11.	197·2 Au 79	200·61 Hg 80	204·39 Tl 81	207·20 Pb 82	209·00 Bi 83	Po 84	85		Rn 222 86			
12.	87	Ra 225·95 88	Ac 89	Th 232·15 90	E Ta***) 91	U 238·17 92						Část periody

*) Typické prvky.   **) Prvky vzácných zemin, nesnadno zařaditelné.   ***) Ekatantal, zvaný též protoaktinium.

Figure 6.5 The Periodic Table of Elements according to Brauner, used in the 1920s and 1930s in most Czech chemistry textbooks. Reproduced from Emil Votoček, *Chemie anorganická, II. doplněné vydání* (Praha: Politika, 1925), 572.

that the new theory became a regular component of the Czech high school curricula since the first decade of the twentieth century.

It is necessary to underline that the Czech Chemical Society financed most editions of the Preis-Votoček and Votoček *Inorganic Chemistry*, as well as Milbauer's manual, while the translation of Reychler's *Physical Chemistry* was supported by the Czech Academy. Also, several high school textbooks were issued or accredited by the Chemical Society. The assistance of the leading Czech scientific institutions suggests that these treatises were officially authorized to serve instruction of chemistry in the Czech Lands together with their chapters on the periodic system.

Although this essay focuses mainly on the Czech chemical community in the Czech Lands, we may also rightly ask, how did the German chemical community in the Czech Lands receive the periodic system? The documentary evidence at our disposal does not allow us to answer this question. We cannot even rely on any textbooks, like in the Czech case. Although there existed before 1945 several linguistically German universities in the Czech Lands (and later in Czechoslovakia), most likely no German chemistry textbook was issued in the Czech Lands prior to 1920, as these universities used textbooks issued in Austria or Germany. In the Prague libraries only one German "textbook" on inorganic/general chemistry is available, published in Prague and written by Brauner's tutor Eduard Linnemann.[118] The volume contains handwritten notes of Linneman's lectures duplicated apparently for students in the winter semester of 1881–82, which are missing any reference to the periodic system. According to Brauner's testimony, his professor of chemistry, Adolf Lieben, had never mentioned Mendeleev in his lectures between 1874 and 1875.[119] Otto Hönigschmid,[120] professor of the German Technical University, was probably the only German chemist in the Czech Lands at the time who was seriously working on the periodic system; however, he did not leave behind any textbook. Due to the firm barrier between the Czech and German scientific communities in the Czech Lands at the turn of the nineteenth century, it is not surprising that we have no indication about any contacts between Brauner and Hönigschmid, although they both were engaged in related problems.

6. CONCLUSIONS

The data in this essay allow us to state that the periodic system of elements had been accepted and appropriated by the Czech chemical community without any serious objections gradually since the 1880s.

As inorganic, physical, or theoretical chemistry were on the periphery of interest of most Czech chemists, the acceptance of an entirely new theoretical system, like the periodic system, was only to a small degree a matter of scholarly discussions, and the process of its appropriation in the Czech Lands was largely conditioned by non-scientific factors. The working and personal contacts of the two Slavic scholars Brauner and Mendeleev were taking place just in the period when the Czech-German antagonism dominated the Czech scientific institutions and the Czech society as a whole. Escalation of this conflict in the last two decades of the nineteenth century led to reorientation of the Czech chemists toward their Slavic counterparts and to focus on Slavic science, which was to replace the previous links to German science. Brauner, himself cosmopolitan by background and upbringing, personified the contradiction between the international character of his research and working contacts, on the one hand, and nationalistic attitudes into which he was pushed by the political and social situation in the Czech Lands, on the other. In any case, his patriotism and nationalism had never become offensive and later in his life, in the times of the afterwar democratic Czechoslovak Republic, faded.

Mendeleev was perceived in the Czech Lands chiefly as a brilliant representative of the ingenious Slavic science and, as such, honored by tributes as no other foreign scientist: he was elected the first foreign honorary member of the Society of Czech Chemists in 1880 (along with other Russian chemists Butlerov and Menshutkin, and the Polish chemist Radziszewski)[121]; he also became in 1891 the first elected non-resident (foreign) member[122] of the 2nd Class of the Czech Academy of Arts and Sciences and in 1894 the first honorary member of the Society for Chemical Industry. The Czech chemical community also expressed its appreciation of Brauner's scholarly achievements and his international fame, and, in spite of his disagreements with some chemists and his solitary position in the Czech chemical community, elected him honorary member of the Czech Chemical Society in 1905. Thus, the fact that the periodic system was the discovery of a Slavic scientist, Mendeleev, promoted on the international level by another internationally accepted Slavic (and Czech!) scientist, Brauner, played a particularly important role in the reception of the periodic system in the Czech Lands. These were the main reasons why Brauner's propagation of the Mendeleev periodic system fell in the Czech Lands on fertile soil and met with only sporadic objections.

Historians of chemistry have shown that the years 1913 and 1914 were of extraordinary significance for the further evolvement of the periodicity principle and some even consider these years as the second discovery of the periodic law.[123] The Bohr theory (1913) of the electronic structure of the atom and other seminal findings of the period, like Moseley's application of X-ray spectra to atomic structure or detection of isotopes, made apparent that the properties of the elements varied periodically with the atomic number and not the atomic weight, as proposed by Mendeleev. These changes were understood and critically recorded by Brauner and the representatives of the young generation of the Czech chemists, like Brauner's pupil and successor Křepelka and the analytical chemist Jaroslav Formánek. Not by chance, during this period also took place the only serious disagreement (relating to a certain extent to the periodic system issue) between Brauner and the Czech chemical community on the Czech chemical nomenclature reform: while the Czech chemistry community officially accepted a national nomenclature, Brauner favored the internationally accepted principles. It is necessary to accentuate that this reform not only concerned the Czech names of the elements or compounds, but also the principles of naming of mostly inorganic compounds, which differed from the rules of the international chemical nomenclature, as they were based on the linguistic peculiarities of the Czech language.[124]

The periodic system gradually penetrated into the Czech educational system (both the university and mid-range one) as systemized knowledge and methodological base to research and instruction since the turn of the nineteenth century, and was fully accepted in the first decade of the twentieth century. In the sphere of education, the appropriation of Mendeleev's ideas was practically unconditional.

The case of the reception of the periodic system in the Czech Lands demonstrates that the acceptance of a scientific discovery in a certain geographic circle is a complex, multistage process that can be conditioned by various factors; aside from the objective ones, like the level of scientific, institutional, and social advancement and preparedness of the scientific community to absorb new ideas and adapt to new paradigms, there also exist political circumstances and even delicate and subjective personal relations.

ACKNOWLEDGMENTS

The paper was supported by the Grant Agency of the Academy of Sciences of the Czech Republic, project No. IAAX00630801.

ABBREVIATIONS

PNP—Archives of the Museum of Czech Literature (PNP-Památník národního písemnictví)

NOTES

1. The role of science in the Czech National Revival until the 1870s was treated in Jan Janko and Soňa Štrbáňová, *Věda Purkyňovy doby* [Science in Purkinje's Time] (Praha: Academia, 1988); for the position of chemists in the later periods, namely in the times of exalted nationalism, see Soňa Štrbáňová, "Patriotism, Nationalism and Internationalism in Czech Science: Chemists in the Czech National Revival," in *The Nationalization of Scientific Knowledge in the Habsburg Empire (1848– 1918)*, ed. Mitchell G. Ash and Jan Surman (Basingstoke: Palgrave Macmillan, 2012), 138–156.
2. The Society has undergone several changes and reorganizations, including various modifications of its name and the title of its journal. For the sake of simplicity, the name Czech Chemical Society is used in this essay. The early history of the Society is treated in Soňa Štrbáňová, "Chapter 3, Czech Lands: Chemical Societies as Multifunctional Social Elements in the Czech Lands, 1866–1919," in *Creating Networks in Chemistry. The Foundation and Early History of the Chemical Societies in Europe*, ed. Anita K. Nielsen and Soňa Štrbáňová (Cambridge: RSC Publishing, 2008), 43–74. For the short history of the Society, see Oldřich Hanč ed., *100 let Československé společnosti chemické její dějiny a vývoj 1866–1966* [100 years of the Czechoslovak Chemical Society, its history and development 1866–1966], (Praha: Academia, 1966), 61.
3. The German chemists participated only in a few associations of chemical, food, or fermentation industry specialists, which were somewhat more open to German professionals, or even bilingual. See Štrbáňová, "Chapter 3, Czech Lands," 49.

4. There exist many sources to Brauner's biography. This essay relies especially on the following ones: Jan S. Štěrba-Böhm, *Bohuslav Brauner* (Praha: Česká akademie věd a umění, 1935) with bibliography of Brauner; Jaroslav Heyrovský, "Professor Bohuslav Brauner died February 15th 1935," *Collection of Czechoslovak Chemical Communications* 7 (1935): 51–56; Gerald Druce, *Two Czech Chemists* (London: The New Europe Publishing Co., 1944) 5–44; Susan G. Schacher, "Brauner, Bohuslav," *Dictionary of Scientific Biography* **1** (1970) 428–430; Soňa Štrbáňová, "Brauner, Bohuslav," in *Lexikon der bedeutenden Naturwissenschaftler 1. Band*, ed. Dieter Hoffmann et al. (Heidelberg-Berlin: Spektrum Akademischer Verlag, 2003), 249–251.

5. A particularly important source of Brauner's biography, namely to his relation with Mendeleev, is Brauner's obituary essay on Mendeleev. It was published several times, originally as Bohuslav Brauner, "Dimitrij Ivanovič Mendelejev," *Pokroková revue* 4 (1907–08), available as a reprint published by the author as Bohuslav Brauner, Dimitrij Ivanovič Mendelejev (Rokycany: Zápotočný, 1907). Another version of the essay with comments was published as part of the Czech edition of Mendeleev's letters to Brauner, *Dopisy Dimitrije I. Mendělejeva českému chemiku Bohuslavu Braunerovi* [Letters of D. I. M. to the Czech chemist B. B.] (Praha: Technicko-vědecké vydavatelství, 1952), 17–70. The abridged and modified English translation of the essay was made by the physical chemist Jaroslav Heyrovský (Nobel Prize 1959), Brauner's pupil; see Bohuslav Brauner "D. I. Mendělěěv as Reflected in His Friendship to Professor Bohuslav Brauner," *Collection of Czechoslovak Chemical Communications* **2** (1930): 219–243. The Czech transcription of Mendeleev's name in this article was proposed by Brauner (see 219). The whole number of the journal was dedicated to Brauner's birthday and the 50th anniversary of his doctorate; it also contains Brauner's bibliography; see "Bibliography of scientific communications published by Bohuslav Brauner," *Collection of Czechoslovak Chemical Communications* **2** (1930): 211–218.

6. For memoirs of the Brauner family written by one of its descendants, see Vladimír Hellmuth Brauner, *Paměti rodu* [The memoirs of the family] (Praha: H&H, 2000). For the relation of Brauner and Mendeleev, see 175–184.

7. Brauner, "D.I. Mendělěěv" (note 5), 230. Ernst Mach was, in the years between 1867 and 1895, professor of experimental physics at Prague University.

8. The *dozent* title (which compares to associate professor or senior lecturer) is used in some European university systems for someone who pursues an academic career, has the qualification of doctor, and passed the procedure of habilitation.

9. In Czech *Ústav pro chemii obecnou, anorganickou a analytickou.*

10. This paragraph is mostly based on Brauner's already cited biographies (note 4 and 5), and where necessary, additional sources will be referred to. Works referred to in notes 4, 5, and 6 repeatedly testify about Mendeleev's personal influence on Brauner's lifelong research project; Brauner's correspondence with Mendeleev is the most important source in this respect. Mendeleev's letters to Brauner are kept at the Archives of the Museum of Czech Literature (PNP-Památník národního písemnictví); see PNP, Brauner Bohuslav, Personal Collection, Correspondence. The letters were translated into Czech and published, as *Dopisy* (note 5); see also Blanka Ondráčková, "Poslední dopis D.I. Mendělejeva Bohuslavu Braunerovi [Mendeleev's last letter to Brauner]," *Dějiny věd a techniky* **7** (1974): 172–175. The PNP archives also contain an undated, incomplete draft of Brauner's letter to Mendeleev; it was most probably written at the beginning of February 1881, as a reply to the first letter of Mendeleev to Brauner dated January 27, 1881, which Brauner received (according to his answer) on January 31. The letters of Brauner to Mendeleev are

kept in the D. I. Mendeleev Museum and Archives of St. Petersburg State University; some of their content is described and examined in Bonifatii Kedrov and Tamara Cheptsova, *Brauner-spodvizhnik Mendeleeva* [Brauner, associate of Mendeleev], (Moskva: Izdatel'stvo Akademii Nauk SSSR, 1955); this book also presents other important data, especially on Brauner's contacts with Mendeleev. See also the chapter on Russia in this book.

11. Brauner "D.I. Mendělěěv" (note 5).
12. Lothar Meyer, *Die modernen Theorien der Chemie und ihre Bedeutung für die chemische Statik* (Breslau: Maruschke u. Berendt, 1864); Brauner probably read the third edition, which was published in 1876. See also Brauner "D.I. Mendělěěv" (note 5), 230. On Meyer and his textbook, see the chapter on Germany and the on Russia in this book.
13. Brauner to Mendeleev, draft letter of February 1881 (note 10).
14. Brauner had in mind apparently the article Dmitri Mendelejeff, "Die periodische Gesetzmässigkeit der chemischen Elemente," *Annalen der Chemie und Pharmacie*, Supplement **8** (1872): 133–229.
15. Brauner, "D.I. Mendělěěv" (note 5), 230–231.
16. Bohuslav Brauner, "O atomech a mocenstvích některých prvků, jakož i o pravidelnostech v číslech atomových [About the atoms and valence of some elements, as well as on the regularities in atomic numbers]," Listy chemické **2** (1877): 30–36, 87–93, 129–137.
17. Bohuslav Brauner, "Ueber das Atomgewicht des Berylliums (I)," *Berichte der Deutschen chemischen Gesellschaft* **11** (1878): 872–874.
18. Lars Fredrik Nilson (1840–1899), Swedish chemist who discovered scandium in 1879.
19. Brauner, "D.I. Mendělěěv" (note 5), 232.
20. Brauner, "D.I. Mendělěěv" (note 5), 227.
21. Pavel Drábek, "Profesor chemie Bohuslav Brauner [Professor of chemistry, Bohuslav Brauner]," *Rozpravy Národního technického muzea v Praze* 203 in *Řada Dějiny vědy a techniky* **15** (2007) 21–24, quotation 21.
22. On Toyokichi Takamatsu, see the chapter on Japan in this book. He wrote one of the first chemistry textbooks to mention Mendeleev's periodic law in Japan.
23. Schacher, "Brauner, Bohuslav" (note 4), 428.
24. Brauner published his key paper on the position of rare earths in the periodic table in August 1902; see Bohuslav Brauner, "Ueber die Stellung der Elemente der Seltenen Erden im periodischen System von Mendelejeff [On the Position of the Rare Earth Elements in Mendeleev's Periodic System]," *Zeitschrift für anorganische Chemie* 32 (1902): 1–30. The article also appeared in Czech and Russian. Early comparison of various types of the periodic table, including Brauner's table, is presented in G. M. Quam and Mary B. Quam, "Types of Graphic Classifications of the Elements," *Journal of Chemical Education* **11** (1934): 27–32; Brauner's table is on p. 29.
25. For references, see note 5.
26. Kedrov and Cheptsova, *Brauner-spodvizhnik* (note 10).
27. Boguslav Fedorovich, "K 150-letiu B. Braunera [150th anniversary of B. Brauner]," *Khimia*, No. 11 (2005), http://him.1september.ru/article.php?ID=200501102; page numbers are not visible.
28. See Bohuslav Brauner and John I. Watts, "On the Specific Volumes of Oxides," *Philosophical Magazine* **11** (1881): 60–65. Brauner mentions the story in Brauner, "D.I. Mendělěěv" (note 5), 233.
29. Kedrov and Cheptsova, *Brauner-spodvizhnik* (note 10), 45–46. Brauner reported in Odessa about the atomic weight of tellurium. His paper was published in Russian

as Boguslav Brauner, Ob atomnom vese tellura, Zhurnal Russkogo Fiz.-Khim. Obshchestva, October 1883; quoted from Brauner, "D.I. Mendělěěv" (note 5), 212.

30. The date 1902 comes from Brauner's own account. See Brauner, "D.I. Mendělěěv" (note 5), 242–243. Kedrov and Cheptsova, *Brauner-spodvizhnik* (note 10), 9 and Fedorovich, "K 150-letiu" (note 24), state that this last meeting took place in 1901.

31. Published in Russian as Boguslav Brauner, "O polozhenii redkozemelnikh elementov v periodicheskoi sisteme Mendeleeva [On the position of rare earth elements in Mendeleev's periodic system]," *Zhurnal Russkogo Fiz.-Khim. Obshchestva* 34 (1902): 142–153.

32. Fedorovich, "K 150-letiu" (note 27).

33. Ibid.

34. Brauner, "D.I. Mendělěěv" (note 5), 243.

35. Brauner to Mendeleev, draft letter of February 1877, see note 10.

36. *Dopisy* (note 5), 69–70, quotation translated by SŠ.

37. Brauner, "D.I. Mendělěěv" (note 5), 243.

38. Ibid., 219; highlighted by SŠ.

39. "Bibliography" (note 5).

40. Brauner, "D.I. Mendělěěv" (note 5), 219.

41. Bohuslav Brauner, "Ueber die Weiterentwicklung des periodischen Systems der Elemente [On the further development of the periodic system of elements]," *Tageblatt der 54. Versammlung Deutscher Naturforscher und Aerzte in Salzburg* 1881, 49–51.

42. Brauner himself testifies about his reading knowledge of Russian in his letter to Mendeleev written at the beginning of February 1881 (note 10): "I can read very well your clear Russian handwriting and understand everything completely in your letter . . . I would be very sorry if you would use other than the Russian language because otherwise the sense of your letters would be significantly distorted. . . .What concerns me, I would still like to use with your permission the German language which unfortunately is the universal language of the Slavs." (Translated by SŠ.)

43. This request was a consequence of Brauner's expression of displeasure in the article published with Watts regarding the unfamiliarity with Mendeleev's *Osnovy khimii* (Principles of Chemistry) in the West.

44. See Francis P. Venable, *The Development of the Periodic Law* (Easton, Pa.: Chemical Publishing, 1896), 233; Bohuslav Brauner, "The Standard of the Atomic Weights," *Chemical News* **58** (1888), 307–308; Bohuslav Brauner, "Die Basis der Atomgewichte," *Berichte der Deutschen chemischen Gesellschaft* **22** (1889), 1186–1192.

45. Gregoire Wyrouboff (Grigoriy Nikolayevich Vyrubov, 1843–1913), Russian inorganic chemist, positivist philosopher, and historian of science, 1903–1913 holder of the chair of history of science at the Collège de France.

46. Information and quotations taken from Druce, *Two Czech Chemists* (note 4), 19.

47. Brauner, "O atomech" (note 14).

48. For instance, Bohuslav Brauner, "On Cerium and Its Compound Nature," *Bulletin International České Akademie* **2** (1895): 1–6; Bohuslav Brauner, "Revise atomové váhy lanthanu [Revision of the atomic weight of lanthanum]," *Rozpravy České Akademie* **11** (1902): 1–41.

49. For instance, Bohuslav Brauner, "Experimentální studie o periodickém zákonu [Experimental Studies on the Periodic Law]," *Listy chemické* **14** (1889), 1–30.

50. Brauner, "D.I. Mendělěěv" (note 5), 231.

51. The issue is treated in detail in, for instance, several essays in Harald Binder, Barbora Křivohlavá, and Luboš Velek, ed., *Position of National Languages in*

Education, Educational System and Science of the Habsburg Monarchy, 1867–1918, vol. 11 of *Práce z dějin vědy* (Praha: VCDV, 2003), and Mitchell G. Ash and Jan Surman, ed., *The Nationalization of Scientific Knowledge in the Habsburg Empire (1848–1918)*, (Basingstoke: Palgrave Macmillan, 2012).

52. See, for instance, Janko and Štrbáňová, *Věda Purkyňovy doby* (note 1) 74–121 and 215–246; Soňa Štrbáňová and Jan Janko, "Uplatnění nového českého přírodovědného názvosloví na českých vysokých školách v průběhu 19. století [Assertion of the new Czech scientific nomenclature at the Czech universities in the 19th century]," in *Position of National Languages in Education, Educational System and Science of the Habsburg Monarchy, 1867–1918*, ed. Harald Binder, Barbora Křivohlavá and Luboš Velek, vol. 11 of *Práce z dějin vědy* (Praha: VCDV, 2003), 297–311.

53. Šafařík's chemical nomenclature was published in *Německo-český slovník vědeckého názvosloví pro gymnasia a reálné školy* [German-Czech Dictionary of Scientific Terminology for Gymnasien and Realschulen], (Praha: Kalvéské knihkupectví Bedřich Tempský, 1853). This dictionary was officially authorized by the Austrian Ministry of Culture and Education.

54. Some facts of Batěk's life and work and his relation to Brauner were taken from Batěk's memoirs, Alexandr Sommer Batěk, *Jak jsem padesát let žil a pracoval* [Fifty years of my life and work], (Praha: Kočí,1925), and the article Petr Slavíček, "Vzpomínka na da Vinciho české chemie [Remembrance of the da Vinci of the Czech chemistry]," *Chemické Listy* 102 (2008): 465–466. Alexander (or Alexandr) Batěk later adopted the maiden surname of his mother "Sommer" and became known as Sommer Batěk.

55. Archiv Univerzity Karlovy, fond Matriky Univerzity Karlovy 1882–2008, inventární číslo 1, Matrika doktorů české Karlo-Ferdinandovy univerzity I. [Archive of Charles University, Registry of the Charles University 1882–2008, inventory No. 1, Registry of the doctors of the Charles-Ferdinand University], 561.

56. Alexandr Batěk, O atomové váze ceria. Archiv Univerzity Karlovy, sbírka disertací č. 192 [Archive of the Charles University, Collection of Dissertations No. 192]. I am indebted for this information to Dr. M. Sekyrková from the Institute of History and Archive of the Charles University.

57. Batěk became later known as writer of popular chemistry and fantasy books, author of various weird inventions, spiritualist and pacifist. He also passionately opposed the atomic theory; see Alexander Batěk, "Fikce atomu a molekuly [The fiction of the atom and molecule]," *Chemické listy pro vědu a průmysl* 7 (1913): 333–335.

58. Alexander Batěk, "Návrh k opravě českého názvosloví chemického [Proposal of the correction of the Czech chemical nomenclature]," *Listy chemické* **24** (1900): 225–226.

59. Batěk, *Jak jsem padesát let žil* (note 54), 95.

60. For details of the reform, history of its acceptance, and the rules of the new Czech chemical nomenclature, see František Mašek, "Nové chemické názvosloví [The new chemical nomeclature]," *Časopis pro pěstování matematiky a fysiky*, **48** (1919): 337–342.

61. In chemistry textbooks and Votoček biographies, Votoček is often considered co-author of the Czech chemical nomenclature, or even its chief proponent.

62. For details of the nomenclature, see Emil Votoček, *Slovník sloučenin anorganických dle názvosloví sjezdového zavedeného officielně roku 1918* [Nomenclature of inorganic compounds introduced officially in 1918], (Praha: Česká chemická společnost, 1919).

63. Popular rumor said that Brauner's opposition against the new chemical nomenclature was also motivated by his lifelong animosity against his former protegé Alexander Batěk.

64. The lecture was published as Bohuslav Brauner, "O rozvoji periodické soustavy [On the Evolvement of the Periodic System]," *Chemické listy pro vědu a průmysl* **10** (1916): 238–242.

65. Nobel Prize winner in 1914 "in recognition of his accurate determinations of the atomic weight of a large number of chemical elements." Richards in his Nobel lecture pointed out that the existence of isotopes would necessarily lead to modifications of the periodic system: "*For theoretical purposes . . . not only must we learn how to separate the isotopes of lead . . . ; we must study likewise all other elementary substances, in order to find out whether they also may have atoms of differing weight. This quest has already been begun by several experimenters. Who knows what modifications our Periodic system of the elements may suffer, and what illumination it may gain, from such experiments?*" The quotation was taken from Richard's Nobel lecture posted on the Internet: http://www.nobelprize.org/nobel_prizes/chemistry/laureates/1914/richards-lecture.html, accessed on November 7, 2013.

66. Brauner, "O rozvoji periodické soustavy" 240. Highlighted by Brauner. It is useful to mention that Brauner was in close friendly and professional relation with Richards, as evidenced by the letter of 1911, where Richards thanks Brauner for his support in the matter of the Nobel Prize: "*My dear friend, I am indeed greatly honoured that you should continue to propose me for the Nobel Prize. That you in particular, who understand better than anyone else the object and value of my work on atomic weights, should wish to propose me, gives me especial pleasure.* (Richards to Brauner, January 5, 1911, PNP Brauner Bohuslav, Personal Collection, Correspondence.)

67. Bohuslav Brauner and Jindřich Křepelka, *Analysa kvalitativní pro posluchače (začátečníky) České university* [Qualitative analysis for students—beginners of the Czech University] (Praha: Česká společnost lékarnická, 1919). Křepelka worked on Brauner's recommendation with Richards at Harvard University in the years 1920 and 1921 and became in 1925 Brauner's successor at the Prague Czech university. Křepelka reported about his internship in Richards' laboratory in the meeting of the Czechoslovak Chemical Society on December 17, 1921; see Jindřich Křepelka, "Rok na Harvardově universitě v Cambridgi [One year at the Harvard University in Cambridge]," *Chemické listy pro vědu a průmysl* **16** (1922): 39–46.

68. Bohuslav Brauner, "Předmluva II," in Brauner, and Křepelka, *Analysa kvalitativní* (note 62), 161–174.

69. Ibid.,172.

70. Ibid., 171. No documents that would allow us to learn about this session and the counterarguments of the listeners are available.

71. Brauner, Dimitrij Ivanovič Mendelejev (note 5), 14.

72. Vojtěch Šafařík, *Rukověť chemie pro vysoké učení české. Díl I. Chemie anorganická* [Handbook of chemistry for the Czech Technical University. Part I, Inorganic chemistry] (Praha: Slavík a Borový,1878).

73. Ibid., 59.

74. Ibid., 58. Quotation translated by SŠ.

75. Ibid.,58. Quotation translated by SŠ.

76. Vojtěch Šafařík, *Počátkové chemie* [The beginnings of chemistry], (Praha: Fuchs, 1884).

77. See Ladislav Niklíček, Irena Manová, and Bohumil Hájek, "Význam Vojtěcha Šafaříka pro počátky české výuky chemie [The significance of Vojtěch Šafařík for the beginnings of Czech instruction in chemistry]," *Dějiny věd a techniky* 3 (1980): 129–143. The paper mentions Šafařík's personal and professional problems and his gradual diversion from chemistry to astronomy since the 1880s, in spite of his professorship

of chemistry at the Czech University. In 1880 Šafařík married his second wife Paulina, an enthusiastic amateur astronomer, who persuaded Šafařík that his scientific mission was astronomy and not chemistry. Šafařík became full professor of astronomy in 1891.

78. Brauner in his letter to Mendeleev dated October 19, 1885, complains about his difficult position at the university, where he stands in fully for Šafařík, who is in poor health and does not do any work. "*Šafařík believes that I should long have been appointed extraordinary professor. But as I am in close relation with his family, he does not want to propose my nomination because other professors could consider this undeserved favoritism.*" Kedrov and Cheptsova, *Brauner-spodvizhnik* (note 10), 48. (Citation translated from Russian by SŠ.) For explanation: Brauner married in 1886 Šafařík's adopted daughter Lidmila. They had two sons and a daughter.

79. Formánek was the author of a five-volume monograph on absorption spectroscopy of organic dyes; Jaroslav Formánek, *Untersuchung und Nachweis organischer Farbstoffe auf spektroskopichem Wege* [Investigation and proof of organic dyes by Spectroscopic Way], (Berlin: Springer, 1908, 1911, 1913, 1926, 1927).

80. Jaroslav Formánek, *Stručný nárys anorganické chemie, 2. úplně přepracované vydání s 29 obrazci* [Short outline of inorganic chemistry, 2nd fully reworked edition with 29 tables], (Praha: Ministerstvo zemědělství ČSR, 1921).

81. Ibid., 267–272.

82. Otto Hönigschmid (1878–1945) was born in Horowitz (Hořovice) in Bohemia, studied at the Prague German University under Guido Goldschmied, worked under Henry Moissan and Theodore Richards, and in the years between 1911 and 1918 was professor of inorganic and analytical chemistry at the Prague Technical University. In 1918 Hönigschmid was appointed professor at the Munich University. He estimated the exact atomic weights of approximately forty elements. For his biography, see Lothar Birkenbach, "Otto Hönigschmid 1878–1945," *Chemische Berichte* **82** (1949): XI–LXV; Heinrich Wieland, "Hans Fischer und Otto Hönigschmid zum Gedächtnis [In memory of Hans Fischer and Otto Höigschmid]," *Angewandte Chemie* **62** (1950): 1–4.

83. We may assume that Formánek expressed these views even earlier—in 1919—in the first edition of his textbook, not available in the Prague libraries.

84. Brauner, "O rozvoji periodické soustavy" (note 63), 241.

85. The following text is based upon a survey of the majority of Czech textbooks of general, inorganic, organic, theoretical, physical, and medical chemistry (approximately twenty titles) published between 1876 and 1925. The date 1876 exemplifies the year when the first information on the periodic table and Mendeleev's research arrived to the Czech Lands. The date 1925 is more or less arbitrary; it delimits the end of the Brauner era (he retired in 1925).

86. Šafařík, *Rukověť chemie* (note 71).

87. Rayman was disciple of F. A. Kekulé, A. Wurtz, and Ch. Friedel, dozent at the Czech Technical University, and professor at the Czech University. He assumed key positions in the organization of Czech science. For more details and further literature, see Štrbáňová, "Patriotism" (note 1).

88. Bohuslav Rayman, *Chemie theoretická* [Theoretical chemistry] (Praha: Borový, 1884).

89. Ibid., 32–51; the periodic system is explained on p. 34–50; the table is on p. 35.

90. Ibid., 43.

91. Karel Preis and Emil Votoček, *Anorganická chemie* [Inorganic chemistry] (Praha: Spolek českých chemiků, 1902).

92. Ibid., 341–350.

93. Karel Preis and Emil Votoček, *Anorganická chemie, 2. vydání* [Inorganic chemistry, 2nd edition] (Praha: Česká chemická společnost pro vědu a průmysl, 1909), 374–384.

94. Ibid., 379.

95. Emil Votoček, *Chemie anorganická* [Inorganic chemistry] (Praha: Česká chemická společnost pro vědu a průmysl, 1922).

96. Ibid., 566–577.

97. Ibid., 572.

98. Albert Reychler, *Les théories physico-chimiques* [The physicochemical theories] (Bruxelles: H. Lamertin, 1897).

99. Albert Reychler, *Chemie fysikálná* [Physical chemistry] (Praha: Česká akademie věd a umění, 1902).

100. Ibid., p. 31–33.

101. Ibid., p. 33–37.

102. Ibid. p, 31.

103. Jaroslav Milbauer, *Cvičení v anorganické chemii* [Practical exercises in inorganic chemistry] (Praha: Česká chemická společnost pro vědu a průmysl, 1912), 152.

104. Ibid., 152.

105. Jan Horbaczewski, *Chemie lékařská, I. díl, Chemie anorganická* [Medical chemistry, vol. I, Inorganic chemistry] (Praha: Grosman a Svoboda, 1904).

106. Ibid., 227–231; the periodic table is on p. 229.

107. Národní pedagogická knihovna J. A. Komenského.

108. *Historické učebnice fyziky (1825–1948); historické učebnice chemie (1828–1948) ve fondu SPKK* [Historical textbooks of physics (1825–1948); historical textbooks of chemistry (1828–1948) in the collections of the State Pedagogic Library] (Praha: Ústav pro informace ve vzdělávání, Státní pedagogická knihovna Komenského, 1999). I am indebted to Ms. Marie Hošková from the National Pedagogic Library for calling my attention to this collection.

109. In the Austrian school system, the so-called *Realschule, Gymnasium*, and *Realgymnasium* type high schools emphasized education in sciences. The *Realschule* was preparing pupils principally for technical studies, while the *Gymnasium* offered more general education with a strong accent on humanities; the *Realgymnasium* offered a gymnasium type education giving prominence to sciences and mathematics.

110. Jan Matzner, *Chemie anorganická pro vyšší školy reální* [Inorganic chemistry for higher Realschule] (České Budějovice: Dubský, 1903).

111. Ibid., 95–99.

112. Jan Matzner, *Chemie anorganická pro vyšší školy reální. Druhé přepracované vydání* [Inorganic chemistry for higher Realschule, 2nd Revised Edition] (Praha: Grosman, 1908).

113. Ibid., 104–107.

114. Ibid., 108–109.

115. František Mašek and Hynek Němeček, *Anorganická chemie pro 5. třídu reálky* [Inorganic chemistry for the 5th grade of Realschule] (Praha: Česká chemická společnost pro vědu a průmysl, 1910).

116. Ibid., 132–136.

117. František Mašek and Hynek Němeček, *Chemie pro reálky. 2. přeprac. vydání s chemickým názvoslovým zavedeným r. 1918 na vysokých školách v Praze a v Brně. Díl I.* [Chemistry for realschule, 2nd revised edition, with the chemical nomenclature introduced in 1918 at the universities in Prague and Brünn] (Praha: Česká společnost pro vědu a průmysl, 1918).

118. *Allgemeine chemie. Anorganische Chemie nach Un. Prof. Dr. Linnemann, Wintersemester 1881-82.* Undated.
119. Brauner, Dimitrij Ivanovič Mendelejev (note 5), 14.
120. See note 81.
121. See Hanč ed., *100 let Československé společnosti* (note 2), 61.
122. This type of membership corresponded to honorary membership.
123. See, for instance, Juriy I. Soloviov, Dimitriy. N. Trifonov, Aleksey N. Shamin, *Istoria khimii* [History of Chemistry] (Moskva: Prosveshchenie, 1984) 37. The book devotes a detailed chapter to this "second discovery"; see pp. 36–56.
124. The main principles of the Czech chemical nomenclature devised in the first two decades of the twentieth century have survived until today. Roughly speaking, the names of the compounds have suffixes that reflect the valence of the element in these particular compounds. Therefore, it is quite simple to deduce the formulas and names of the inorganic compounds in the Czech language, but because the principles differ from the Western convention, the Czechs may encounter difficulties in using the international nomenclature. I appreciate the comments of Dr. G. Palló, who reminded me of the broader significance of the nomenclature case, which shows, among others, that not only the names of some anciently known elements like iron, carbon, and so on can have national names, but also that the "grammar" (how to form the names of compounds) could be dependent on local culture.

PART IV

Response in the Northern
European Periphery
(Scandinavian Countries)

CHAPTER 7

⚬⚭⚬

When a Daring Chemistry Meets a Boring Chemistry

The Reception of Mendeleev's Periodic System in Sweden

ANDERS LUNDGREN

1. INTRODUCTION

The reception of Mendeleev's periodic system in Sweden was not a dramatic episode. The system was accepted almost without discussion, but at the same time with no exclamation marks or any other outbursts of enthusiasm. There are but a few weak short-lived critical remarks. That was all. I will argue that the acceptance of the system had no overwhelming effect on chemical practice in Sweden. At most, it strengthened its characteristics. It is actually possible to argue that chemistry in Sweden was more essential for the periodic system than the other way around. My results might therefore suggest that we perhaps have to reevaluate the role of Mendeleev's system in the history of chemistry.

Chemistry in Sweden at the end of the nineteenth century can be characterized as a classifying science, with chemists very skilled in analysis, and as mainly an atheoretical science, which treated theories at most only as hypothesis—the slogan of many chemists being "facts persist, theories vanish."[1] Thanks to these characteristics, by the end of the nineteenth century, chemistry in Sweden had developed into, it must be said, a rather boring chemistry. This is obviously not to say that it is boring to study such a chemistry. Rather, it gives us an example of how everyday science, a part of science too often neglected but a part that constitutes the bulk of all science done, is carried out. One purpose of this study is to see how a theory, considered to be important in the history of chemistry, influenced every-day science. One might ask what happened when a daring chemistry met a boring chemistry. What happened when a theory, which had been created by a chemist who has been described as "not a laboratory chemist,"[2] met an atheoretical experimental science of hard laboratory work and, as was said, the establishment of facts? Furthermore, could we learn something about the role of the periodic system per se from the study of such a meeting?

Mendeleev's system has often been considered important for teaching, and his attempts to write a textbook are often taken as the initial step in the chain of thoughts that led to the periodic system. I will therefore start by looking at how textbooks in chemistry in Sweden structured their material, before and after Mendeleev. Thereafter, I will shortly say something about the atheoretical attitude, before going into the system's effects on laboratory work.

2. TEXTBOOKS

2.1 Before Mendeleev

The structure of a textbook obviously depends on its pedagogical aims. With a good structure it should be possible to present chemical knowledge in a clear and easily understood way, and to give the students a firm basis from which to learn chemistry. Did the periodic system influence or change the structure of textbooks? The answer is, no.

The pedagogical aim of a textbook was, and is, obviously and evidently, to teach chemistry—a truism if there ever was one, and an important part of teaching chemistry, for many the most important part, involved making the student learn the characteristics of many different substances. To learn chemistry was to learn lots of facts, and the textbook authors had to facilitate the reaching of that goal by presenting chemical knowledge in the best possible way. Unsurprisingly, the presentation of such an amount of facts could be done in many different ways.

Since the elements were considered the smallest units in chemistry (more on this later), it was natural that the main structure of a textbook was determined by the characteristics of these elements. The fundamental division found in all textbooks was that of metalloids and metals. For further structuring within each of these two groups, different qualitative criteria were chosen. Among metalloids gases could constitute one group, halogens another. Alkali and alkaline metals could also constitute two different groups among the metals. Smaller subgroups such as haloids, the noble metals, or the triad S, Se, and Te were held together by similar chemical qualitative properties. Many of these groups were later also to be recognized as groups in Mendeleev's system. Such qualitatively determined groups constituted the smallest units, chapters, paragraphs, and so on in the textbooks. The order in which these different subgroups were presented could vary, but in the main the different subgroups were the same in all textbooks, regardless of how or where they were presented in the textbook. If there was a leading pedagogical principle, it was to treat the metalloids in the first chapters, since they could combine with almost all the other elements, and to discuss the other groups in the subsequent chapters.

2.2 After Mendeleev

It is easily seen in the textbooks that Mendeleev's system quickly became known in Sweden, as well as that the textbook authors themselves considered it as fruitful for chemistry and an important step in its development. Hence, the fact that the system entered the textbooks without changing them might seem paradoxical. The system was simply mentioned, although positively, more or less in passing, or in a few pages, sometimes at the end of the textbook, and it never became pedagogically important in the teaching of a science, in which uniqueness and the special property of every substance was a central issue. To structure such a teaching according to Mendeleev's system—or to any other system—did not bring any pedagogical advantages compared to already existing ways of presentation. In fact, it could make pedagogy worse. One chemist, J. O. Rosenberg (1840–1925), professor of chemistry at Chalmers technical school in Gothenburg, was generally very pleased with the system, but added that it could not be used in teaching, since that would create "great practical disadvantages."[3] Following the system would mean that some of the most frequent elements in nature would not be discussed in their proper place—the beginning of the text.

There are other examples of the non-speculative and non-dramatic way of presenting Mendeleev. In his chemical dictionary from 1882, the professor of chemistry at Uppsala University, Per Teodor Cleve (1840–1905), mentioned the system in only two pages, under the entrance "element" [grundämne]. Though he agreed that it was an important contribution to chemistry, in his textbook from 1886 Cleve presented the system in just a few pages without even mentioning Mendeleev.[4] In 1886, when Hjalmar Berwald (1848–1930), teacher in mathematics at the Royal Institute of Technology, published a textbook in chemistry for gymnasiums, Mendeleev's system was mentioned only as an addendum at the end of the book.[5] The gymnasium teacher David Kempe (1864–1949) published a small textbook, also intended for gymnasiums in 1896, in which he presented the periodic system on one and a half pages (of which one page was the table) out of 270 pages.[6] Another gymnasium teacher, Wilhelm Abenius (1864–1956), published an elementary textbook in 1903, in which he presented Mendeleev on three pages (of which one was occupied by the table), and with the very positive judgment that Mendeleev had given the key to "a more *natural system* [ett mer *naturligt system*]."[7] But as Kempe, Abenius never used it as a pedagogical tool. More examples could easily be given.

Even if the general structure of these textbooks continued to differ, also on a lower hierarchical level, the same qualitatively determined subgroups (alkaline earths, noble metals, and so on) as before Mendeleev were still discernible. Mendeleev's system presented a way to systematize the elements without changing the structure of the textbooks. Hence, the old ways to structure the books according to subgroups kept their pedagogical value after the periodic system had been introduced.

2.3 The System of C. W. Blomstrand

It should be added that other attempts to systematize the elements suffered the same fate. For example, the professor of chemistry in Lund, Christian Wilhelm Blomstrand (1826–97), presented a system of the elements in 1870 in his heavily revised third edition of the much used textbook by Nils Johan Berlin (1812–91). He also discussed this system in his own textbook from 1873 and in a more popular version in 1875.[8]

Blomstrand was the internationally best-known chemist in Sweden during these years, and he had a strong position in the chemical community in Sweden. He is also of special interest here, since Mendeleev himself once is said to have argued that Blomstrand had "theilweise [. . .] die Priorität meines Systems streitig machen [to some extent [. . .] has contested the priority of my system]."[9] Blomstrand searched for what he called a natural system of classification of elements (the classification in metals and metalloids was not natural), and he found one by using the concept of atomicity, and the electrochemical properties of the elements.[10] His point of departure was atomicity and he distinguished between elements, with atomicity expressed in odd numbers, starting with one, and in even numbers, starting with two. Since hydrogen was its simplest element, the first group was called the hydrogen group, with an atomicity that varied with uneven numbers: 1, 3, 5, and 7. The second group was the oxygen group, since oxygen was its simplest element with an atomicity that varied with even numbers: 2, 4, and 6.

With respect to electronegative properties, he constructed a subgroup of "combustories" [kombustorer]. Here he identified a one-atomicity group of haloids (Fl, Cl, Br, I) and a two-atomicity group of "amphides" [amfider] (O, S, Se, Te). Among the electropositive alkaline metals he also identified one one-atomicity group of "true alkali metals" (Cs, Rb, K, Na, Li), and one two-atomicity group of "alkaline earth metals" (Ba, Sr, Ca, Mg). Between these four groups he added two electrochemically indefinite groups, the nitrogen group (N, P, As, Sb), with atomicity 3, 5, and the coal group (C, Si, Ti, Sn), mostly with atomicity 4, in rare cases 2. Putting all these groups beside each other in a system consisting of groups and rows, he could create a table similar to the periodic system (see Figure 7.1).

A remaining problem was the metals, the place of which Blomstrand himself was uncertain: "their natural groups are still in many cases vague," he had declared, even though he discerned certain groups also among the metals.[11]

Blomstrand's system was initially constructed using atomicity and the electrochemical properties of the elements. Having done so, he became aware of, in his own words, an extraordinary regularity when looking at the atomic weights—for example, the fact that in some groups, determined by atomicity, the positive electrochemical character increased with increasing atomic weight and that in some groups the negative electrochemical character decreased. He also noted that some

TABELLARISK ÖFVERSIGT AF DE VIGTIGARE GRUPPERNA:

	I.	II.	III.	IV.	V.	VI.	VII.
				^{I} H 1			
	^{II} Be (beryllium) 9.4	^{III} B (bor) 11	^{IV} C (kol) 12	$^{III, V}$ N (kväfve) 14	^{II} O (syre) 16	^{I} Fl (fluor) 19	
Li (lithium) 7	Mg (magne-sium) 24	Al (alumi-nium) 27.4	Si (kisel) 28	P (fosfor) 31	$^{II-VI}$ S (svafvel) 32	$^{I-VII}$ Cl (klor) 35.5	
Na (natrium) 23	Ca (calcium) 40		Ti (titan) 50	As (arsenik) 75	Se (selen) 78	Br (brom) 80	
K (kalium) 39	Sr (strontium) 87	Y (yttrium) 89	Zr (ziroc-nium) 89.4	Sb (antimon) 122	Te (tellur) 128?	I (jod) 127	
Rb (rubidium) 85	Ba (barium) 137	La (lanthan) 139	Sn (tenn) 118	Bi (vismut) 210			
Cs (cæsium) 133	Pb (bly) 207						
Tl (tallium) 204							

Figure 7.1 Blomstrand's table of the most important element groups, *Naturens grundämnen i deras inbördes ställning till hvarandra* (Stockholm: Klemmings, 1875), 36.

properties were regularly repeated in the system, noticing the same phenomenon as Mendeleev had pointed out. The differences from Mendeleev were, of course, firstly that the atomic weights had not been a prerequisite for finding the system but only later became part of it, and secondly that since the metals were not included, the system was not complete. It is therefore only with utmost difficulties, if at all, that Blomstrand's system could be seen as a "forerunner" to Mendeleev, in spite of their superficial similarities. Rather, it was one of many contemporary attempts to classify the elements.

In later textbooks, Blomstrand's system, if mentioned at all, suffered the same fate as Mendeleev's—it was described in few pages, and never used to structure the textbooks. Cleve referred to it in his textbook from 1872, without mentioning Blomstrand, but for practical reasons did not use it.[12] Had he used Blomstrand's system to structure the book, oxygen as a two-atomicity element would have been treated after the one-atomicity elements (chlorine, bromine, iodine, fluorine). However, from a pedagogical point of view, oxygen was best treated immediately after the first one-atomicity element, hydrogen, since that would give a possibility of discussing the importance of water early in the text.[13] Nevertheless, Blomstrand himself had a strong belief in his system and lectured on it at least as late as 1888.[14]

The parallel between the fate of Blomstrand's system and that of the periodic system is obvious. In the chemical pedagogical tradition reigning in Sweden, no system or structure was better (or worse) than any other. I have not been able to trace any competition or discussion about pros and cons concerning Blomstrand's system versus Mendeleev's. Pedagogical considerations were much more important than theoretical commitments.

3. BEING ATHEORETICAL

Though theoretical discussions were never prominent among scientists in Sweden, Blomstrand was an exception.[15] Recall the slogan "facts persist, theories vanish"; to avoid theories and speculations was an outspoken ideal among chemists in Sweden. Carl Gustaf Mosander (1797–1858), pupil to Berzelius and professor in chemistry and pharmacy at the Pharmaceutical Institute in Stockholm, as well as discoverer of four new elements, almost bragged of his lack of interest in theory.[16] Some chemists privately claimed to have taken an "agnostic standpoint,"[17] implying that the question of why classification looked as it did was of no interest to natural scientists, but rather a metaphysical problem. At the same time, chemists in Sweden used philosophically complex concepts such as atoms, elements, and simple bodies, but did so in an intuitive, unreflective, "naïve" way. The ontological/philosophical debate on these concepts was almost nonexistent. The practical result of this attitude was that the simplest substance found by analysis determined what kind of atoms there were, and therefore also what atomic weights should be determined, regardless of theoretical demands and the philosophical status of these elements. This attitude was facilitated by the complete consensus on which substances were simple and elementary; "for the moment there is no uncertainty with respect to the division between simple and compounds bodies."[18] Probably Blomstrand expressed a very common attitude when he said that he used the atomic concept "in its most simple material meaning and freed from the nimbus of philosophical speculation in which it [had] been wrapped."[19] If the question of smaller constituents were to be discussed, such a discussion would "under all circumstances be outside of the field of chemistry."[20]

However, this methodological attitude also gave rise to the only instance of critical response to Mendeleev that I have found. This response came from one of the most advanced analytical chemists in Sweden—a "pure" laboratory chemist, who experienced all systems as too hypothetical. When the associate professor of chemistry at Uppsala University, Lars Fredrik Nilson (1840–99), determined the atomic weight of beryllium, his results did not fit the supposed place of the element in the system. He doubted not his own results but rather the system, which he would not take for granted, and certainly not "only as a mere doctrine."[21] At the very least he wanted the system to be better experimentally verified, a claim scientists in Sweden routinely demanded from any theory. Soon thereafter, chemists in Sweden could give such an experimental support.

4. IN THE LABORATORY

The results of chemical analysis were crucial for the fate of the periodic system. Discoveries of new elements could contribute to verifying the system, as could determination of atomic weights. Both these fields were strong in Sweden, and both involved extensive laboratory work.

4.1 New Elements

Chemists in Sweden had since the eighteenth century routinely taken part in the discovery of new elements. Mosander became one of the most known chemists during this time thanks to his isolation of many of the rare earth metals, even though he did not contribute with anything else to chemistry.[22]

During the nineteenth century, discoveries of many new elements were reported and seriously discussed, including elements not considered as such today. Berzelius's discovery of Gahnium almost went into print, and he also reported on Vestaeium.[23] Blomstrand referred, although not fully convincing, to the existence of two new possible elements, niobium and pelopium.[24] Johan Fredrik Bahr (1805–75), teacher in chemistry at Uppsala University, had found wasium in no less than three different minerals.[25] During a meeting with The Royal Academy of Science (KVA), Nilson orally reported on the new research by his German colleague Gerhard Krüss (1859–95), "according to which cobalt and nickel, in the so far *purest possible state reached,* both contain a foreign metal, not possible to identify with any so far known element" (my italics), and which they named gnomium.[26] The hope of finding new elements never died. In 1891 A. E. Nordenskiöld (1832–1901) also orally reported in KVA on "a pyrite which seems to contain a new simple substance."[27] However, the promised article was never published, indicating too much optimism concerning the existence of this simple substance.

Many more examples could be given.[28] The belief in the existence of more elements is easy to understand. If Mosander had found no less than four new elements, why should there not be many more? In addition, the pleasure of being a "discoverer" must have been tempting.[29] To explain the many "false" elements that were reported simply by dismissing the discoverers as bad chemists would certainly be too simple. True, the average chemist in Sweden during the second half of the nineteenth century was certainly not a Berzelius, a Kekulé, a Laurent, or a Ramsay, but he was generally a good analytical chemist of an average standard, and that standard was high in Sweden.[30]

4.2 Hard Laboratory Work and Atomic Weights

Chemical analysis was a hard job. It took time, certain operations had to be carried out hundreds of times, and, said Cleve, it wasn't until "after many years of work I [. . .] finally succeeded to isolate the real erbium earth."[31] The otherwise unknown chemist J. A. Alén from Uppsala University remembered that "after $2^{1}/_{4}$ year the salt was still completely intact."[32] Nilson started with 10 kilos of the rare mineral euxenite, in order to produce 20 grams of ytterbium, and he needed 68 decomposing series, which was "trying and time-consuming."[33]

It should come as no surprise that chemists in Sweden showed great admiration for the experimental work of Jean Servais Stas (1813–91), who dedicated a tremendous amount of time (and money) to an enormous amount of analyses, which sometimes meant, said Rosenberg; "*60 hours of unbroken cleaning of a precipitation in a dark room*, to which Stas adds that he 'did not often repeat this analysis, which almost went beyond the strength of a human being'" (emphasis in original).[34] The professor of chemistry at Lund University, Nils Johan Berlin, said in 1860 that only "with the outmost pain" was it possible to separate lanthanoxide and didymoxide.[35] Such work led to the discoveries of new elements, and was also necessary if chemists wanted to determine the chemical characteristics of these new elements. Obviously, one of these characteristics was their atomic weights, and to be able to determine them as exactly as possible, pure substances were needed. To produce such substances was a job of the able analyst—and still, the problem of exact weighing remained.

Focus on experimental work and on atomic weight determinations meant that theory stayed in the background. Also, it strengthened the consensus during the second half of the nineteenth century on what substances were to be considered elements and what were their atomic weights. This consensus was a prerequisite for Mendeleev, not one of his results. New elements were reported without reference to the periodic system, as Nilson's announcement of the discovery of scandium.[36] The similarity between scandium and Mendeleev's eka-boron was instead later pointed out by another chemist (Cleve).[37] Nilson himself did not until somewhat later point out that his results were "the more interesting since they coincide with the atomic weight Mendeleev calculated for the predicted element eka-boron = 44."[38] His belief in the system was at the same time strengthened by the news of the discovery of gallium, which he considered "the best confirmation of the validity of these speculations by the Russian chemist."[39] Mendeleev saw it the same way, calling Nilson a "true corroborator."[40] Mendeleev also wrote a well-known letter to Nilson saying that the latter's determination of the atomic weight of beryllium, and his work on scandium, was a proof "dass sie sich auf den Standpunct des periodischen Systems gestellt hatten [that he had accepted the periodic system],"[41] even if Nilson's intention never was to verify the periodic system. Looking from the analyst's point of view it would be possible to argue that it was Mendeleev who verified the analysis, not that the analysis verified Mendeleev.

Analytical methods did not change by the coming of Mendeleev's system; rather, there was a strong continuity in research in Sweden. This is obvious from the many examples given by Cleve in his various analyses of the rare earth metals, from the beginning of the 1870s to the beginning of the 1880s.[42] In 1887 Nilson, together with Krüss, published a long series of articles on thorium, niobium, and other rare earths, all very descriptive, but at the same time accepting the periodic system, without making too much of it.[43] In this matter there actually are some noteworthy similarities between Nilson's discovery of scandium and Clemens Winkler's of germanium. The latter had a background in mineralogy, was an accomplished analyst, and, like Nilson, did not obtain his results after having consulted Mendeleev. During certain periods Winkler actually cooperated with Nilson, and sent him germanium material for further analysis.[44] Winkler also initially thought that the new element wasn't the predicted ekasilicon but rather the predicted ekastibium.[45]

It is important to stress that there were no conflicts between the periodic system and the analytical tradition. Rather, they complemented each other. In relation to the analytical tradition, the acceptance of Mendeleev was so smooth that in 1892, when Cleve reported on the most important recent research in inorganic chemistry, he did not mention the periodic system, but conveyed the analytical results in a way that indicates that the system by now was part of standard knowledge in chemistry.[46]

5. TEXTBOOKS, ATHEORETICAL CHEMISTRY, AND ANALYTICAL CHEMISTRY

The structure of textbooks, the atheoretical attitude, and the advanced laboratory technique were effected in similar ways to Mendeleev's system: they all continued unbroken as they had been carried out before the coming of the system. The periodic system did not change the structure of Swedish textbooks, and in this Sweden was not unique.[47]

The atheoretical attitude shown by chemists in Sweden could also, it has been argued, be seen as shared by Mendeleev, since he did not formulate a hypothesis as to why his system looked the way it did, and since he was skeptical toward ideas

such as Prout's hypothesis.[48] Neither did he think it was necessary to have a theoretical understanding of the causes behind the system prior to its being put to
use.[49] I cannot but think that Mendeleev's ideas here paralleled the attitude toward
theory among chemists in Sweden and that the common disregard for philosophical and theoretical questions, even if it led to some small objections, basically facilitated the acceptance of the system. Actually, a closer study of Mendeleev's thoughts
might show another attitude, but my interest here is the way in which chemists in
Sweden could read Mendeleev. And one such possible reading was to consider him
as uninterested in philosophy.

Practical analytical work in the laboratories continued as before, and it was
exactly this continuity that facilitated an acceptance of Mendeleev's system.
Furthermore, it was this analytical skill that converted the discoveries of the
eka-elements to empirical facts and thereby converted the last remaining doubts
concerning the periodic system into acceptance. Once the periodic system had been
verified by analysis, the system could actually be used as a standard against which
experimental results could be measured. If experimental work had given two considerably different atomic weights for one element, the place of the element in the
periodic system could be used as an arbiter when choosing between them.

Thus, in Nilson and Krüss's work on thorium, the possibility that thorium carried the valence 2 was discarded because in that case thorium "with difficulties
could be placed in the natural system of the elements."[50] According to Blomstrand,
if there were difficulties in determining the formula of an oxide, and thus its atomic
weight, the verdict could come from the periodic system; "the periodic system has
decided the matter."[51] Åke Gerhard Ekstrand (1846–1933), in a late memorial of
Nilson, explicitly remembered that, "when studying the atomic weight of new elements one therefore had a certain support in the periodic or natural system."[52]

Finally, there is also the possibility that the acceptance of the periodic system
was facilitated by the fact that Mendeleev's first publications were in the field of
mineralogy. Hence, they can be said to have combined classification and analytical chemistry, in a way that was familiar to all chemists in Sweden. According to
Masanori Kaji, Mendeleev's work in mineralogy showed a "talent for compiling and
systematizing large amounts of data."[53] As a talent for systematizing was endemic
in Linnaean Sweden, chemists there recognized such a capacity when seeing it,
even if first-hand knowledge of Mendeleev's mineralogical work was probably rare.

6. CHEMISTRY, PHYSICS, AND SPECTRAL ANALYSIS

The acceptance of the periodic system in Sweden was also closely linked with developments in physics. First of all, the features of chemistry in Sweden—the wish
to classify, the analytical skill, and the atheoretical attitude—were shared by
other sciences in Sweden during the late nineteenth century. Thus, the mapping
of physical phenomena such as spectral lines, stars, and magnetic inclination was
one main preoccupation of physicists in Sweden, and it can be said to correspond to
the wish to classify in chemistry, and the use of advanced techniques for precision

measurements in physics corresponded to the minute analysis among chemists, even if the instruments used by chemists were not as complicated or delicate as those used by physicists. Finally, many physicists philosophically embraced the same ideas about theories and hypotheses as the chemists, and remained essentially atheoretical.[54]

An important part of the story of the periodic system in Sweden relates to the development of spectral analysis in physics. Since every element had a unique spectrum, spectral analysis became an important means to find, identify, observe, and distinguish among different elements as well as to identify new ones. For Nilson, spectral analysis was a strong argument for initiating the research that led to the discovery of scandium, since it "showed some lines, which could not be found for any known substances."[55] When N. A. Langlet (1868–1936), assistant to Cleve, analyzed the gases evolving from the mineral cleveite and among them identified helium, he used spectroscopy.[56] Johan Bahr had used spectral analysis to argue for the new element wasium (see earlier).

The physicists themselves also contributed to the field, especially the professor of physics in Uppsala, Robert Thalén (1827–1905), with whom Cleve regularly worked. [57] Thalén published and wrote articles on the spectral lines of many of the rare earth metals, and Nilson, Cleve, and others almost always referred to him in their articles. But all publications in the field by both chemists and physicists remained descriptive and almost clinically free from outspoken theoretical statements. They were simply presented as the result of long work and many observations that had been difficult to carry out—sometimes extremely difficult.[58]

An important similarity between physics and chemistry was the technical difficulties and the tedious work in the laboratory. Reading spectral lines was a skill, which required extensive practice and training of how to judge the intensity of certain lines, as well as the possibility of distinguishing between different lines. Thalén had "no end of trouble to decide if these two lines coincide[d] or not," and he mentioned that differences in wavelengths are so small "that measurements, done the usual way, could not answer the question with complete certainty."[59] This is not to even mention a mundane problem such as light—it happened that the weakest lines in the spectra could not be seen "because of lacking strong sunlight."[60] Furthermore, the more-than-able physicist from Lund University Janne Rydberg (1854–1919) experienced the insufficient exactness of spectral analysis. He could not say if argon was a unique element or not, because when trying to see the supposed double spectral line of argon, "the precision of the measurements did not suffice for the present purpose."[61]

There was (is?) nothing like a simple and smooth reading of the result of a chemical analysis or of a spectrum; rather, the observer's judgment had to be trusted. But despite these problems, basic trust in the ultramodern technique of spectral analysis remained strong and it became an important complement to traditional chemical analysis. Spectral lines could, as the periodic system in itself, be used as standards. The proof that a sample was pure could be that it "had been investigated [by Thalén] through spectral analysis."[62]

Such a use of spectral analysis in chemistry did not in itself decisively contribute to the acceptance of the periodic system among chemists in Sweden. But it was instrumental in the sense that it could verify, or at least disprove, the existence of most of the new elements reported by chemists, and that it could demonstrate the pureness of a certain sample, which in turn was important when its atomic weights and other characteristics were to be determined. In Sweden spectral analysis and chemical analysis existed in a fruitful symbiosis at the end of the nineteenth century.

7. CONCLUSIONS: OR WHAT DID THE PERIODIC SYSTEM DO IN SWEDEN?

If the periodic system were that easily accepted, it should be legitimate to ask for an eventual effect. But from what has been said, not much can be found. The most important contribution of the periodic system in Sweden was rather to reinforce the way chemistry already was done. It reinforced, but certainly did not create, the work for a more exact analysis. Such analysis, as well as more exact atomic weights, was part of the description of the unique properties of chemical elements and substances, which were considered a goal for chemistry.[63] Fulfilling this goal could now also be motivated by a wish to make Mendeleev's system as complete as possible. Otto Pettersson (1848–1941) and Gustaf Ekman (1852–1930) argued that since the atomic weights of oxygen and sulphur had been carefully determined, it was now time for more exact determination of the values of the other elements in "the group of elements to which selenium belongs."[64] This group had certainly been considered a group before, but its place in a natural system increased motivation to extend research to cover all its members.

Also, the search for new elements to fill out the gaps was helped in the sense that the search for such a new element could be directed toward the minerals where other elements in the same group had been found.

Another possible side effect of the acceptance of the periodic system was that the search for new elements became more focused, since the periodic system set a limit to the amount of elements that could be discovered. A marked decline of reports on new elements can be seen toward the end of the century. None of the earlier systems, including Blomstrand's, had given such a limit to the number of new elements. This "no-limit attitude" partly explains the many earlier reports of new elements, which later turned out not to be elements. Among some chemists spectral analysis at first also seemed to have contributed to an increased belief in the possibility of finding new elements. Nilson and Krüss referred to this method when they, admittedly according to a rather late source, declared that it was in principle possible to find more than twenty new elements, using spectral analysis.[65]

Of course, the fact that many new elements so often turned out to be mixtures of already known ones made the chemists themselves suspicious of such reports. But it also seems reasonable to conclude that the periodic system, by allowing only a definite amount of new elements, restricted the search for new elements, or at

least made chemists more careful to announce new elements. Under all circumstances, the still most commonly used argument against the reports of new elements was more and better analysis.

The periodic system was compatible with the chemistry done in Sweden. That
the predictions of Mendeleev were corroborated by empirical analysis, both in
chemistry and in physics, contributed to its acceptance. The more difficult empirical
tests a theory could withstand, the more it could be accepted as a theory—such was
the view of the atheoretical scientists in Sweden. Mendeleev's system passed that
test relatively easily, but chemists in Sweden never did, thanks to their atheoretical
attitude, ask for an explanation as to why regularities occurred in Mendeleev's system. To give such an explanation would be to propose a hypothesis. To the empirical chemists in Sweden the system remained a description of empirical facts, and a
description that did not have to connect with any unproved metaphysical hypothesis on the ultimate structure of matter.

The system never became relevant for much of what was going on in chemistry
in Sweden. The writing of textbooks continued as before. In fact, to use a natural
system for textbooks is rarely a good pedagogical idea, since it would make descriptions much more complex. Neither did the periodic system help the chemists in one
of their most important pedagogical tasks: to describe and to distinguish between
different elements. Analytical work in the laboratory continued as earlier, and at
most the system had the not very dramatic consequence of contributing to a decline
in the search for new elements. To predict chemical properties of a certain element
was one thing; but it also had to be shown analytically that the predictions were
correct, and that meant advanced handicraft work within chemistry.

It was the extreme carefulness of daily analytical work that strengthened the
periodic system, rather than the other way around. Furthermore, analytical results
obtained by chemists in Sweden could be used to argue for the system in other
places, countries, laboratories, and so on. Perhaps one could say that Mendeleev's
periodic system transited chemistry in Sweden—with its classifying, analytical,
and atheoretical features—and that the system thereby became strengthened in its
future travels, even in its travels back to Mendeleev.

Could one conclude something about the general importance of Mendeleev's
system from its smooth and unproblematic acceptance in Sweden? Are we entitled
to draw any conclusion about Mendeleev and the significance of the system for the
development of chemistry in general from this case? It is generally said that the
periodic system and its history and development was a most important discovery.[66]
It is also a view shared by most chemists since the introduction of the system. But
in what way was/is it important? If a "big" chemical theory meets a "small" and
"boring" chemistry, and the theory becomes accepted but nothing happens—how
"big," then, is the theory? The short-lived and yet not very dramatic "fight" over the
atomic weight of beryllium cannot change this view, especially since that dispute
quickly was solved within existing academic frames.[67] It is indeed difficult to point
to any new essential chemical work brought about by the acceptance of the periodic
system—to find a new element can hardly be considered as something essentially
novel. In Sweden the periodic system was certainly not considered revolutionary to

the everyday laboratory worker, but how unique was Sweden in this case? It would be surprising if it·was very unique.

If the periodic system did not change much in Sweden, and if Sweden is not unique, which undoubtedly must be the case, would it be possible to suggest that the system was not that important at all? Could it be that its prominent place on the walls in every lecture hall for chemistry (or almost any other science lecture hall) has made us believe it was more important to chemistry than it actually was when it was presented? The Swedish historian Yvonne Hirdman has once, in an attempt to open new fields for historical research in gender studies, talked about "the tyranny of definite events [de bestämda händelsernas tyranni]."[68] That is, we have with hindsight determined what events were important in history, and concentrated our historical efforts on those. But considering the little effect Mendeleev's system had on textbooks, the little effect it had on laboratory work, and the fact that it did not initiate new substantial scientific work, what would happen if we looked beside the periodic system, and considered it an event that has "tyrannized" the history of chemistry? Could it be that its main role became to legitimize already existing chemical knowledge rather than to create new knowledge? With all certainty, many will disagree with such a conclusion, which, until we have more comparable material, still has to be tentative. But I think the question is worth asking, and even if I cannot rule out the possibility that a future answer to such a question could sharpen the arguments in favour of the importance of Mendeleev's system, I sincerely doubt that would be the case. Historical studies on the periodic system have often, perhaps too often, discussed its philosophical aspects, instead of the relation between everyday chemistry and Mendeleev's system.

NOTES

1. This paragraph is based on my earlier studies of nineteenth-century chemistry in Swedish. In the references, the abridgment *Öfversigt KVA* will stand for *Öfversigt af Kongl. Vetenskaps-Akademiens Handlingar* [Survey of the Transactions of the Royal Academy of Science] (Stockholm); *KVA Bihang* for *Bihang till Kongl. Svenska Vetenskapsakademiens Handlingar* [Supplement to the Transactions of the Swedish Royal Academy of Science] (Stockholm).
2. Michael D. Gordin, *A Well-Ordered Thing: Dmitrii Mendeleev and the Shadow of the Periodic Table* (New York: Basic Books, 2004), xviii.
3. J. O. Rosenberg, *Lärobok i oorganisk kemi* [Textbook in Inorganic Chemistry] (Stockholm: Norstedts, 1888), 126; 2:a uppl. [2nd ed.], 1892, 128; 3:e uppl. [3d ed.], 1903), 139. The only difference among the editions is that in the third edition the noble gases have been added. Here and thereafter all the translation from Swedish is mine (Anders Lungren).
4. Per Teodor Cleve, *Kemiskt hand-lexikon* [Chemical Dictionary] (Stockholm: Seligmann, 1883), 146–148; idem, *Lärobok i kemiens grunder*, 3. omarbetade uppl. [Textbook in the Foundation of Chemistry, 3d Revised Edition] (Stockholm: Geber, 1886), 178–180.
5. Hjalmar Berwald, *Lärobok i oorgansik kemi för läroverken* [Textbook in Inorganic Chemistry for Secondary Schools] (Stockholm: Norstedts, 1886), 134–136.
6. David Kempe, *Lärobok i oorganisk kemi för allmänna läroverken* [Textbook in Inorganic Chemistry for Secondary Schools] (Stockholm: Norstedts, 1896), 234–236. The reading is the same in the second and third editions (1900 and 1903).

7. Wilhelm Abenius, *Elementär lärobok i oorganisk kemi* [Elementary Textbook in Inorganic Chemistry] (Stockholm: Bille, 1903), 32–34 (italics in original).

8. N. J. Berlin, *Elementar-lärobok i oorganisk kemi*, 3:e upplagan bearbetad af C. W. Blomstrand [Elementary Textbook in Inorganic Chemistry, 3d ed. revised by C. W. Blomstrand] (Lund: Gleerup, 1870), 597–602; C. W. Blomstrand, *Naturens grundämnen i deras inbördes ställning till hvarandra* [The Elements of Nature and their Mutual Relationship] (Stockholm: Klemmings, 1875). I have here mostly used the last item.

9. Quoted from J. W. Spronsen, *The Periodic System of Chemical Elements: A History of the First Hundred Years* (Amsterdam: Elsevier, 1969), 343. I have not found any such claims on behalf of Blomstrand, who in *Kort lärobok i oorganisk kemi*. 4:e uppl. [Short Textbook in Inorganic Chemistry. 4th edition] (Lund: Gleerup, 1897), 253 called Mendeleev the founder of the periodic system.

10. Blomstrand, *Naturens grundämnen* (note 8), 21–32.

11. Blomstrand, *Naturens grundämnen* (note 8), 31. Such groups could be the Fe-group (Fe, Mn, Cr, Ni, Co), the W-group (W, Mo), the Ta-group (Ta, Nb), and so on.

12. Cleve, *Lärobok i oorganisk kemi* [Textbook in Inorganic Chemistry] (Stockholm: Seeligmann, 1872), 32.

13. Cleve, *Lärobok* (note 12), 35.

14. "Anteckningar från prof. C.W. Blomstrands offentliga föreläsningar öfver oorganiska kemin H.T. 1884 af S[imon] Bengtsson" [Notes from prof. C. W. Blomstrand's public lectures in inorganic chemistry during autumn 1884, by S[imon] Bengtsson], 10, 26 (my pagination), Ms Blomstrand, Lund University Library. These lecture notes contain material also from the period after 1884.

15. This paragraph draws on my unpublished Ms "Looking at the atomic theory from below: Atomism in chemical textbooks during the 19th century."

16. A. Lundgren, "Carl Gustaf Mosander," *Svenskt biografiskt lexikon* 25 (Stockholm: Svenskt biografiskt lexikon, 1987), 739–742.

17. Adolf Nordenskiöld to Carl Wilhelm Blomstrand, 10.2.1897, Ms Archives of The Royal Academy of Science, Stockholm.

18. G. Swederus, *Natur-vetenskaperna populärt afhandlade med tillämpning på industrien, m.m.* [A Popular Treatise on the Sciences and their Application in Industry, etc.] (Stockholm: Fahlstedt, 1869), 377. This text was a revised edition of a French original.

19. C. W. Blomstrand, *En blick på våra dagars naturforskning* [A Look at the Sciences of Today] (Lund: Lunds universitet, 1874), 8.

20. Blomstrand, *Naturens grundämnen* (note 8), 64.

21. L. F. Nilson & Otto Pettersson, "Om Berylliums atomvigt och väsendtliga egenskaper [On the Atomic Weight and Essential Properties of Beryllium]," *Öfversigt KVA* 37:6 (1880), 42; on the beryllium "fight," see Spronsen, *The Periodic System* (note 9), 300–302.

22. A. Lundgren, "Carl Gustaf Mosander" (note 16).

23. J. Berzelius, *Lärbok i kemi*, vol3 [Textbook in Chemistry] (Stockholm, 1818), 464. Cf James L. & Virginia R. Marshall, "Reinvestigating Vestium: One of the Spurious Platinum Elements," *Bulletin for the History of Chemistry* 35:1 (2010), 33–39.

24. C. W. Blomstrand, "Om metallsyrorna af tantalgruppen, samt några mineralier, hvari dessa syror ingå [On the Metallic Acids in the Tantalum Group, and on Some Minerals which Contain Them]," *Öfversigt KVA* 21 (1864), 541–558.

25. J. Bahr, "Om en ny metalloxid [On a New Metallic Oxide]," *Öfversigt KVA* 19 (1862), 415–423. Ten years earlier, however, Bahr had been critical toward the claim by

Clemens Ullgren (1811–68) to have found the new metal aridium, "Något om den förmodade metallen Aridium [On the Supposed Metal Aridium]," *Öfversigt* KVA **9** (1852), 161–173.

26. "Öfversigt af sammankonstens förhandlingar onsdagen den 13 feb. 1889 [Minutes from the Meeting February 13, 1889]," *Öfversigt* KVA **46** (1889), 40.

27. "Öfversigt af sammankonstens förhandlingar onsdagen den 10 juni. 1889 [Minutes from the Meeting June 10, 1889]," *Öfversigt* KVA **48** (1891), 359.

28. See the German contribution, this volume.

29. The national motive for finding gave rise for many national combats on the discovery of hafnium. The French chemist Georges Urbain (1872–1938) wanted it to be called "celtium," whereas George von Hevesy (1885–1966) and Dirk Coster (1889–1950), working at Niels Bohr Institute in Copenhagen, originally proposed "danium" before being satisfied with "hafnium." Helge Kragh, "Anatomy of a Priority Conflict: The Case of Element 72," *Centaurus* **23** (1980), 275–301.

30. Needless to say, not only chemists in Sweden discovered new elements. One of the most well-known cases is Thomson's discovery of junonium. Thomas Thomson, "Ein neues Metal, Junonium [A New Metal, Junoium]," *Annalen der Physik* **42** (1812), 115–116.

31. P. T. Cleve, "Om Erbinjorden [On Erbium Earth]," *Öfversigt* KVA **37**:7 (1880), 3.

32. J. E. Alén, "Etylsvafvelsyrade salter [Salts of Etylic Sulphuric Acid]," *Öfversigt* KVA **37**:8 (1880), 39.

33. L. F. Nilson, "Om Ytterbiums atomvigt [On the Atomic Weight of Ytterbium]," *Öfversigt* KVA 37:6 (1880), 5.

34. J. O. Rosenberg, *Den kemiska kraften framställd i dess förnämsta verkningar* [The Chemical Force Described through Its Most Important Effects] (Stockholm: Fahlkrantz, 1887), 231.

35. N. J. Berlin, *Elementar-lärobok i oorganisk kemi,* 2.uppl [Elementary Textbook in Inorganic Chemsitry, 2nd ed.] (Lund: Gleerup, 1860), 348.

36. L. F. Nilson, "Om Scandium, en ny jordmetall [On Scandium, a New Rare Earth Metal]," *Öfversigt* KVA **36**:3 (1879), 47–51.

37. P. T. Cleve, "Om Skandium [On Scandium]," *Öfversigt* KVA **36**:7(1879), 3–10.

38. L. F. Nilson, "Om Scandiums atomvigt och några karakteristiska Scandium-föreningar [On the Atomic Weight of Scandium and some Typical Scandium Compounds]," *Öfversigt* KVA **37**:6 (1880), 19.

39. Nilson, "Om Scandiums atomvigt," (note 38), 30.

40. Stephen Brush, "The Reception of Mendeleev's Periodic Law in America and Britain," *Isis* **87** (1996): 595–628, p. 610.

41. Mendeleev to Nilson 18/4 1884, Ms Archives of The Royal Academy of Science, Stockholm. Mendeleev also asked Nilson for pieces of scandium and ytterbium to be able to study "der specifischen Gewichte von Salzlösungen [the specific gravity of its salt solution]."

42. P. T. Cleve & Otto M. Höglund, "Om Yttrium- och Erbium-föreningar [On Yttrium and Erbium Compounds]," KVA *Bihang* **1**:8 (1873); idem, "Bidrag till jordarts-metallernas kemi I. Torium [Contribution to the Chemistry of the Rare Earth Metals. I. Thorium]," KVA Bihang 2:6) (1874); idem, "Bidrag till jordartsmetaller-nas Kemi II. Lantan [Contribution to the Chemistry of the Rare Earth Metals. II. Lanthanum]," KVA *Bihang* **2**:7 (1874); idem, "Bidrag till jordartsmetallerna kemi III. Didym [Contribution to the Chemistry of the Rare Earth Metals. III. Didymium]," KVA *Bihang* 2:8 (1874); P. T. Cleve, "Om Yttriums atomvigt [On the Atomic Weight of Yttrium]," *Öfversigt* KVA **39**:9 (1882), 3-7; idem, "Om Lantans atomvigt [On

the Atomic Weight of Lanthanum]," *Öfversigt* KVA **40**:2 (1883), 15-21; idem, "Om Didyms atomvigt [On the Atomic Weight of Didymium]," *Öfversigt* KVA **40**:2 (1883), 23–26; idem, "Om Samarium [On Samarium]," *Öfversigt* KVA **40**:7 (1883), 17–26.

43. L. F. Nilson and G. Krüss, "Om Thoriums equivalent- och atomvigt [On the Equivalent- and Atomic Weight of Thorium]," *Öfversigt* KVA **44**:5 (1887), 251–265; idem, "Om jordarterma och niobsyran i fergusonit [On the Rare Earths and Niobic Acid in Fergusonite]," ibid., 267–285; idem, "Om produkten af niobfluorkaliums reduktion med Natrium [On the Product of Niobfluorpotash with Sodium]," ibid., 287–297; idem, "Om kaliumgermanfluorid [On Potassiumgermaniumfluoride]," ibid., 299–303. Krüss was staying in Sweden during the autumn 1886.

44. Nilson & Krüss, "Om kaliumgermaniumfluorid," (note 43), 299.

45. Eric Scerri, *The Periodic Table: Its Story and Its Significance* (Oxford: Oxford University Press, 2007), 138.

46. P. T. Cleve, "Öfversigt öfver de senare årens forskningar på den oorganiska kemiens område [Survey of the Progress in Inorganic Chemistry during the Last Years]," *Svensk kemisk tidskrift* [The Swedish Chemical Journal] **4** (1892), 33–39, 62–70.

47. See other chapters in this volume.

48. Bernadette Bensaude-Vincent, "Mendeleev's Periodic System of Chemical Elements," *The British Journal for the History of Science* **19** (1986): 3–17.

49. Spronsen, *The Periodic System* (note 9), 61.

50. G. Krüss & L. F. Nilson, "Om Thoriums eqvivalent—och atomvigt," (note. 43), *Öfversigt* KVA **44**:5 (1887), 252.

51. C. W. Blomstrand, *Kort lärobok i oorganisk kemi,* 4:e uppl. [Short Textbook in Inorganic Chemistry, 4th ed.] (Lund: Gleerup, 1897), 262.

52. Å. G. Ekstrand, *Lars Fredrik Nilson*, Levnadsteckningar öfver Kongl. Vetenskapsa kademiens Ledamöter [Biographies of Fellows of the Royal Academy of Science] 6:1 (Stockholm: Kungl. Svenska Vetenskapsakademien, 1921), 18.

53. Masanori Kaji, "Mendeleev's Discovery of the Periodic Law: The Origin and the Reception," *Foundations of Chemistry* **5** (2003): 189–214, p.191.

54. For Swedish physics in general at the end of the nineteenth century, see Sven Widmalm, *Det öppna laboratoriet: Uppsalafysiken och dess nätverk 1853–1910* [The Open Laboratory: Uppsala Physics and its Networks], (Stockholm: Atlantis, 2001).

55. Nilson, "Om Scandium," (note 36), 47.

56. N. A. Langlet, "Om förekomsten af Helium i cleveit [On the Presence of Helium in Cleveit]," *Öfversigt* KVA **52**:4 (1895), 211–213; idem, "Om Heliums atomvigt [On the Atomic Weight of Helium]," *Öfversigt* KVA **52**:6 (1895), 371–377; idem, "Profning af kolm på Helium [Testing Kolm for Helium]," *Öfversigt* KVA **53**:9 (1896), 663f. (Kolm is a mineral rich in uranium and found in mid-Sweden).

57. On Thalén, see Widmalm, *Det öppna laboratoriet* (note 54), 77–130, especially 106–113. Thalén's atheoretical commitment became clear during his negative (although understandable) reaction toward Arrhenius's theory of electrolytical dissociation.

58. Robert Thalén, "Spektralundersökningar rörande Skandium, Ytterbium, Erbium och Thulium [Spectral Analysis on Didymium, Ytterbium, Erbium and Thulium]," *Öfversigt* KVA **38**:6 (1881), 13–21; idem, "Om de lysande spektra hos Didym och Samarium [On the Luminous Spectra of Didymium and Samarium]," *Öfversigt* KVA **40**:7 (1883), 3–16.

59. Thalén, "Spektralundersökningar" (note 58), 14.

60. Thalén to Nilson 11.3.1879, letter quoted in Nilson, "Om Scandium," (note 36), 49.

61. "On the Constitution of the Red Spectrum of Argon," *Astrophysical Journal* **6** (1897), 338. Rydberg, however, came to accept the idea that argon was a single element. The

theoretically most advanced treatment of the periodic system in Sweden came from Rydberg. Since he, however, by this time, remained an outsider in Swedish science, and since this essay is on the general acceptance of the periodic system, I have chosen not to discuss him, however interesting his attempt ever may be from the theoretical point of view. For Rydberg's position in Sweden, see Paul Hamilton, "Reaching out: Janne Rydberg's Struggle for Recognition," in S. Lindqvist (ed), *Center on the Periphery: Historical Aspects of 20th-Century Swedish Physics* (Canton, Mass.: Science History Publications, 1993), 239–292.

62. Cleve, "Bidrag till jordartsmetallerna Kemi II. Lantan," (note 42), 5.
63. It should be noted that J. S. Stas started his exact determinations of precise atomic weights before the advent of Mendeleev's periodic table.
64. O. Pettersson & G. Ekman, "Om Selens atomvigt [On the Atomic Weight of Selenium]," *Öfversigt KVA* **33**:6 (1876), 57.
65. Ekstrand, *Lars Fredrik Nilson* (note 52), 44.
66. See, for example, Aaron Ihde, "Introduction," in Spronsen, *The Periodic System* (note 9)—or any standard work in the history of chemistry.
67. On the beryllium "fight," see Spronsen, *The Periodic System* (note 9), 300–302.
68. Yvonne Hirdman, "Att skriva historiska synteser—en liten moralism," [Writing Historical Syntheses—a Story with a Moral], *Historisk tidskrift* [Historical Journal] 2003, 215–224.

CHAPTER 8

༠ᐯᢒ

Reception and Early Use of the Periodic System

The Case of Denmark

HELGE KRAGH

In this essay I examine how the periodic system or table was introduced in Denmark in the late nineteenth century, how it was used in chemical textbooks, and the way it was developed by a few of the country's scientists. Danish chemists had in the period an international orientation, which helped them in getting acquainted with Mendeleev's system and appreciating its strength. The main reason they felt the system to be attractive was its predictive force, especially its prediction of new elements and ability to accommodate new chemical knowledge. I pay particular attention to the work of Hans Peter Jørgen Julius Thomsen (1826–1909), which is an important example of "neo-Proutean" attempts to understand the periodic system in terms of internally structured atoms. Moreover, I direct attention to Mendeleev's connection to Danish science by way of his membership in the Royal Danish Academy of Sciences and Letters.

Thomsen's speculations of composite atoms as the ultimate cause of the periodicity of the elements were vindicated by the new developments in atomic theory. A semi-quantitative explanation was offered by Niels Bohr (1885–1962) in 1913, and in subsequent refinements of his atomic model he came close to an explanation of the entire periodic system. The essay briefly considers Bohr's work on the periodic system in its local context, including its relation to the earlier ideas of Thomsen.

1. THE DANISH CHEMICAL COMMUNITY, 1870–1920

In order to appreciate how the periodic system of the elements was received in Denmark, it will be helpful to provide some basic information of the country's chemical landscape.[1] In the period here considered, approximately 1870–1920, Denmark was a small country, scientifically and culturally almost completely dominated by its capital, Copenhagen. As far as chemical research and education was concerned, the most important institutions were the University of Copenhagen, the Polytechnical College, the Royal Veterinary and Agricultural College, and the Pharmaceutical College, all located in Copenhagen. Although the number of chemists grew rapidly during this period, only a few of them were trained at the University and even fewer had an interest in the more theoretical aspects of the chemical sciences. University-trained chemists were not only outnumbered by chemical engineers, trained at the Polytechnical College, but also by chemists with a background in medicine and pharmacy. The professionalization of chemistry manifested itself locally with the foundation in 1879 of the Danish Chemical Society, the first such society in Scandinavia. The Danish Chemical Society was broadly composed, appealing not only to professional chemists, but to all "men with an interest in chemistry."[2]

The research interests of most Danish chemists had a practical orientation, either connected to chemical engineering or dairy products and the fermentation industry. At approximately the turn of the century, many chemists worked in the biochemical and biotechnological sectors, where a central institution was the Carlsberg Laboratory, established in 1875. Leading chemists such as Johan Kjeldahl (1849–1900) and Søren P. L. Sørensen (1868–1939) worked at this laboratory, which was an integrated part of the Copenhagen chemical network. Whether working with applied or pure chemistry, Danish chemists had a strong international orientation. They had typically spent some time abroad, mostly at German universities, and kept abreast of the international literature. Moreover, the large majority of them published one or more of their research articles in German or other foreign-language journals. Because of the small size of the population, and also because the local Chemical Society did not publish its own journal, Danish chemists were forced to adopt an international attitude.

At about the time of the foundation of the Danish Chemical Society, the total number of regular academic positions amounted to only two full professors and three associate professors. These professors taught and did research at both of the twin institutions, the University and the Polytechnical College. (In addition, the Agricultural College had a professor of chemistry, Christen T. Barfoed (1815–89)). Because of their small number and central positions, they were of great importance with regard to introducing and disseminating new ideas and theories from abroad. From approximately 1870 to the early years of the new century, academic chemistry was much dominated by two professors and powerful personalities, Julius Thomsen and Sophus M. Jørgensen (1837–1914). Thomsen served as a professor between 1866 and 1901, and Jørgensen during roughly the same period, from 1871 to 1908. During the first decade of the twentieth century, a much-needed generation shift occurred in Danish chemistry, followed by an increased interest in theoretical and physical chemistry. With the appointment of Johannes Brønsted (1879–1947) as professor in a new chair at the University in 1908, and Niels Bjerrum (1879–1958) at the Agricultural College in 1914, a new era started in Danish chemistry.[3] Other chemists of relevance to the subject of this essay, the periodic system, will be mentioned later.

Several chemists ended up in teaching positions, either at higher institutions (such as the Military Academy) or in the "learned" gymnasium schools attended by students who wanted to proceed to a university education. The system of "learned schools" or "Latin schools" went back to the Middle Ages, and in the seventeenth century some of them were named gymnasia. Education in these elite schools was originally dominated by Latin and other classical learning, but with the reform of 1871 it became possible also to graduate in a branch focused on mathematical and scientific subjects.

Realizing the need for strengthening the scientific subjects compared to humanistic studies and classical languages, in 1871 an educational reform was implemented. According to this reform, chemistry would be an obligatory part of the gymnasium curriculum, if only in modest doses and in combination with physics. As a result of this and other reforms, several textbooks in elementary chemistry were published, either by schoolteachers or academic chemists. The most widely used of these books were Hannibal Jespersen's *Kortfattet Lærebog i Uorganisk Kemi* [Brief Textbook in Inorganic Chemistry] (1874) and S. M. Jørgensen's *Kemiens Begyndelsesgrunde* [Introductory Chemistry] (1876) and *Mindre Lærebog i Uorganisk Chemi* [Smaller Textbook in Inorganic Chemistry] (1888).[4] Another major reform followed in 1903, and according to the ministerial instruction of this reform the students should not only be taught descriptive chemistry but also elementary theoretical chemistry. The periodic system was not mentioned explicitly, which meant that teachers and textbook writers could choose to mention it or not. Some did.

2. EARLY DISCUSSIONS OF THE PERIODIC SYSTEM

The first published recognition of the periodic system among Danish chemists that I have come across dates from 1880. However, there is little doubt that many of the chemists were aware of the classification of either Mendeleev or Lothar Meyer, or both, at an earlier date. Thomsen had dealt with the groupings of the chemical elements according to their atomic weights as early as 1865, in a work that may well be counted as one of the many incomplete anticipations of the periodic system.[5] Thomsen's aim was not so much to establish a natural chemical classification as to defend the Proutean hypothesis—so named after the English chemist and physician William Prout (1786–1850)—that the elements are really composite bodies made up of more elementary entities. Some twenty years later, he would return to this kind of reasoning and develop it in detail (see section 4 of this chapter). I have not found any references in Danish chemical literature to earlier versions of the periodic system, such as those proposed by John Newlands, William Odling, and Gustavus Hinrichs.[6]

In an article of 1880 in a popular science journal, the young chemist Odin T. Christensen (1851–1914), at the time an assistant at the laboratory of the Polytechnical College, reviewed the recent discoveries of chemical elements, including gallium and scandium. In this connection, he discussed the place of the new elements in "the system," such as predicted by Mendeleev in the form of the hypothetical elements eka-aluminium and eka-boron. He concluded that gallium and scandium "provide strong support in favor of the view of Mendeleev, namely, that the properties of the elements and the constitution of their compounds are periodic functions of the atomic weights of the elements."[7] Christensen further noted that the discovery of gallium by Emile Lecoq de Boisbaudran (1838–1912) had taken place wholly independently of Mendeleev's prediction, which he found to be further confirmation of the essential truth of the periodic system.

Christensen was evidently impressed by the agreement between the predictions of Mendeleev and the metals discovered by Boisbaudran and Lars Fredrik Nilson (1840–99). When Clemens Winkler (1838–1904) some years later discovered germanium and identified it with Mendeleev's eka-silicium, he was no less impressed: "One can scarcely think of a more striking proof of the theory of the periodicity of the elements than Mendeleev's prediction of the properties of eka-silicium, such as realized in the discovery of germanium and its compounds."[8] Yet another triumph of Mendeleev's law was the outcome of the controversy concerning the correct classification of beryllium, as a homologue of either magnesium or aluminum. Based upon the law of Dulong and Petit, its atomic weight came out as approximately 14, a value indicating that beryllium was a tervalent element, in disagreement with Mendeleev's conclusion. Only in approximately 1880 was the element's atomic weight determined as 9.1, which largely settled the controversy. As Christensen saw it, the problem had been solved, with the new atomic weight being "proof of the great significance that must be ascribed to Mendeleev's periodic law."[9]

Thomsen and Christensen were not the only Danish chemists who paid tribute to the periodic system in the 1880s. A twenty-three-year-old graduate student in chemistry, Rudolph Koefoed (1862–1924), published in 1885 an extensive survey article on what he called the periodic law and in which he referred to Mendeleev's as well as Lothar Meyer's works of 1869–71. Like Christensen, he assigned much significance to the successful predictions of gallium and scandium. So much, in fact, that he suggested that now the chemists were on their way to establishing their science on a principle nearly as universal and reliable as Newton's law of gravitation was for the astronomers. Ironically, adding to Koefoed's confidence in the periodic system was that it—apparently—resulted in atomic weights in agreement with recent measurements. For example, it was well known that tellurium's atomic weight of 128 conflicted with the periodic system, which required a value of about 125. However, as Koefoed was happy to report, recent determinations made by Bohuslav Brauner (1855–1935) gave just this value and "thus confirm Mendeleev's prediction."[10]

3. THE SYSTEM IN TEXTBOOKS AND EDUCATION

While the periodic system seems to have been well known among Danish chemists by the mid-1880s, naturally it took some time until it percolated to the level of education and became part of the teaching of chemistry students. Julius Thomsen never wrote a textbook on chemistry, such as did his younger colleagues S. M. Jørgensen and Emil Petersen (1856–1907). Most university lectures in inorganic chemistry were given by Jørgensen, who however chose to ignore the periodic system. Although not opposed to the atomic hypothesis, the arch-empiricist Jørgensen used to warn his students that atoms and molecules should primarily be conceived as convenient means of representing empirical data. Likewise, although he may have appreciated the predictive power of the periodic system, he seems to have conceived it as somewhat speculative and neither as a necessary nor fundamental classification of the elements.

In his textbooks on chemistry, which exerted great influence on a generation of Danish chemists, Jørgensen did not so much as mention either Mendeleev or his periodic system of the elements. The system was absent from both the first and the second editions of his textbook on inorganic chemistry, published in 1888 and 1896, respectively.[11] Nor was the system to be found in his 1902 textbook on general chemistry, which was widely used and translated into several languages.[12] In this work Jørgensen included some of the more recent developments, such as the Thomsen-Berthelot theory of thermochemistry, le Chatelier's theorem of chemical equilibria, Arrhenius's and Ostwald's ideas of ionic dissociation, and Ramsay's discovery of the noble gases—but not the periodic system. He listed the chemical elements and their atomic weights alphabetically, without any indication of relations between them. The discoveries of gallium and scandium were duly mentioned, but again without stating their relations to Mendeleev's system. This is all the more remarkable in light of the fact that Jørgensen also disregarded the periodic system in the second edition of 1913, where he mentioned such novelties as the liquefaction of helium, the radioactive transmutation of elements, and the electron theory of atomic constitution. Some of these novelties were also included in the English translation of 1908.[13]

The absence of the periodic system was noted in an otherwise positive review in the German periodical *Naturwissenschaftlicher Rundschau*: "Neither the periodic system of the elements nor the related question of a primary matter is mentioned in the book. The reviewer is unaware of the author's reason for this reservation, but it seems to him [the reviewer] that this question—which possibly goes deeper into the philosophical foundation of chemistry than any other subject—might well have fitted into the book."[14]

The first university-level textbook to incorporate the periodic system, written by Odin Christensen in connection with his lectures at the university, appeared in 1890 and ran through four editions. Without mentioning Mendeleev by name, he introduced his system in the form of an appendix, not as an organizing principle for treating the properties of the elements.[15] Using "periodic system" interchangeably with "periodic law," his main justification for the classification was its ability to predict new elements in accordance with later experiments.

Another advocate of the periodic system was Emil Petersen, who after studies in Paris and Leipzig had taken up the new physical chemistry of the Ostwald school. In 1889 he gave a lecture series at the University of Copenhagen on the rare elements, with special emphasis on the problem of their places in the periodic system. By that time he was convinced of the basic truth of the system and also that it reflected an underlying unity of matter.[16] After having been appointed professor of chemistry in 1901, he wrote a textbook in inorganic chemistry in which he included a fairly detailed account of the system. According to Petersen, the periodic system was a useful classification, yet "it is far from a perfect expression of the facts [and] . . . many deficiencies are attached to it." Among these deficiencies he mentioned the Ar-K and Te-I atomic weight inversions, and he also found it problematic that copper and mercury ("which chemically are so analogous") were placed in different groups. On the other hand, he was convinced of the importance of the periodic system, not least because "in several cases the existence of elements and their main properties were predicted in advance, many years before they were actually discovered." Rather than mentioning the classic cases of gallium, scandium, and germanium, he called attention to the new element radium, "which is very similar to barium and, with an atomic weight of 225, fits nicely into the system."[17]

Some of the features of the textbooks of Christensen and Petersen can also be found in the university textbooks of the next generation of Danish chemists: Although the periodic system was now included, it played no great role and did not function as a principle for organizing the discussion of the elements and their compounds. The two new professors of the 1910s, J. Brønsted and N. Bjerrum, each wrote a textbook in inorganic chemistry, intended to supplement their lectures at the University and the Agricultural College, respectively. Whereas Brønsted's book of 1916 still based the periodic system on atomic weights, Bjerrum's work of the following year incorporated the recent developments in atomic physics.[18] This was the first time in Danish chemistry that the atomic number appeared as an ordering parameter for the elements, and also the first time that the Rutherford-Bohr nuclear model was introduced as a way of explaining the periodic system in terms of atomic structure. But apart from this novelty, Mendeleev's system played no prominent part in the book.

In the period under consideration, the custom in Denmark was to use textbooks written by local authors, in most cases the professors. Textbooks translated from other languages were not, or only very rarely, used either at the University or elsewhere. Nor were German or other papers from abroad on the periodic system translated into Danish.

Among the elementary textbooks intended for the gymnasium schools that appeared in the early part of the twentieth century, some referred to or made use of the periodic system. This was the case with a book written by Julius Petersen (1865–1931), a polytechnically trained chemist and former assistant of S. M. Jørgensen, who in 1908 was appointed professor of chemistry at the University of Copenhagen. Petersen followed the tradition by emphasizing the successful predictions of elements based on Mendeleev's system, and at the same time, much like his colleague and namesake Emil Petersen, pointed to its incompleteness and problems, such as the Ar-K and Te-I atomic weight anomalies. Another book for the gymnasium, written by the teacher Hans Rasmussen (1869–?), is noteworthy because it presented the periodic system in the unconventional form suggested by Julius Thomsen, with vertical groups and horizontal periods.[19] The pedagogical value of the system was not always appreciated, and some teachers suggested that it, being too theoretical, should not be part of the curriculum.[20] It took until 1958 before the periodic system became a formally required part of the Danish gymnasium education system.

4. SPECULATIONS ON THE COMPLEXITY OF ATOMS

A pioneer of thermochemistry, Julius Thomsen was first and foremost an experimentalist. Yet he also had an interest in chemical theories, and he was the only Danish scientist who, until Bohr in 1913, actively examined and contributed to the understanding of the periodic system. As mentioned, ever since the 1860s he entertained the heterodox view that the atoms of chemistry are complex particles and that this is revealed by regularities in their atomic weights. Of course, he was far from the only neo-Proutean of his time, but he was one of the most distinguished and articulate advocates of the idea of a basic unity of matter. In a work of 1887 he connected for the first time this idea with the periodic system, undoubtedly inspired by an address that William Crookes (1832–1919) had given the year before to the British Association for the Advancement of Science.[21] Another likely inspiration was the British astronomer Joseph Norman Lockyer (1836–1920), whose work on the cosmic evolution of the elements had a great deal of similarity with the views expounded by Thomsen.[22]

The Danish chemist was particularly concerned with the question of why only some atomic weights are realized in nature, while other possible weights seem to be missing. An ardent advocate of so-called inorganic Darwinism, he thought that the answer was to be found in the slow evolution of elements from simple to more complex structures. "The right of the fittest has manifested itself and only allowed the formation of atoms with a structure firm enough for a continuous existence," he wrote.[23] As to Mendeleev's system, he praised it for its ability to identify missing elements and predict their properties, such as had been the case with gallium, scandium, and germanium. Contrary to Mendeleev and most other chemists, he was convinced that the system was a key to understanding the complexity of the elements and that it would eventually be possible to represent it as a mathematical function of the atomic weight. The version of the periodic system he presented in 1887 was fairly orthodox, not differing significantly from Mendeleev's. Like the Irish chemist Thomas Carnelley (1854–90) had done the year before, Thomsen suggested an analogy between the chemical elements and the hydrocarbon radicals.[24]

The questions addressed by Thomsen were taken up also by Emil Petersen, who in 1890 discussed the nature of the chemical elements and the idea of a basic unity of matter such as considered by Crookes and others.[25] Evidently in sympathy with the idea, he suggested that it received support from the periodic system. "It is hardly to doubt," he wrote, "that in this way we will eventually get insight into the unity that lies behind the varied diversity of the elements." Referring to Mendeleev's recent Faraday lecture, he admitted that the dream of a primary matter was somewhat speculative, but he nonetheless found the dream worthy of pursuit.[26] Whether in Mendeleev's or Meyer's version, Petersen thought highly of the periodic law, which he summarized in the formula "The properties of the elements stand in a periodic relationship to the atomic weight." He explained that there were two major reasons for accepting the truth of the law, the one relating to its unifying power and the other to its predictive power:

> It is the merit of the periodic law that it has arranged all known elements—and in some cases also unknown elements—in one coherent system and demonstrated the intimate mutual relationship between their properties. It has assigned the right place for some elements whose relationships to other elements were doubtful. For some of the less well known elements it has proved possible, by means of the table, to correct their atomic weights such as found experimentally. . . . These and other applications of the system are of considerable scientific importance. Another application of the system is less important, but on the other hand more striking and amazing, namely, its ability to predict as yet undiscovered elements—to predict their existence and most important properties years before they were actually discovered and manufactured.[27]

In other words, according to Petersen, the scientific value of the periodic law was primarily its ability to arrange the elements into a coherent system, whereas he gave lower priority to its predictive power. No other Danish chemist expressed a similar view.

To return to Thomsen, in a memoir of 1894 published by the Royal Danish Academy of Sciences, he offered a detailed examination of the atomic weights and their significance. His purpose was to establish that they, if only properly interpreted, revealed that "the so-called atoms of our elements have evolved out of combination of particles of a common basic substance."[28] He did not on this occasion discuss the relation to the periodic system, but this is what he did the following year, in a paper in which he proposed a new classification of the elements (Figure 8.1).[29] From a formal point of view, Thomsen's innovation was merely to reverse periods and groups, which was not entirely original since versions of this kind had been proposed earlier, first by Thomas Bayley in 1882 and again by Carnelley in 1886.[30] However, in 1894 Thomsen was unaware of these two systems, such as he stated in a letter to the American chemist Francis Venable (1856–1934), who in a book of 1896 described Thomsen's system in some detail.[31]

Thomsen designed his version of the periodic system in such a way that it immediately suggested a common origin of the elements, that is, an evolutionary interpretation. Irrespective of such an interpretation, it included several novel features and indicated the existence of possible new elements. For example, it was the first

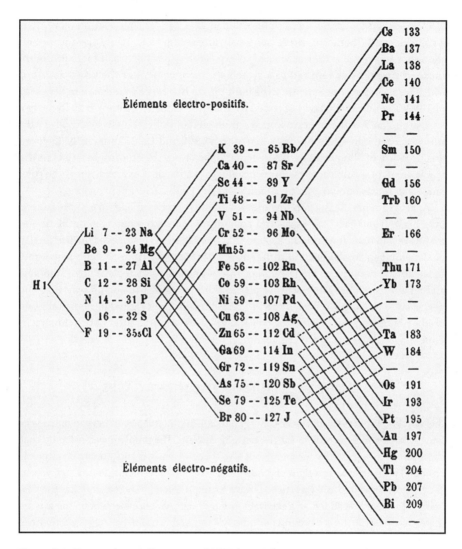

Figure 8.1 Thomsen's periodic system of 1895 (note 29).

version of the periodic system that included the correct number of rare earth metals, namely fourteen, and placed this group between cerium and an unknown element of atomic weight 180 with chemical properties analogous to those of zirconium. This hypothetical element—later identified as hafnium—also implicitly appeared in Mendeleev's original periodic system of 1869, but it was only with Thomsen that it was given explicit attention and placed outside the rare earth group.

Another feature of Thomsen's brief paper deserves mention, namely the "curious fact" that the number of elements in the periods is 1, 7, 17, and 31. These numbers, Thomsen pointed out, can be written as $1, 1+2\times3, 1+2\times3+2\times5$, and $1+2\times3+2\times5+2\times7$. Expressed slightly differently, the number of elements follows the expression

$N = 2n^2 - 1$, or, if the inert gases are included, $N = 2n^2$. "Is this relation more than a coincidence?" Thomsen asked, cautiously answering that, "Only the future will show, but I have nevertheless wished to expose the possibility of a more profound cause."[32] He probably referred to a systematic arrangement of the proto-atoms of which he assumed the elements to be built up, such as he had indicated in his essay of 1887. The numerical law suggested by Thomsen came to be known as Rydberg's rule, after the Swedish spectroscopist Janne (Johan Robert) Rydberg (1854–1919) who proposed it in different forms in works of 1906 and 1913. Apparently, Rydberg was unaware of Thomsen's earlier speculation, as were the atomic physicists in the tradition of quantum theory who eventually gave a rational explanation of the rule, namely in terms of quantum mechanics and the Pauli exclusion principle.[33]

Apart from his inclination toward numerology, Thomsen had no good reason to take his "curious fact" seriously and apparently soon lost whatever confidence he may have had in it. The periodic system that he used in his lectures at the University of Copenhagen in about 1898 differed in some respects from the published one, especially by having the long period of thirty-one elements replaced by three new and smaller periods. Moreover, he placed the inert gases in such a way that the new system no longer revealed the $2n^2$ relationship. The original periodic table, as used by Thomsen in 1898 and for several years by other lecturers of chemistry in Copenhagen, is preserved at the Technical Museum in Elsinore, Denmark.

5. THE POSITION OF THE INERT GASES

It is well known that the discovery of argon in late 1894, and also of helium half a year later, caused a major problem for the periodic system. The problem was not only that there was no natural place for argon, but also that the new gas appeared to be monoatomic and with an atomic weight of 39.9, greater than the one of potassium.[34] However, the crisis disappeared and was turned into a triumph when it was realized that the new inert gases could be added as a separate group of zero-valence elements. This was an important test for the still young periodic system, and it has been suggested that the successful incorporation of the inert gases was of no less importance for the authoritative status of the system than the earlier predictions of metallic elements.[35]

The issues that emerged with the discovery of argon were known among Danish chemists and reflected in their works. It was these problems that induced Thomsen to "publish some ideas, with which I have been occupied for years, but which I have wished not to publish until now, because I would not encumber science with unverifiable hypotheses."[36] The ideas he referred to were probably mathematical relations between the electrochemical character of the elements and their atomic weights. From such considerations Thomsen argued that there supposedly existed a new group of chemical elements that were electrochemically indifferent and possessed zero valence. Moreover, based on his new and still unpublished periodic system, he suggested that the atomic weights of the elements—of which only argon was known at the time—were 4, 20, 36, 84, 132, and 212.[37] For the seventh period he proposed that it would end with a noble-gas element of atomic weight 292. Like

several other scientists at the time, Thomsen searched for a mathematical representation of the periodic system, and he thought that his new extension of the system
was a step in the right direction.

In his essay of 1887, Thomsen speculated that the hypothetical solar element,
helium, might be a subhydrogenic primary element (Crookes did the same in 1886).
When he read his paper on the inert gases to the Royal Danish Academy on April
19, 1895, William Ramsay (1852–1916) had not yet announced his discovery of
helium in terrestrial sources. Helium initially raised questions with regard to its
place in the periodic system, but after a couple of years it was realized that it and
argon belonged to a new group, in agreement with Thomsen's proposal. Thomsen
kept an interest in the inert gases, and in 1898 he succeeded in detecting helium
in a red fluorite mineral in Greenland. In the same year he gave an address to the
15th Scandinavian Meeting of Natural Scientists, held in Stockholm, in which he
emphasized the scientific importance of what appeared to be a new group of gases
belonging to the periodic system.[38]

Thomsen was not the only Danish chemist with an interest in the new gases. In a
survey article addressed to Scandinavian pharmacists of June 1895, Emil Petersen
discussed the sensational discovery made by Ramsay and Lord Rayleigh. In agreement with the "two distinguished British chemists"—to his dismay, Rayleigh was
often thought to be a chemist—he concluded that the evidence spoke in favor of argon
being monoatomic and with an atomic weight close to 40. He was confident that there
was no fundamental disagreement between argon and Mendeleev's system:

> As soon as a new element is discovered and a determination of its atomic weight
> has been obtained, what is usually done is to look at Mendeleev's well known
> periodic system. All known elements have been secured a place in this system, as
> determined by their atomic weights and in agreement with the element's physical
> and chemical properties.[39]

After having discussed various solutions to the problem of argon's place in the system, he ended with suggesting that the standard version of Mendeleev's system
was probably incomplete. Later the same year, the delicate question was reviewed
in detail by S. P. L. Sørensen, Thomsen and Petersen's colleague at the Carlsberg
Laboratory and later famous for his invention of the pH scale. Sørensen expressed
strong support of Thomsen's view of the periodic system and its "convincing argument for the existence of a group of elements of an inactive character."[40]

6. FROM THOMSEN TO BOHR

At about the time when Mendeleev and Thomsen passed away (in 1907 and 1909,
respectively), there was increasing evidence that the periodic system was a manifestation of the internal structure of atoms, such as Thomsen and other neo-Prouteans
had speculated. This was an important feature of the atomic model by J. J. Thomson
(1856–1940), according to which atoms were conglomerates of electrons structured

in concentric circles and moving in an imponderable positive charge of atomic dimension. Indeed, as early as 1897, in the paper in which he announced the discovery of the electron, Thomson explicitly referred to rings of electrons as an explanation of Mendeleev's system.[41] However, by 1910 the Thomson model had run out of power, to be replaced a few years later by the highly successful quantum theory of the nuclear atom. Nonetheless, the general idea that the periodicity of the elements reflected the configurations of the electron survived the demise of the Thomson model. It is worth pointing out that according to Thomson the periodicity was due to similar configurations of internal rings of electrons, not a similarity of the outer configurations close to the surface of the atom.

Niels Bohr was well acquainted with general chemistry, including the periodic system in the versions of Mendeleev, Meyer, and Thomsen, which he had been taught in lectures in inorganic chemistry held at the University of Copenhagen in 1905.[42] The lecturer was the young chemist Niels Bjerrum, who was familiar with the recent attempts to explain the periodicity of the elements in subatomic terms. In an article of 1907, Bjerrum reviewed these ideas, including the view that the atomic weights reflect the internal composition of the atoms. He concluded that "the law-like connection between the properties of the elements and their atomic weights, such as expressed in the periodic system, can hardly be explained without assuming an internal constitution of the atom."[43] This was not an original observation, but at the time it was unusual for chemists to relate the periodicity of the elements to their internal structure.

Although Bohr's great work of 1913 focused to a large extent on the hydrogen atom, he also dealt with the electron structures of more complex atoms. As he wrote in a letter of February 1913 to George von Hevesy (1885–1966), his still unpublished theory would include a "very suggestive indication of an understanding of the periodic system of the elements."[44] Contrary to earlier physicists and chemists, Bohr could make use of the very recent introduction of the atomic number by Dutch amateur physicist Antonius van den Broek (1870–1926)— corresponding to the charge of the atomic nucleus—as the ordinal number of the periodic system. Taking advantage of this new definition of a chemical element, and also of the periodic variation of the atomic volume of the elements, he ventured to suggest electron configurations for the first twenty-four elements, from hydrogen to chromium. For example, for the first three alkali metals he assigned the configurations Li = (2,1), Na = (8,2,1), and K = (8,8,2,1). In this way, he explained the chemical similarity between elements of the same group as due to the same number of electrons in the outermost ring. However, he cautiously avoided explicitly identifying the electron structures with definite chemical elements.[45]

Bohr's 1913 explanation of the periodic system was incomplete and wrong in its details, but nonetheless a great progress compared to earlier attempts. It was a first step toward the much fuller and more detailed theory he composed between 1921 and 1923, still based on the semi-classical so-called old quantum theory. In this important theory, he made use of a slightly modified version of Thomsen's table with vertical periods and horizontal groups (Figure 8.2). Thus, in an address

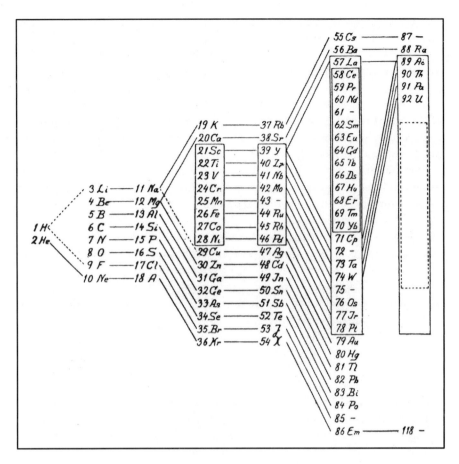

Figure 8.2 The Thomsen-Bohr system, as Bohr discussed it in his Nobel lecture in Stockholm in December 1922.

of 1921 on the periodic system, he said that he preferred the graphic version "proposed more than twenty years ago by Julius Thomsen, . . . [because it] is more suited for comparison with theories of atomic constitution."[46] Likewise, when he gave his Nobel lecture in Stockholm in December 1922, he used the occasion to pay tribute to the Danish chemist. Like Thomsen, Bohr ambitiously suggested an extension of the periodic system to cover transuranic elements, offering a full electronic configuration of the element Z = 118, supposed to be an inert gas homologous to radon.[47] As mentioned, Thomsen had speculated that the same hypothetical element had an atomic weight of about 292.

7. CONCLUSIONS

Although dating from 1869, Mendeleev's periodic system was only explicitly noticed by Danish chemists about a decade later. By the mid-1880s it seems to have been broadly known and also accepted as a useful classification by many chemists. Compared with other small countries, this delay in the reception was not unusual. The system or law was considered of interest only by relatively few chemists, whereas it tended to be ignored by the majority who worked within the more practical fields of chemistry, such as related to engineering, pharmacy, and dairy products. Generally speaking, one should be careful not to confuse lack of references in the literature with ignorance: Although the periodic system was not mentioned very frequently by Danish chemists, this does not mean that it was unknown or considered unimportant.

One indication of the status that Mendeleev, and by implication the periodic system, enjoyed in Denmark is that on April 5, 1889, the Russian chemist was elected a foreign member of the Royal Danish Academy of Sciences and Letters.[48] Julius Thomsen served at the time as president of the Academy, and it appears to have been on his initiative that Mendeleev was invited to become a member of the prestigious society going back to 1742. (The slightly older Royal Swedish Academy of Sciences elected Mendeleev a member in 1905.) The letter of motivation was written by Thomsen and signed jointly by him and S. M. Jørgensen. Given Jørgensen's lack of appreciation of the periodic system, one may assume that the proposal was actually due to Thomsen. At any rate, Denmark's two leading chemists, both of them of international repute, motivated their proposal as follows:[49]

> During many years, Prof. Mendeleev has conducted a great number of excellent investigations, in part of a general chemical nature and in part of a physico-chemical nature, and they have all been characterized by a superior mind. It would be too long-winded to recount the subjects of all these investigations, but we would like to emphasize his great works on the dependence of gases on temperature and pressure. However, Mendeleev's name has become even more generally known by his brilliant work on the theory of how the chemical and physical properties of the elements depend on their atomic weights—the so-called periodic system. In this way he has opened a wide field for a philosophical discussion of the most important chemical phenomena; his theories have several times been remarkably confirmed by the discovery of elements whose existence and most important properties he had predicted as a consequence of the system. Objections can indeed be raised against the full justification of the system, such as can be done against many other theories; but the system has, to a very high degree, advanced chemistry as a science, and for this reason Mendeleev's name will forever be inscribed among the first in the history of chemistry.

It is not unreasonable to assume that the reference to Mendeleev's work on gases reflected Jørgensen's view of the relative significance of the Russian chemist's work. During the 1870s Mendeleev conducted extensive investigations on the compressibility of gases, which led him to suggest a generalization of the ideal gas laws. This work, completely overshadowed by his research on the periodic system, was not seen as particularly important, but apparently Thomsen and Jørgensen thought that it was.

Mendeleev quickly responded to the invitation, expressing how great an honor it was for him to become a foreign member of the Royal Danish Academy. He was pleased to accept this sign of "the scientific brotherhood of the peoples," which he considered a manifestation of "the sympathy which unites the Danes and the Russians."[50]

As seen from the perspective of Danish chemists, the periodic system was of importance primarily because of its successful predictions of new elements. It was this feature that provided the system with a measure of credibility and authority. Because the predictions were associated with Mendeleev and his version, rather than the versions of Meyer and others, the periodic system was invariably associated with the name of the Russian chemist. Whereas the periodic system did not appear prominently in Danish academic textbooks in chemistry between 1880 and 1900—and in some cases did not appear at all—it was introduced in elementary textbooks at a relatively early date. By 1910, most Danish students in the gymnasium schools would have encountered the system, if only in its most rudimentary form. On the other hand, both in university- and gymnasium-level textbooks, it typically appeared isolated from the systematic description of the elements and their properties.

Education apart, only one Danish chemist took an active scientific interest in the periodic system of the elements. During the 1890s Julius Thomsen did important work on its interpretation, which to him was to be found in terms of the complex structure of atoms. Although Thomsen's contributions to this area of speculative chemistry were well known internationally, and although they to some extent served as an inspiration for Bohr's later work, they did not make much of an impact on Danish chemistry. Emil Petersen shared some of his ideas, but on the whole Thomsen was a lone figure in his advocacy of neo-Proutean speculations. In approximately 1900, the view of most Danish chemists may have been something like this: Sure, the periodic classification of the elements is an interesting hypothesis with a certain predictive power, but scarcely more than that; it is probably not of fundamental importance, nor is it necessary for understanding inorganic chemistry; in any case, it has little to do with what most chemists are occupied with, namely practically oriented experiments.

Given the vast difference in the amount of consulted sources, whether textbooks or articles, it is problematic to compare the case of Denmark with Stephen Brush's much more detailed study of the reception in the United States and Britain. Nonetheless, I think two comments may be appropriate. First, among Danish chemists the prediction of new elements was generally given more attention than the correlation between the physicochemical properties and atomic weights. This is contrary to what Brush found in his survey. Second, Brush observes that in chemistry textbooks the periodic system was "not as a rule introduced at the beginning or used as an organizing principle for those books."[51] This conclusion fully agrees with my more limited study of the Danish case.

NOTES

1. For a general account of the development of science in Denmark, see Helge Kragh, Peter Kjærgaard, Henry Nielsen, and Kristian H. Nielsen, *Science in Denmark: A Thousand-Year History* (Aarhus: University of Aarhus Press, 2008), which includes sections on the chemical sciences. More details and references can be found in Helge Kragh, "Out of the Shadow of Medicine: Themes in the Development of Chemistry in Denmark and Norway," in *The Making of the Chemist: The Social History of Chemistry in Europe 1789–1914*, ed. David Knight and Helge Kragh (Cambridge: Cambridge University Press, 1998), 345–364 and Anita K. Nielsen, "The Chemists: Danish Chemical Communities and Networks, 1900–1940" (PhD diss., Aarhus University, 2000). See also Stig Veibel, *Kemien i Danmark, II: Dansk Kemisk Bibliografi* [Chemistry in Denmark, II: Chemical Bibliography] (Copenhagen: Nyt Nordisk Forlag, 1943), a valuable bibliography of works by Danish chemists.

2. Statute of October 1879; see Kragh, "Out of the Shadow" (note 1), 246. For details about the early phase of the Danish Chemical Society, see Nielsen, "The Chemists" (note 1), 92–102 and Anita K. Nielsen, "Denmark: Creating a Danish Identity in Chemistry between Pharmacy and Engineering, 1879–1914," in *Creating Networks in Chemistry: The Founding and Early History of Chemical Societies in Europe*, ed. Anita K. Nielsen and Sona Strbanova (Cambridge: RSC Publishing, 2008), 75–90.

3. Thor A. Bak, "The History of Physical Chemistry in Denmark," *Annual Review of Physical Chemistry* **25** (1974): 1–10. Anita K. Nielsen and Helge Kragh, "An Institute for Dollars: Physical Chemistry in Copenhagen between the Wars," *Centaurus* **39** (1997): 311–331.

4. On chemistry education in the Danish gymnasium schools, see Børge Riis Larsen, *Otte Kapitler af Kemiundervisningens Historie* [Eight Chapters in the History of Chemical Education] (Copenhagen: Dansk Selskab for Historisk Kemi, 1998).

5. Julius Thomsen, "Om de saakaldte Grundstoffers Natur [On the nature of the so-called elements]," *Tidsskrift for Physik og Chemi* **4** (1865): 65–115. On Thomsen's speculations on composite atoms, see Helge Kragh, "Julius Thomsen and 19th-Century Speculations on the Complexity of Atoms," *Annals of Science* **39** (1982): 37–60.

6. On the work of these precursors, see Eric Scerri, *The Periodic Table: Its Story and Its Significance* (Oxford: Oxford University Press, 2007), 72–92. The Danish-American (Holsteinian born) scientist Gustavus Detlef Hinrichs studied at the University of Copenhagen until 1861 and probably knew Thomsen. After he immigrated to the United States, he seems to have had no connections to Danish chemists. He did however refer to Thomsen's determinations of atomic weights: Gustavus D. Hinrichs, *Absolute Atomic Weights . . . and the Unity of Matter* (St. Louis: C. G. Hinrichs, 1901), 261. On Hinrichs, see Carl A. Zapffe, "Gustavus Hinrichs, Precursor of Mendeleev," *Isis* **60** (1969): 461–476 and Jan W. Van Spronsen, "Gustavus Detlef Hinrichs Discovered, One Century Ago, the Periodic System of the Chemical Elements," *Janus* 56 (1969): 47–62.

7. Odin T. Christensen, "Nogle i de Senere Aar Opdagede Grundstoffer [Some recently discovered elements]," *Tidsskrift for Populære Fremstillinger af Naturvidenskaben* 27 (1880): 417–429, on 421. In 1895 O. T. Christensen was appointed professor of chemistry at the Agricultural College, where he served until his death in 1914. On the significance of Mendeleev's predictions for the general acceptance of his system, see Stephen G. Brush, "The Reception of Mendeleev's Periodic Law in America and Britain," *Isis* **87** (1996): 595–628.

8. Odin T. Christensen, "Meddelelser om Germanium [Communication concerning Germanium]," *Tidsskrift for Physik og Chemi* **25** (1886): 256–263, on 257.

9. Odin T. Christensen, "Berylliums Atomvægt [The atomic weight of beryllium]," *Tidsskrift for Physik og Chemi* **23** (1984): 310–311, on 311. On the beryllium anomaly, see Jan W. Van Spronsen, *The Periodic System of Chemical Elements: A History of the First Hundred Years* (Amsterdam: Elsevier, 1969), 300–302.

10. Rudolph G. Koefoed, "Den Periodiske Lov [The periodic law]," *Tidsskrift for Physik og Chemi* 24 (1885): 161–174, on 172, and Emil Petersen "Grundstoffernes Natur [The nature of the elements]," *Naturen og Mennesket* **4** (1890): 13–32. In fact, Brauner found in 1883 Te = 127.6 but convinced himself (and apparently others) that the value was too high and due to impure tellurium. The anomaly remained a problem, clarified only by the introduction of isotopy and the new definition of elements based on the notion of the atomic number. On the tellurium-iodine inversion problem, see Van Spronsen, *The Periodic System* (note 9), 238–240 and Scerri, *The Periodic Table* (note 6), 130–131. Koefoed was employed as an assistant at the Carlsberg Laboratory from 1886 to 1890 and subsequently worked for various brewing companies.

11. Sophus M. Jørgensen, *Mindre Lærebog i Uorganisk Chemi* [Smaller Textbook in Inorganic Chemistry] (Copenhagen: Gad, 1888). Jørgensen, who was internationally renowned for his important work on complex metal compounds, received in 1906 the Lavoisier gold medal from the Académie des Sciences in Paris and in 1907 he was nominated for the Nobel Prize in chemistry. See details in Helge Kragh, "S. M. Jørgensen and His Controversy with A. Werner: A Reconsideration," *British Journal for the History of Science* **30** (1997): 203–219.

12. Sophus M. Jørgensen, *Kemiens Grundbegreber Oplyste ved Exempler og Simple Forsøg* [The Concepts of Chemistry Informed by Examples and Simple Experiments], (Copenhagen: Gad, 1902), with translations into German (1903), Greek (1904), Italian (1904), and English (1908). The English translation, an extended version of the German translation, was Sophus M. Jørgensen, *The Fundamental Conceptions of Chemistry* (London: Society for the Promotion of Christian Knowledge, 1908). It was critically reviewed in *Nature* **79** (1908), 218.

13. It is curious to read in Jørgensen's account of the new electron theory of matter that, "The atom is, in fact, now considered to be a nucleus of positive electricity, around which negative electrons rotate with immense velocities in definite paths, like the planets in the solar system." Jørgensen, *Fundamental Conceptions* (note 12), 26. A similar formulation appeared in the Danish edition of 1913, except that he spoke of "a nucleus made up of positive electrons." This looks very much like the Rutherford-Bohr model, but is written three years before Rutherford's nuclear atom and five years before Bohr's model! The most likely explanation is that Jørgensen was aware of the planetary atomic model that the Japanese physicist Hantaro Nagaoka (1865–1950) published in 1904, and that he mistakenly thought that this kind of model was generally accepted. In fact, by 1908 the only atomic model that enjoyed wide recognition was J. J. Thomson's "plumcake" model where the electrons moved in an extended positive charge of atomic dimension. The suggestion receives support from a passage in the Danish (but not the English) text on p. 30: "Thus, in atoms of considerable weight, such as uranium with a weight of about 240 times that of hydrogen, one must assume the existence of several hundred thousand electrons." Jørgensen's comment is of some interest because it may be the first time that the positive central charge was called a "nucleus." Neither Nagaoka nor Rutherford in 1911 used this term.

14. Review by "R.M." in *Naturwissenschaftlicher Rundschau* **19** (1904), 271.

15. Odin T. Christensen, *Grundtræk af den Uorganiske Kemi* [Elements of Inorganic Chemistry] (Copenhagen: Wilhelm Priors Forlag, 1890). Later editions appeared in 1896, 1902, and 1908.

16. Petersen "Grundstoffernes Natur" (note 10). On his series of lectures, see Kai A. Jensen, "Kemi [Chemistry]," in *Københavns Universitet 1479–1979*, vol. 12, ed. Mogens Pihl (Copenhagen: Gad Jensen 1983), 427–580, on 511.

17. Emil Petersen, *Lærebog i Uorganisk Kemi for Begyndere* [Textbook in Inorganic Chemistry for Beginners] (Copenhagen: Schubotheske Forlag, 1902), 317–321. A second edition of the book appeared in 1906. Pierre and Marie Curie's determination of the atomic weight of radium dated from 1902.

18. Johannes N. Brønsted, *Grundrids af den Uorganiske Kemi* [Elements of Inorganic Chemistry] (Copenhagen: Gjellerup, 1916). Niels Bjerrum, *Lærebog i Uorganisk Kemi* [Textbook in Inorganic Chemistry] (Copenhagen: J. Jørgensen, 1917), 214–215.

19. Julius C. Petersen, *Kemi for Gymnasiet* [Chemistry for High Schools] (Copenhagen: Gjellerup, 1907). Hans Rasmussen, *Kemi for Gymnasiet* [Chemistry for High Schools] (Copenhagen: Gyldendal Rasmussen, 1912).

20. Niels Berg, "Nogle Bemærkninger vedrørende Kemiundervisningen i Gymnasiet [Considerations concerning the teaching of chemistry in high schools]," *Fysisk Tidsskrift* **37** (1939): 149–155.

21. Julius Thomsen, *Om Materiens Enhed* [On the Unity of Matter] (Copenhagen: J. H. Schultz, 1887). William Crookes, "Address, President of Section on Chemical Science," *Report, British Association for the Advancement of Science* (1886), 558–577. Although Thomsen did not refer to Crookes's address, it is evident that he knew about it.

22. J. Norman Lockyer, *Inorganic Evolution* (London: Macmillan, 1900). William H. Brock, *From Protyle to Proton: William Prout and the Nature of Matter, 1785–1985* (Bristol: Adam Hilger, 1985), 180–194.

23. Thomsen, *Om Materiens Enhed* (note 21), 37. On Thomsen's neo-Prouteanism and inorganic Darwinism, see Kragh, "Julius Thomsen" and Helge Kragh, "Uorganisk Darwinisme: Udviklingstanken i de Fysiske Videnskaber [Inorganic Darwinism: Evolutionism in the physical sciences]," *Slagmark: Tidsskrift for Idéhistorie* no. 54 (2009): 63–76.

24. Thomsen, *Om Materiens Enhed* (note 21), 22. Thomas Carnelley, "Suggestions as to the Cause of the Periodic Law and the Nature of the Chemical Elements," *Chemical News* **53** (1886): 169–172.

25. Petersen, "Grundstoffernes Natur" (note 10). Contrary to Thomsen, Petersen referred explicitly and in great detail to Crookes's address.

26. Mendeleev's paper in the *Journal of the Chemical Society*, in which he rejected the hypothesis of a unity of matter and denied that it was supported by the periodic law, has been reprinted several times. See, e.g., William B. Jensen, ed., *Mendeleev on the Periodic Law: Selected Writings, 1869–1905* (Mineola: Dover Publications, 2002), 162–188.

27. Petersen, "Grundstoffernes Natur" (note 10), 22–23.

28. Julius Thomsen, "Relation Remarquable entre les Poids Atomiques des Éléments Chimiques," *Royal Danish Academy of Sciences and Letters, Oversigt* (1894): 325–343, on 334.

29. Julius Thomsen, "Classifications des Corps Simples," *Royal Danish Academy of Sciences and Letters, Oversigt* (1895): 132–136. The paper also appeared in German, in *Zeitschrift für anorganische Chemie* **9** (1895): 190–193, and was translated into English in *Chemical News* **71** (1895): 89–91.

30. Thomas Bayley, "On the Connexion between the Atomic Weight and the Chemical and Physical Properties of Elements," *Philosophical Magazine* **13** (1882): 26–37. Carnelley, "Suggestions" (note 24).

31. Francis P. Venable, *The Development of the Periodic Law* (Easton, Pa.: Chemical Publishing Company, 1896), 209 and 271–276.

32. Thomsen, "Classifications" (29), 136.
33. At the Rydberg centennial conference in Lund in 1954, the physicist Wolfgang Pauli, referring to a paper by Rydberg of 1897, erroneously stated that, "At that time no sufficient attention had been paid to Rydberg's claim and only later the work of Julius Thomsen and others on the periodic system of the elements followed." Wolfgang Pauli, "Rydberg and the Periodic System of the Elements," in *Pauli: Writings on Physics and Philosophy*, ed. Charles P. Enz and Karl von Meyenn (Berlin: Springer-Verlag, 1994), 73–76. On Rydberg's elaborated and controversial version of the periodic system, see Johannes R. Rydberg, *Elektron, der erste Grundstoff* [Electron, the First Element] (Lund: Håkan Ohlssons Buchdruckerei, 1906). In her valuable study of Rydberg, Sister St. John Nepomucene states, also erroneously, that Rydberg in 1906 was the first to recognize the $2n^2$ relationship. St. John Nepomucene, "Rydberg: The Man and the Constant," *Chymia* **6** (1960): 127–145, on 141.
34. On this problem, see Richard F. Hirsh, "A Conflict of Principles: The Discovery of Argon and the Debate over Its Existence," *Ambix* 28 (1981): 121–130 and Giunta Carmen, "Argon and the Periodic System: The Piece that Would Not Fit," *Foundations of Chemistry* 3 (2001): 105–128.
35. Scerri, "The Periodic Table" (note 6), 156.
36. Julius Thomsen, "Über die mutmassliche Gruppe inaktiver Elemente [On the Presumed Group of Inactive Elements]," *Zeitschrift für anorganische Chemie* **8** (1895): 283–288, on 283, with English translation in *Chemical News* **77** (1895): 120–121. According to the English chemist Edward Thorpe, the paper was of great importance because it foreshadowed the discovery of the congeners of argon. Edward Thorpe, "Thomsen Memorial Lecture," *Journal of the Chemical Society* **97** (1910): 161–172, on 170.
37. Thomsen's figures were not wide off the mark, cp. that He = 4, Ne = 20, Ar = 40, Kr = 84, Xe = 131, and Rn = 222.
38. Julius Thomsen, "Über die Abtrennung von Helium aus einer natürlichen Verbindung unter Licht- und Wärmeentwicklung [On the Separation of Helium from a Natural Compound under the Evolution of Light and Heat]," *Zeitschrift für physikalische Chemie* 25 (1898): 112–114. Julius Thomsen, "Nogle Resultater af de seneste Aars Naturforskning [Some results from recent science]," *Förhandlingar vid det 15de Skandinaviska Naturforskarmötet* (1898): 66–67.
39. Emil Petersen, "Om Argon [On argon]," *Nordisk Farmaceutisk Tidskrift* **2** (1895): 233–40, on 238.
40. Søren P. L. Sørensen, "Om Argon og Helium [On argon and helium]," *Nyt Tidsskrift for Fysik og Kemi* **1** (1896): 1-18, on 17.
41. On this and other early explanations of the periodic system, see Helge Kragh, "The First Subatomic Explanations of the Periodic System," *Foundations of Chemistry* **3** (2001): 129–143.
42. The role of chemistry in Bohr's atomic theory is discussed in Helge Kragh, *Niels Bohr and the Quantum Atom: The Bohr Model of Atomic Structure 1913–1925* (Oxford: Oxford University Press, 2012).
43. Niels Bjerrum, "Nyere og Ældre Anskuelser om Grundstoffernes Natur [New and old views about the nature of the elements]," *Fysisk Tidsskrift* 6 (1907): 71–85, on 77.
44. Quoted in Léon Rosenfeld, ed., *Niels Bohr. Collected Works*, vol. 2 (Amsterdam: North-Holland, 1981), 530.
45. Niels Bohr, "On the Constitution of Atoms and Molecules," *Philosophical Magazine* 26 (1913): 476–502.
46. Quoted in J. Rud Nielsen, ed., *Niels Bohr. Collected Works*, vol. 4 (Amsterdam: North-Holland, 1977), 272. A similar reference to Thomsen's periodic table appeared

in Bohr's article on "Atom" in the fourteenth edition (1929) of the *Encyclopedia Britannica*, reprinted in Finn Aaserud, ed., *Niels Bohr. Collected Works*, vol. 12 (Amsterdam: Elsevier, 2007), 42–48.

47. Element 118, provisionally named ununoctium, was produced in nuclear reactions in 2006. Curiously, its calculated structure agrees with Bohr's electron configuration of 1922. Clinton Nash, "Atomic and Molecular Properties of Elements 112, 114 and 118," *Journal of Physical Chemistry* **A 109** (2005): 3493–3500.

48. The invitation was confirmed at the next meeting of the Academy, on April 26, 1889. See Asger Lomholt, *Det Kongelige Danske Videnskabernes Selskab 1742–1942* [The Royal Danish Academy of Sciences 1742–1942], vol. 1 (Copenhagen: Munksgaard, 1942), 407, and *Oversigt over det Kongelige Danske Videnskabernes Forhandlinger* (1889): 44–46. Mendeleev was awarded the Davy Medal by the Royal Society of London for 1882, and the next year the same honor was bestowed on Thomsen (in both cases they shared the prize, with Lothar Meyer and Marcellin Berthelot, respectively). While Mendeleev was elected a foreign member of the Royal Society in 1892, Thomsen was elected a member ten years later.

49. Letter in Thomsen's handwriting to the Academy's secretary of February 25, 1889. In Danish, author's translation. Archive of the Royal Danish Academy of Sciences.

50. Mendeleev to Hieronymus G. Zeuthen, secretary of the Royal Danish Academy, of April 14, 1889. In French. Archive of the Royal Danish Academy of Science. Mendeleev's mention of the special Danish-Russian relationship was probably a reference to Princess Dagmar, the daughter of the Danish king Christian IX and, as Empress Maria Feodorovna, the wife of Russia's tsar since 1881, Alexander III. Mendeleev was a loyal and appreciated consultant of the tsar's administration. See Michael D. Gordin, *A Well-Ordered Thing: Dmitrii Mendeleev and the Shadow of the Periodic Table* (New York: Basic Books, 2004), chapter 6.

51. Brush, "The reception of Mendeleev's periodic law" (note 7), 612.

CHAPTER 9

ᥫᩚ

Ignored, Disregarded, Discarded? On the Introduction of the Periodic System in Norwegian Periodicals and Textbooks, c. 1870–1930s

ANNETTE LYKKNES

It is . . . the usual complaint from beginners in chemistry, that they cannot understand interrelationships before the whole book has been taught. And for this reason, chemistry has been taken to be a subject for which you have to "cram"—which, in my opinion, is groundless. . . . from the beginning one should skip the presentation of the elements and compounds and instead emphasize much more demonstrations and explanations of the chemical properties of the substances.

P. K. Hustad, *Lærebok i kemi* [Textbook of chemistry], 2nd ed., 1913.[1]

The above quote from P. K. Hustad (1878–?), a textbook author and teacher at an agricultural school in mid-Norway, might at first glance be taken as an argument for the use of the periodic system in the teaching of chemistry, as opposed to introducing element by element as was the tradition before the periodic system was presented and used in textbooks. Hustad, however, did not mention the periodic system at all in his text. As I will demonstrate in this chapter, Hustad's book was not exceptional in this respect: Even though university professor Thorstein Hallager Hiortdahl (1839–1925) introduced the periodic system in his textbook in 1888 and some textbook authors continued this tradition into the 1890s, others ignored the periodic system completely as late as the 1920s and '30s. This contrasts sharply with Stephen Brush's conclusion that by the late 1880s the periodic system was "widely accepted" in America and Central Europe, and that most textbooks in America and Britain after this time discussed the periodic system.[2] In fact, the periodic system received little attention in Norway, not only compared to the United States and Central Europe, but compared to the rest of Scandinavia as well.[3]

The aim of the present volume is to compare how the periodic system was *received* in different countries. Answering such a comprehensive question, is, of course, challenging given the limited selection of printed sources. Quite often the sources that could shed light on the kinds of discussions that took place between relevant actors are either lost or inaccessible. As a consequence, I have chosen to look at how the periodic system was *presented* in Norwegian periodicals and to what degree—if any—it was *introduced* and/or *used* in Norwegian chemistry textbooks during the years between 1870 and the 1930s. I have based my investigation on textbooks available in the Norwegian national library database[4] and other texts of which I was aware, as well as on the most common Norwegian chemistry and science periodicals.

In the traditional view—although there are examples to the contrary—textbooks are the place to look for established consensus, while journal articles can tell us about what was regarded, in a particular context, as cutting-edge research.[5] Information on what was presented to students as "accepted" knowledge along with news of research that was presented to peers in journals are both useful sources that help in understanding interest in the periodic system—or lack thereof—among Norwegian chemists after the publication of Dmitrii Ivanovich Mendeleev's system in 1869. Interestingly, the periodic system in Norway was introduced in a textbook more than a decade before its first mention in a Norwegian periodical. I will argue that the system attracted great interest in the scientific community only when seen in light of new developments such as radioactivity and atomic theory. Furthermore, I will demonstrate that even when a theory, or as in this case, an organizing principle, is widely known and accepted in the community of researchers, this does not mean it will readily make its way into teaching. In the case of Norway, the long-standing monopoly held by one twentieth-century textbook author in particular hindered the teaching of the periodic system in Norwegian gymnasiums. Before looking more closely at Norwegian periodicals and textbooks from the 1870–1930s, I will provide some background information on the Norwegian chemical community.

1. A SMALL CHEMICAL COMMUNITY

With a population of less than one million at the turn of the nineteenth century, Norway was a small country fighting for its independence.[6] After almost four hundred years as a province of Denmark-Norway, Danish rule was replaced in 1814 by a union with Sweden that lasted until 1905.[7] The country's first university, the Royal Frederik University of Christiania (currently the University of Oslo),[8] was established in 1811, tied to the effort to achieve political liberation from Denmark. Almost a hundred years later, in 1910, it was joined by the Norwegian Institute of Technology in Trondheim (currently the Norwegian University of Science and Technology). During the period of interest here, the University and the Institute were the only institutions that offered advanced training in chemistry in Norway.[9] There were thus few chemistry positions in institutions of higher education. Until 1872 there was only one professor of chemistry (who served in Christiania); at the end of the period investigated here, there were three chairs in Christiania and four in Trondheim.

The size of the chemistry community was to some extent compensated for by the travels of its members to well-equipped and topic-specific laboratories abroad, typically to Germany, where Norwegian researchers were able to establish international connections. Upon their return, Norwegian chemists typically published extensively in a foreign language—mostly in German. It was also probably perceived as a natural part of their work as chemists and educators for the professors to keep abreast of the latest developments in their field by reading international periodicals and texts.[10] For example, English and German translations of Mendeleev's *Principles of Chemistry*, both from 1891, are still found in the chemistry library at the University of Oslo. It is thus believable that Mendeleev's work found its way to Christiania at around that time and was read by the two Norwegian professors who were employed at the University in Christiania, the internationally recognized Peter Waage (1833–1900), famous for his work on the law of mass action (1864) and Thorstein Hallager Hiortdahl, who authored a textbook in chemistry for university students that ran to seven editions. As mentioned, Hiortdahl introduced the periodic system to his readers in the 1888 edition of his book.

Apart from the few chairs available in academia, Norwegian chemists typically worked as pharmacists and officers, (in the later part of the period) in the chemical industry, and some entered teaching positions at technical schools and gymnasiums.[11] From its modest beginnings in 1893 with thirty members, the Norwegian Chemical Society's membership grew to between 250 and 300 in the 1930s, which testifies to the increase in number of educated chemists during the period of investigation.[12] Few of these chemists had received their advanced training in Norway; only between seven and thirty-one chemical engineers were trained each year at the time at the Institute of Technology in Trondheim,[13] and even fewer at the University's chemistry laboratory, which mainly educated pharmacists and medical doctors.[14] Some might have been trained at institutes of technology in other countries, however: at the turn of the century more than two hundred Norwegians studied at a German *Technische Hochschule*.[15]

The University's study program in mathematics and the natural sciences for budding teachers had been established in 1851. An educational reform in 1896 that made chemistry compulsory for all first-year gymnasium students must have prompted a need for chemistry teachers; however, before 1914, only ten candidates with a *major* in chemistry had graduated from the teacher training program. Hence, there may have been many teachers with limited knowledge of chemistry.[16]

Between 1896 and 1935, when lower secondary school (realskole) was established in Norway, the chemistry course at the gymnasium spanned four to six periods (hours) per week.[17] One thousand three hundred students followed the chemistry course around 1900; by 1920 the number of students who graduated from a Norwegian gymnasium had reached 4,200.[18] One textbook especially dominated the market for the gymnasium chemistry course for more than fifty years. This was Sverre Bruun's *Lærebok i kjemi for gymnaset* [Textbook of chemistry for the gymnasium], which appeared regularly in new editions, but with few substantial changes. The periodic system was not mentioned explicitly in the curriculum from 1896, yet the plan was quite detailed both on the conceptual level and regarding which of the elements and their compounds should be taught.[19] We will soon have a closer look at the extent to which the periodic system was mentioned in chemistry textbooks. But first let us consider the introduction of the periodic system in the most common academic publication channels in Norway.

2. IGNORED BY SCHOLARS?

If one is to judge from the papers published in Norwegian during the first decades following Mendeleev's system from 1869, the conclusion must be that the periodic system received little attention in Norway. The first mention that I have found dates back to 1900. None of the publications listed in the indexes of the communications from the country's two learned societies—the Royal Norwegian Society of Science and Letters (hereafter the Royal Society), established in Trondheim in 1760, and the Norwegian Academy of Science and Letters (hereafter the Academy), founded in Christiania in 1858—were concerned with the periodic system between 1870 and the 1930s.[20] In fact, chemistry did not appear as a separate category in the index of the Royal Society until 1926, which testifies to the sparse attention given to this field in Trondheim during the period of interest here.[21] The contributions in chemistry presented at the Academy in Christiania spanned from the study of physiological processes and reactions of organic compounds to photochemistry and quantitative-analytical methods.[22] Communications in chemistry in Norwegian journals in the period investigated dealt with, for instance, metal winning, noble gases, radioactivity, organic chemistry reactions, and acids. Very few articles on the periodic system were published in the general science journal *Archiv for mathematik og naturvidenskab* [Archive for mathematics and science], and in the chemistry journals *Pharmacia* [Pharmacy] and *Tidsskrift for kemi og bœrgvesen* [Journal for chemistry and mining].[23] There were, however, fourteen articles in the popular science journal *Naturen* [Nature] that mentioned the periodic system during the time of interest. None appeared in *Nyt magazin for naturvidenskaberne* [New magazine for the natural sciences].[24] In many cases the periodic system was treated as a part of historical accounts on the development of chemistry, but separate notices on the discoveries of some of Mendeleev's eka-elements appeared as well. Nevertheless, these new discoveries do not seem to have prompted any further publications on the periodic system.[25]

As noted, most of the articles written on the periodic system were published in the popular journal *Naturen*, and it is in this journal that the very first mention of the system appeared, in 1900. In an article entitled "How one can work out the existence of unknown chemical elements and predict their physical and chemical properties," Einar Simonsen, a chemistry teacher in Christiania and one of the founders of Norwegian Chemical Society, explained the periodic system of elements and its basis in historical context, crediting J. W. Döbereiner, A.-E. B. de Chancourtois, J. A. R. Newlands, L. Meyer, and Mendeleev.[26] He argued that many elements remained to be discovered, and that Mendeleev had left vacant places for them in the periodic system. The discoveries of scandium, gallium, and germanium as well as the predictions of their physical and chemical properties by Mendeleev were described in detail. Simonsen argued that, "the history of the discovery of scandium and gallium quite clearly illustrates the justification for and the correctness of the periodic system."[27] He characterized it as "one of the finest developments of mankind"—which he eagerly shared with non-expert readers.[28] The reason for this, he argued, was that "[t]he old arrangement, which, partly out of convenience, still persists in numerous textbooks, the separation into metals and non-metals or metalloids, has proved to be scientifically unsatisfactory," because quite a few elements had properties from both groups of elements.[29] As a teacher at the technical school in Christiania and later at a commercial college, he wrote various chemistry textbooks. It is interesting, therefore, to notice that in his small introductory chemistry book *Indledning til chemien* [Introduction to chemistry] from 1906, Simonsen did not devote any attention to the periodic system, although atomic and equivalent weights and valency were addressed.[30]

3. NEW CONTEXTS—RENEWED INTEREST?

Apart from historical articles, Norwegian journals mainly addressed the periodic system in articles on radioactivity, and only some decades after the system had been published by Mendeleev and Meyer. The discovery of more than thirty new radioactive elements evidently became a puzzle to chemists; before the concept of isotopy was introduced by Frederick Soddy (1877–1956) in 1913, chemists were worried about how to fit all the new "elements" into the periodic system. An instructor at the technical school in Christiania, Haavard Martinsen (1879–1967), conveyed this challenge in an article in *Pharmacia*, several years before many of the radioelements were recognized as isotopes of known chemical elements.[31] Martinsen, who had spent the previous summer working in William Ramsay's laboratory in London, acknowledged the advantages of Mendeelev's periodic system. However, faced with the evidence that radioelements such as radium emit helium, chemists were, in Martinsen's opinion, left with two options: "Either to maintain the old established definition of the concept of element and according to this perceive radium and similar substances as common chemical compounds, or to throw the old definition overboard and admit the divisibility of the elements."[32]

The chemist who would most actively disseminate information about radioactivity, including the challenge posed by finding a place for all the new radioelements in the periodic system, was Ellen Gleditsch (1879–1968, figure 9.1). Gleditsch had worked in Marie Curie's laboratory in Paris between 1907 and 1912 and became the

Figure 9.1 Ellen Gleditsch (in the middle) in the chemistry laboratory at the University in 1929, with her assistants Ernst Føyn (left) and Ruth Bakken. Photo courtesy Norsk Farmasihistorisk Museum.

sole authority on radioactivity in Norway thereafter.[33] In 1916 she was appointed associate professor of radiochemistry at the University in Christiania, and radioactivity remained her main research and teaching interest until she took up a professorship in the more comprehensive field of inorganic chemistry in 1929. Through numerous articles and books on radioactivity she shared the most recent problems and results from international research in radioactivity with her peers and the general public in Norway. The concept of isotopy and the elements' atomic weights occupied her in particular, as she herself had conducted atomic weight determinations on some of the elements in the radioactive series.[34]

In a long article on Rutherford's atomic model and the radioactive elements in *Archiv for mathematik og naturvidenskab* in 1915, Gleditsch presented aspects of the history of the periodic system, paying special tribute to periodic systems by Chancourtois, Mendeleev, and Meyer.[35] She also described the periodicity of the system and thoroughly explained the concept of isotopy and its consequence for understanding what a chemical element is. In another contribution, published in *Naturen* the following year, Gleditsch explained how research in radioactivity had led to "interesting deductions" concerning the structure of the atom.[36] Referring to Fajan-Soddy's group displacement law, Gleditsch demonstrated how the positions of the elements in the periodic system were linked to the "electrical state" of the elements' atoms. Finally, she described one of the most recent developments in this area, the introduction of atomic number to replace atomic weight as the atom's most important constant. For this, she cited her colleague, associate professor of physics Lars Vegard (1880–1963), who had contributed a most innovative account. Vegard had studied under Joseph John Thomson (1856–1940) at the Cavendish laboratory in Cambridge and was thus well acquainted with recent developments in the study of atoms.

Gleditsch had referred to Vegard's article entitled "X-rays and atomic structure," which appeared in *Naturen* as early as 1915.[37] This was only one out of a series of papers by Vegard on atomic structure that appeared between 1915 and 1923 in British and German journals as well as in popular form in Norwegian.[38] As recent investigations had shown that the properties of the chemical elements were not dependent on atomic weights, Vegard argued that it was necessary "to seek another quantity, of such a nature that the properties are unambiguously determined by this quantity," namely, the atomic number.[39] Vegard set out to prepare an atomic model that would explain the spectra and X-rays emitted by different substances, thus establishing the connection between atomic structure and chemical and physical properties. This work culminated in a long paper that was published in *Philosophical Magazine* in 1918.[40] According to Helge Kragh, who has studied Vegard's contribution in detail,[41] this "highly ambitious" work is "the first full account of the periodic system in terms of atomic theory and for this reason alone merits recognition."[42]

4. FIRST MENTION OF THE PERIODIC SYSTEM IN TEXTBOOKS

Thorstein Hallager Hiortdahl (figure 9.2) was, to my knowledge, the first textbook author who referred to the periodic system. As noted in the introduction, he mentioned the system in 1888, in the fourth edition of his *Kortfattet lærebog i anorganisk chemie* [short textbook of inorganic chemistry],[43] intended for students at the university, technical schools, or in the chemistry courses offered in gymnasiums.[44] However, he did not use the system as an organizing principle.

Hiortdahl's text contained an introductory chapter that dealt with aggregation state, crystallization, and specific weight, including a table of the elements with atomic weights. However, this section made no reference to a periodic system. The elements were thereafter divided into two groups, metals and non-metals, and presented in twenty-two chapters, dealing with occurrence, chemical compounds, physical and chemical properties, and chemical reactions of each element. The structure of the book is consistent with that used in his first edition of 1870[45]—except that the section on organic chemistry had been developed into a separate volume.[46] The periodic system appeared toward the end of the book, in a separate chapter, yet it contained references to comments on periodical properties in two other parts of the text.

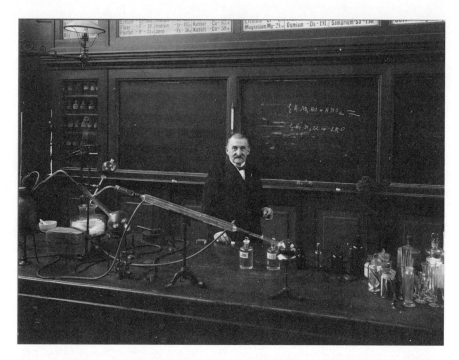

Figure 9.2 Thorstein Hallager Hiortdahl in his lecture hall in April 1909. Photo by *Tidsskrift for kemi, farmaci og terapi,* Photo courtesy Museum for university and science history, University of Oslo (MUV).

The chapter on the periodic system was just three pages and was among the smallest chapters in the book—the number of pages for other chapters averaged between seven and eight. In this chapter, Hiortdahl explained the principle of the periodicity of the system, and that the elements were listed by increasing atomic weight. Although he concluded that there was no doubt that when the periodic system had been further developed it would provide "the most rational classification of the elements," he also highlighted the uncertainties of the contemporary version of the system.[47] Among the problems Hiortdahl named was the position of beryllium in the system; if one were to consider the oxide Be_2O_3, this would yield an atomic weight, and hence a position, that was inconsistent with the chemical properties of the element. He also pointed out the challenge of incorporating metals such as iron, cobalt, and nickel and heavy elements such as gold, thallium, and lead into the system in a satisfactory way. On the positive side, Hiortdahl referred to the predictions of chemical properties of unknown elements given by Mendeleev, and how the discoveries of gallium, scandium, and germanium in the subsequent decade demonstrated that his predictions were correct.

The fifth edition of Hiortdahl's text, from 1893, included adjustments to some few atomic weights, but the text itself remained essentially unchanged.[48] The sixth edition appeared only in 1917, and included Gleditsch as a co-author.[49] In this edition, the account of the periodic system had been completely rewritten and extended by a few pages (now totalling six); it addressed chemical and physical properties as functions of atomic weight, radioactive elements, isotopes, and atomic numbers. As noted, Gleditsch was the authority on radioactivity in Norway. She had authored papers on the periodic system and had been teaching new generations of chemistry students for years; thus, her experience and competency was well suited to this task.[50]

Hiortdahl died in 1925, by which time no new edition of the textbook had been published. Gleditsch decided to revise it once again and published a seventh edition in 1928. She stated in the preface that she wished to revise the book according to the "new basis for chemistry through recent years' work on the constitution of the atom and the arrangement of the elements in the periodic system," but she had found no time for this.[51] Many people had urged her to publish a revised version, so she decided to publish it with only minor changes. Many years later, in 1940, with her assistant Einar Jensen (1907–63?), she published a textbook for students in medicine and pharmacy, whom she had been teaching for eleven years. Not surprisingly, she emphasized that this was a modern book based on the new principles of chemistry: "The book is based on our current understanding of the constitution of the atom and the arrangement of the elements in the periodic system," Gleditsch wrote in the preface.[52] Furthermore, she encouraged the readers "from the very beginning to make use of the table of the periodic system, chart I at the very end of the book, and get accustomed to always think about an element in connection with the others."[53] References to the system were made throughout the book and the elements were presented according to the groups to which they were assigned in the periodic system. This appears to be one of the first textbooks in which the periodic system was, indeed, *used*, not only *introduced*, in a text, more than seventy years after the system was published by Mendeleev and Meyer.

5. TEXTBOOKS AFTER HIORTDAHL'S 1888 EDITION

Hiortdahl's text apart, at least four different Norwegian textbooks in chemistry for gymnasiums and/or higher institutions appeared in the 1890s, although few ran through several editions over many years.[54] Most were published by professors and school teachers and it seems plausible to assume that these were the most commonly used, although we also find Swedish and Danish texts in Norwegian libraries at that time.[55] Furthermore, Henry Roscoe's *Kurzes Lehrbuch der Chemie* was available in a Danish edition from 1877,[56] and an 1898 Norwegian translation of the eighth edition of Rudolf Arendt's text could be found as well.[57] The preface to Arendt's text was written by an associate professor at the agricultural college in Ås, John Sebelien (1858–1932), who probably used the text for his chemistry classes there (in 1910 he authored his own textbook, which ran through two editions).[58] Sebelien argued that the teaching of chemistry at more elementary levels should not be mere abridged versions of the college curricula, but at the same time he questioned textbooks that organized their material after principles rather than elements.[59] His statement bears witness to an attitude clearly shared by many textbook authors at the time, and which, as we shall see, came to be dominant in Norwegian texts up until the 1970s.

Let us now turn to the textbooks originally published in Norwegian. Like Hiortdahl, pharmacist N. Davidsen (*Kursus i anorganisk kemi*, Course in inorganic chemistry, 1894), who drew on texts by J. B. Richter, C. Arnold, H. E. Roscoe and C. Schorlemmer, and O. Dammer, presented the periodic system in a separate part of his book—a section of five pages—where periodic tendencies such as specific weight, metallic character, valency, and affinity for oxygen were explained. Davidsen concluded his account by assuring that even though the system was not yet perfect, "it cannot, though, be assumed that all the peculiar relations that the arrangement shows, are a matter of coincidence."[60] The discovery of gallium, germanium, and scandium were presented to support this claim.

More interesting, however, is the textbook authored by Knut T. Strøm (*Lærebog i uorganisk kemi*, Textbook of inorganic chemistry, 1893). For the sake of overview and because it was consistent with the periodic system, as Strøm explained in the preface of the book, he chose to present the elements by groups. What particularly distinguishes this book from other Norwegian texts, however, is that he provided a summary of the common properties of each group before each element was described in detail. The halogens, as he explained, have high affinities for most other elements, are univalent, form strong acids in water, and do not combine with oxygen directly. As was true of the other textbooks named here, the presentation of the periodic system in Strøm's text was provided in a separate section at the end of the book, although it was not as detailed and explanatory as the one provided by Davidsen. The discovery of the three eka-elements and their prediction by Mendeleev was given some attention at the very end.

Few of the textbooks that appeared in the 1890s were published in new editions in the twentieth century, so what happened next? I have already mentioned the modest revisions of Hiortdahl's text and, more importantly, Gleditsch and Jensen's new book of 1940, which took into account the periodic system and modern theories of the atom. Meanwhile, new authors came forward, especially for the gymnasium market, and we will see that some of them would become more dominant than others.

6. TWENTIETH-CENTURY TEXTBOOKS FOR GYMNASIUMS

As noted, between 1896 and 1935 all gymnasium students had to follow a course in chemistry. Two chemistry textbooks intended for gymnasiums had large market shares and lasted for many decades; the most widespread was *Lærebok i kjemi for gymnaset* (Textbook of chemistry for the gymnasium, figure 9.3) by Sverre Bruun (1886–1987), which was published regularly from 1914 until the end of the 1970s, totalling 144,000 copies before 1964.[61] The other was *Lærebog i kemiens elementer: for gymnasiene* [Textbook of chemistry: for the gymnasiums] by Ole

Figure 9.3 Cover of one of Sverre Bruun's chemistry textbooks for the gymnasium.

Johannesen (1846–1917) and Carl Nicolaysen from 1897, which ran in twelve editions over almost forty years (until 1936).[62] Another book that was published at about the same time was *Lærebok i kjemi* (Textbook of chemistry), authored by O. Lindstad and Johs. Lindeman (1893–1963), which appeared in seven editions between 1926 and 1949. Common to each of the three is the limited attention given to the periodic system. The periodic system is only named in the last edition of Johannesen and Nicolaysen's text, in 1936. None of the textbooks used the periodic system as an organizing principle. I will concentrate here on Bruun's book.

Bruun's text was organized in the traditional manner, where each element and its chemistry were treated separately.[63] The first editions included a brief notice that there was such a thing as a periodic system, discovered by Mendeleev and Meyer, but the system was not connected to the main part of the book. The fact that such a connection was missing was noted in 1914 by two reviewers appointed by the Ministry who approved the book for use in the gymnasium. In general, they were pleased with the book, its thorough and precise descriptions, and its emphasis on experiments, which was in keeping with the emphasis in the curriculum. However, in a statement circulated to schools, the reviewers presented some suggestions for improvements to the text. For example, the metals were too briefly mentioned, and the presentation of the dissolution of salts in water too cursory; moreover, "it would have been useful and natural if the book had pointed to the fact that there are metals, which through common properties are reciprocally connected, even if one could not bring the presentation to a systemizing of the elements."[64] But for decades to come, the author, who also authored textbooks for secondary schools and higher education, continued along the same path.[65] In the 1941 edition, the brief notice on the periodic system was completely omitted and no *periodic* table was shown. A cursory glance at later editions reveals that the periodic system did not have a central place in chemistry, at least according to Sverre Bruun.[66] His reputation as an inspiring teacher did not prevent him from authoring a dull textbook: It has been argued that the way chemistry was presented until the 1960s and '70s, mainly through Bruun's textbooks, gave generations of Norwegians the idea that chemistry is an incomprehensible subject for which one can only "cram,"[67] just as Hustad states in the introductory quote, which dates from before the first edition of Bruun's text. Whenever new knowledge was included in Bruun's chemistry text, it appeared separately in an appendix, with no restructuring of the book itself. At best, this choice was made "out of convenience," as another author, Simonsen, put it; at worst, Bruun disregarded, or even discarded, the periodic system as a pedagogical tool.

In 1970 a new author published a Norwegian chemistry textbook. The book by Tor Brandt (1937–1991), *Kjemi* (Chemistry), stands as something completely fresh in the Norwegian chemistry textbook tradition, with its emphasis on the micro level.[68] Brandt introduced atomic structure, electron shells and the periodic system, chemical bonding, and the structure of matter. This was taken as a basis for the introduction of the elements and their compounds at the macro level and for the explanation of chemical reactions.[69]

7. SOME CONCLUSIONS

The periodic system of elements clearly did not attract much attention in Norwegian scientific journals from the 1870s until the early part of the twentieth century. The first mention of it in a Norwegian periodical appears to date from 1900. Moreover, most of the papers that did appear were published in the popular science journal *Naturen*, which was aimed at the general public, not the chemical community, although chemistry teachers probably read the journal as well.

If the news of a periodic system had been ignored in periodicals during the first decades following Mendeleev and Meyer's publications, this was not entirely true for textbooks for university students. Quite counterintuitive to what might be expected, the first mention of the periodic system in Norway occurred in what may be described as the "boring" context of established consensus, in Hiortdahl's text of 1888. However, compared to other countries this was not particularly early. The order of appearance in Danish communication channels was opposite of what I found for Norway—it reached periodicals in 1880—however, the system reached its first Danish *text* the same year as in Norway.[70] By the time Hiortdahl introduced the periodic system in his book, it had already been discussed in most textbooks in America and Britain.[71] Let us first consider the reasons for its absence from academic publication channels.

The fact that the periodic system was not debated in academies and periodicals does not mean chemists were not aware of it. Noted Norwegian scientists were clearly familiar with international scientific literature. We may argue that since the periodic system was a theoretical framework more than a practical tool for most chemists, it may not be surprising that the system was not what interested them most.[72] Norway's university tradition was also quite young and small, with only two professors of chemistry in the whole country from 1872. In contrast, chemists in Sweden, for example, took part in the discovery of elements as early as the eighteenth century and therefore were probably eager to discuss recent developments in the study of elements.[73] As the periodic system neither resulted in any revolutionary insights nor overturned accepted experimental results, it probably did not add much to the daily work in the laboratories at the university in Christiania. The chemistry professors at the Norwegian Institute of Technology appeared late (1910) on the Norwegian scene and were dedicated to the study of industrial chemical processes, not teaching tools and organizing principles.[74]

The most active contributors to the dissemination of the idea of elements and the periodic system in the twentieth century were Haavard Martinsen and Ellen Gleditsch, both interested in and taking their departure from the study of radioactivity. Research on radioactivity raised new questions and revived the relevance of the concept of chemical element and the periodic system; indeed, it seems that the periodic system attracted renewed interest when seen in the light of new developments. Norwegian periodicals were especially interested in the problem of finding a place for all the newly discovered radioelements in the periodic system and the new developments regarding the constitution of the atom. Although the labels "established consensus" and "cutting-edge research"—used earlier to denote the kind of knowledge

traditionally presented in textbooks and journals, respectively—are not easily defin-
able terms, they may be useful as broad categories. Even if the applicability of the periodic system was, to a certain extent, questioned by textbook authors in the nineteenth
century, we may argue that the news of the system was disseminated as factual and
uncontroversial information, not as an effort that was subject to the scrutiny of peers.
New developments in atomic science during the first decade of the twentieth century
changed the status of the periodic system. Arguably, it had become part of the dis-
course of current, hot research; perhaps we can even say that it reached the status
of "cutting edge"—but not on its own terms. In Norwegian periodicals, the periodic
system instead was found useful and secured its place in academic channels in the role
of a supplier of a theoretical framework, or as a servant to another master.[75]

What role did the periodic system play in the education of chemists, then? Did it
change the texts in which it appeared? It may have been regarded as uncontroversial,
but in nineteenth- and early twentieth-century Norway it certainly did not change
textbooks and teaching practice. To my knowledge it was only in the 1940s that the
periodic system prompted a reorganization of a text, counteracting the traditional
way of teaching elements in Norwegian universities. The periodic system was not
only ignored or disregarded; as a pedagogical instrument it was, in fact, discarded.

For the gymnasiums, however, the effect was even more delayed. What particularly distinguishes the implementation of the periodic system in Norwegian gymnasium texts is the monopoly held by Bruun in the gymnasium market. Although
reviewers asked for a more systematic approach to the presentation of the elements
after the first edition in 1914, Bruun never changed his book in this direction,
and it would take a hundred years from Mendeleev's published system until it was
appreciated as a novel teaching tool that transformed the way of teaching chemistry in Norway. Because of Bruun's large market share, therefore, the incorporation
of the periodic system in Norwegian textbooks was exceptionally late. Hustad's
words in the beginning of this chapter could as well have been uttered fifty years
later; chemistry was, indeed, perceived by many as a subject that required "cram-
ming," because students were not given the benefit of overarching principles with
which to understand its details. While Norway was on Europe's geographical and
scientific periphery for many decades, it was even farther out on the periphery in
the teaching of chemistry.

ACKNOWLEDGMENTS

I am grateful to Lise Kvittingen and anonymous reviewers for their valuable comments and suggestions on an earlier draft of this essay, and to Tim Collins and
Nancy Bazilchuk for linguistic help.

NOTES

1. P. K. Hustad, *Lærebok i kemi: Til bruk i skoler for landmænd* (Trondhjem: Bruns
   Boghandels forl., 1913). The quote is from the preface to the book; the same wording
   is used in the preface of the first edition from 1911. Original text: "Det er . . . den

sedvanlige klage fra begyndere i kemi, at de ikke kan forstaa sammenhængen, før hele boken er gjennomgaat. Og derfor er kemien blit holdt for at være et 'puggfag'— noget som efter min mening er uberettiget . . . man [bør] i begyndelsen springe over fremstillingen av de fleste grundstoffer og forbindelser og til gjengjæld lægge saa meget vegt paa at faa vist og forklaret stoffenes kemiske egenskaper." All quotes in this chapter are translated from the Norwegian by the author unless otherwise indicated.

2. Stephen G. Brush, "The Reception of Mendeleev's Periodic Law in America and Britain," *Isis* **87** (1976): 595–628, on 614.

3. See Helge Kragh's chapter on the Danish case and Anders Lundgren's on the Swedish case, this volume.

4. http://ask.bibsys.no/ask/action/resources. Some of the titles are available online from Norwegian IP addresses. These are linked to from the Bibsys website. See also http://www.nb.no.

5. John Hedley Brooke, "Introduction: The Study of Chemical Textbooks," in *Communicating Chemistry. Textbooks and Their Audiences, 1789–1939*, ed. Anders Lundgren and Bernadette Bensaude-Vincent (Canton, Mass.: Science History Publications, 2000), 1–18, on 5.

6. More accurately, the population totalled 881,499 in January 1800. See Statistics Norway. Befolkning: http://www.ssb.no/histstat/tabeller/3-3-13t.txt; accessed February 13, 2011.

7. For contributions on Norwegian and Scandinavian science around 1905, see Reinhard Siegmund-Schulze and Henrik Kragh Sørensen, eds., *Perspectives on Scandinavian Science in the Early Twentieth Century* (Oslo: Det Norske Vitenskaps-Akademi, 2006).

8. From 1624 the capital of Norway was called Christiania (from 1877, also spelled Kristiania) after the Danish-Norwegian King Christian IV (1577–1648). In 1925 it was renamed Oslo, its original name. The university changed its name to the University of Oslo in 1939.

9. The Agricultural College (Norges Landbrukshøiskole, established in 1897) at Ås outside Christiania offered studies in chemistry as well. For a recent history of the Norwegian Institute of Technology, see Thomas Brandt and Ola Nordal, *Turbulens og tankekraft: Historien om NTNU* [The anniversary history of the NTNU](Trondheim: Pax, 2010).

10. Articles in German appeared both in national and international periodicals. Based on a survey of the publications listed in the annual reports of the Norwegian Institute of Technology for 1910–1950, Aslaug Mølmen reports that the Chemistry Laboratory at the Institute published 259 papers in international periodicals and 414 papers in national ones. This was 42.2 percent and 25.8 percent, respectively, of the total number of publications from the Institute. According to Mølmen, most papers from the chemistry laboratory that were written in a foreign language were written in German. I am not aware of a similar report for the University in Christiania, although a cursory glance at the publications listed in the annual reports of the university supports the observation that this pattern was also true of the university's chemistry laboratory. See Aslaug Mølmen, "Rapport over antall publikasjoner ved Norges Tekniske Høgskole i tidsrommet fra 1910–1950." Report prepared for the University anniversary project in 2010, not published. See Brandt and Nordal, *Turbulens og tankekraft* (note 9), 494–495.

11. Bjørn Pedersen, "NORWAY: A Group of Chemists in the Polytechnic Society in Christiania. The Norwegian Chemical Society, 1893–1916" in *Creating Networks in Chemistry: The Founding and Early History of Chemical Societies in Europe*, ed. Anita

Kildebæk Nielsen and Soňa Štrbáňová (Cambridge: The Royal Society of Chemistry, 2008), 223–235; Annette Lykknes and Joakim Ziegler Gusland, *Akademi og indus-tri. Kjemiutdanning og –forskning ved NTNU gjennom 100 år* [Anniversary history of the chemistry departments at NTNU] (Trondheim: Fagbokforlaget, forthcoming), chapter 1, 4.

12. "Norsk Kjemisk Selskap 1893–1943," *Tidsskrift for kjemi, bergvesen og metallurgi*, No. 5 (1943), 23. See also Sven G. Terjesen, *Norsk Kjemisk Selskap: 1893–1993* [The Norwegian Chemical Society 1893–1993] (Oslo: Tidsskriftforlaget, 1993); Pedersen, "NORWAY" (note 11).

13. Aslaug Mølmen, "Statistics of students at the Norwegian Institute of Technology, sorted by faculty." Data sheet prepared for the University anniversary project in 2010, not published.

14. Annette Lykknes and Ola Nordal, "Trondheim or Kristiania? An early 20th century debate on the education of industrial chemists in Norway." Paper presented at the 6th meeting of STEP (Science and Technology in the European Periphery), Istanbul, Turkey, June 18–22, 2008.

15. Brandt and Nordal, *Turbulens og tankekraft* (note 9), 90.

16. *Det Kongelige Fredriks Universitet 1811–1911. Festskrift* (Kristiania: H. Aschehoug & Co, 1911), vol. 1: 263–266; Pedersen, "NORWAY" (note 11). See also John Peter Collett, *Universitetet i Oslo 1811–1970: Universitetet i nasjonen* [Vol. 1 of the anniver-sary history of the University of Oslo] (Oslo: Unipub, 2011), 502–503.

17. *Gymnasiet. Lov om høiere almenskoler. Reglement for de høiere almenskoler. Undervisningsplan. Eksamensreglement* [Curriculum for the Norwegian gymnasium] (Kristiania: Grøndahl & Søn, 1915), 113.

18. Statistics Norway. Gymnas og middelskole 1875–1956: http://www.ssb.no/histstat/aarbok/ht-040230-189.html; accessed February 13, 2011.

19. *Gymnasiet* (note 17), 112–116.

20. Schmidt-Nielsen, Brynjulf, *Fortegnelse over Selskapets skrifter 1760–1910* [List of publications from the Royal Norwegian Society of Science and Letters 1760–1910] (Trondhjem: Adresseavisens boktrykkeri, 1912); Olav Flo, *Det Kongelige Norske Videnskabers Selskab, Skrifter 1911–1925* [Publications from the Royal Norwegian Society of Science and Letters 1911–1925] (Trondheim: Selskapet, 1977); John Arnsteinsson, *Det Kongelig Norske Videnskabers Selskab, Forhandlinger og Skrifter 1926–1936* [List of publications and discussions from the Royal Norwegian Society of Science and Letters 1926–1936] (Trondhjem: F. Bruns Bokhandel, 1937); *Register til forhandlinger og skrifter utgitt av Vitenskapsselskapet i Kristiania 1858–1924* [List of publications and discussions from the Norwegian Academy of Science and Letters 1858–1924] (Oslo: Jacob Dybwad. 1925); *Register til forhandlinger og skrifter utgitt av Det Norske Videnskaps-Akademi i Oslo 1925–1939* [List of publications and discussions from the Norwegian Academy of Science and Letters 1925–1939] (Oslo: Jacob Dybwad, 1940).

21. Chemistry did, however, appear in the index for the period 1760–1910 as a collec-tive category that also included astronomy, physics, and mechanics, but all of these fields were absent from the index for 1911–1926. In both of these periods, not sur-prisingly, fields traditionally belonging under the Museum of Natural History and Archaeology in Trondheim—an institution that was indeed founded by the Royal Society—were well represented. The new emphasis on chemistry and other natu-ral sciences from 1926 coincided with the revival of the Society in which Sigval Schmidt-Nielsen, a professor of technical-organic chemistry in Trondheim, was cen-tral. See Kristoffer Lund Langlie "'Lærde folk er irrable genus': Omorganiseringen av Det Kongelig Norske Videnskabers Selskab 1910–1926" [On the reorganization

of the Royal Norwegian Society of Science and Letters] (Master thesis, Norwegian University of Science and Technology, 2008).

22. *Register til forhandlinger og skrifter 1858–1924* (note 20); Arnsteinsson, *Det Kongelig Norske* (note 20), 57–58.

23. The two journals merged in 1920 to become *Tidsskrift for kemi* (Journal of Chemistry), which is now the official publication of the Norwegian Chemical Society.

24. While this journal mostly covered the biological and geological sciences, Hiortdahl, who was co-editor for many years, nevertheless contributed an article from time to time on mineral occurrence, mineral analysis, or the history of chemistry and chemical industry in Norway. One of his specialties was mineral composition and crystallography. See Ove Kjølberg, "Hiortdahl, Thorstein Hallager" in *Norsk biografisk leksikon*, ed. Jon Gunnar Arntzen (Oslo: Kunnskapsforl, 2001), vol. 4: 283.

25. Another article on scandium (including some historical data) was published when the German R. I. Meyer had been granted a patent on the production of scandium oxide. See L. Andersen Aars, "Om Scandium," *Tidsskrift for Kemi, Farmaci og Terapi (Pharmacia)*, no. 20 (1908): 301–303.

26. Einar Simonsen, "Hvorledes kan man beregne eksistense af ukjendte kemiske grundstoffer og forutsige deres fysiske og kemiske egenskaper," *Naturen* 1900: 65–79. All quotes in this paragraph are taken from Simonsen's article.

27. Simonsen (note 26), 77.

28. Ibid., 79.

29. Ibid., 66.

30. Einar Simonsen, *Indledning til chemien* (Kristiania: A. W. Brøgger, 1906).

31. Haavard Martinsen, "Begrepet element eller grundstof i den modern kemi [The notion of element in modern chemistry]," *Tidsskrift for kemi, farmaci og terapi (Pharmacia)* (1906): 6–13. This was also the topic of an article the following year, but the article made no reference to the periodic system; see Haavard Martinsen, "De kemiske grundstofs avbygning [The decomposition of the chemical elements]," *Tidsskrift for kemi, farmaci og terapi (Pharmacia)* no. 17 (1907): 241–243.

32. Martinsen, "Begrepet element eller grundstof," (note 31) 10–11.

33. See Annette Lykknes, Helge Kragh, and Lise Kvittingen, "Ellen Gleditsch: Pioneer Woman in Radiochemistry," *Physics in Perspective* **6** (2004): 126–155.

34. See Annette Lykknes, Lise Kvittingen, and Anne Kristine Børresen, "Ellen Gleditsch: Duty and Responsibility in a Research and Teaching Career, 1916–1946," *Historical Studies in the Physical and Biological Sciences* **36** (2005): 131–188.

35. Ellen Gleditsch, "Rutherfords atommodel og de radioaktive grundstoffe [Rutherford's atomic model and the radioactive elements]." *Archiv for mathematik og naturvidenskab* **B.XXXIV**, no. 6 (1915): 3–28.

36. Ellen Gleditsch, "Om de radioaktive grundstof-serier," *Naturen* 1916: 234–250.

37. L. Vegard, "Røntgenstraaler og atomstruktur," *Naturen* 1915: 3–51.

38. Helge Kragh, "The Reception of the New Physics among Norwegian Physicists" in *Perspectives on Scandinavian Science*, ed. Siegmund-Schultze and Kragh Sørensen (note 7), 25–44.

39. Vegard, "Røntgenstraaler og atomstruktur," 4.

40. L. Vegard, "The X-Ray Spectra and the Constitution of the Atom," *Philosophical Magazine* **35** (1918): 293–326.

41. See Helge Kragh, "Niels Bohr's Second Atomic Theory." *Historical Studies in the Physical Sciences* **10** (1979): 123–186; Helge Kragh, "An early explanation of the

periodic table: Lars Vegard and X-ray spectroscopy," Cornell University Library 2011: http://arxiv.org/abs/1112.3774v1; accessed February 13, 2012.

42. Kragh. "The Reception of the New Physics" (note 38), 34.

43. Hiortdahl also wrote about the periodic system in the first of his two volumes on the history of chemistry: Th. Hiortdahl, *Fremstilling af kemiens historie* (Christiania: Jacob Dybwad, 1906–1907), 2 vols., vol. 1, 119.

44. Th. Hiortdahl, *Kortfattet lærebog i kemi*, 4th ed. (Kristiania: Cammermeyer, 1888).

45. Th. Hiortdahl, *Kortfattet lærebog i kemi*, 1st ed. (Christiania: Johan Dahl, 1870).

46. Th. Hiortdahl, *Kortfattet lærebog i organisk kemi* (Kristiania: Cammermeyer, 1889).

47. Hiortdahl, *Kortfattet lærebog i kemi*, 4th ed. (1888), 199.

48. Hiortdahl, *Kortfattet lærebog i anorganisk kemi*, 5th ed. (1889), 219–222.

49. Th. Hiortdahl and E. Gleditsch, *Kortfattet lærebok i anorganisk kemi*, 6th ed. (Kristiania: Cammermeyer, 1917).

50. Editing a textbook with Hiortdahl turned out to be a trying experience for Gleditsch, though, as Hiortdahl was very determined. There did not seem to have been any dispute on the mention of the periodic system, but Hiortdahl insisted on mentioning his "friends [Lars Fredrik] Nilson and [Per Teodor] Cleve." See Letter collection no. 456, National Library of Norway, manuscript section, and Lykknes, Kvittingen, and Børresen, "Ellen Gleditsch" (note 34).

51. Th. Hiortdahl and E. Gleditsch, *Kortfattet lærebok i anorganisk kemi*, 7th ed. (Oslo: Cammermeyer, 1928).

52. Ellen Gleditsch and Einar Jensen, *Lærebok i uorganisk kemi* (Oslo: Cammermeyer, 1940), 1.

53. Ibid., 1.

54. It is not always clear who the intended audiences actually were. Hiortdahl assumed a wide readership from university to gymnasiums, whereas other authors specifically wrote for students at technical schools or students of pharmacy. Sometimes no specification was given at all. The books of which I am aware that appeared in the 1890s were J. Hougen, *Anorganisk kemi i kortfattet fremstilling* [Brief introduction to inorganic chemistry] (Kristiania: P. T. Malling Bokhandels forlag, 1891); Knut T. Strøm, *Lærebok i uorganisk kemi* [Textbook of Inorganic Chemistry] (Kristiania: Cammermeyer, 1893); N. Davidsen, *Kursus i anorganisk kemi* [Course in inorganic chemistry] (Kristiania: P. Omtvedts Forlag, 1894); and Ole Johannesen and Carl Nicolaysen, *Lærebog i Kemiens Elementer: For Gymnasierne* [Textbook of chemistry: for the gymnasiums] (Kristiania: Aschehoug, 1897). Hougen's book was intended for technical evening schools, Strøm's for "beginners as well as advanced students," and Davidsen's for pharmacy students, whereas Johannesen and Nicolaysen's book was directed at the gymnasium market.

55. See Kragh, this volume, for similar results for Denmark.

56. Henry E. Roscoe, *Kemi. Efter 4de Udg. ved Sv. Mørk-Hansen* [after the 4th ed. By Sv. Mørk-Hansen]. (København: Woldike, 1877).

57. Rudolph Arendt, *Kortfattet lærebog i kemi* [Short Textbook of Chemistry] (Kristiania: Mallings, 1898).

58. John Sebelien, *Lærebok i uorganisk kemi: til bruk for studerende* [Textbook of inorganic chemistry] (Kristiania: Aschehoug, 1910).

59. Sebelien's own textbook included a separate chapter on "the interrelationships between the elements" at the end of the book, in which Mendeleev and Meyer were mentioned, and periodic tendencies with respect to affinity, valency, and atomic volume were presented. Prout's theory, radioactivity, and isomorphy (isomorphism) were dealt with as well.

60. Davidsen, *Kursus i anorganisk kemi* (note 54), 257.
61. *Sverre Bruun og Olaf Devik—et lærebokjubileum* (Oslo: Olaf Norlis forlag, 1964).
62. The other editions bore the title *Lærebok i kemi for gymnasiet* (Textbook of chemistry for the gymnasium).
63. See, e.g., Sverre Bruun, *Lærebok i kemi for gymnaset* (Kristiania: Olaf Norlis forlag, 1914).
64. Aa. Bryggessaa and J. Hougen, "Circular from the Ministry of Church and Education to Gymnasiums" in *Sverre Bruun og Olaf Devik—et lærebokjubileum*, 23–26, on 25.
65. Bruun even co-authored textbooks in physics, so he clearly reached a broad audience.
66. As late as 1960, the periodic system appeared as a separate chapter after the elements had been presented in depth. See Sverre Bruun, *Lærebok i kjemi for naturfaglinjen* [Textbook of chemistry for the natural science course](Oslo: Olaf Norlis forlag, 1960).
67. Vivi Ringnes, "Lærebøker i kjemi gjennom 100 år" in *Science Didactics—Challenges in a period of time with focus on learning processes and new technology*, ed. P. A. Åstad (Porsgrunn: Høgskolen i Telemark, 2003). I also thank Lise Kvittingen for sharing her views on, and experiences with, Bruun's texts.
68. Tor Brandt, *Kjemi for gymnaset* (Oslo: Aschehoug, 1970).
69. Ringnes, "Lærebøker i kjemi gjennom 100 år" (note 67).
70. See the chapter on Denmark by Kragh, this volume.
71. Brush, "The Reception of Mendeleev's Periodic Law" (note 2).
72. It was perceived, however, as a practical guide for chemists searching for missing elements; see Brigitte Van Tiggelen and Annette Lykknes, "Ida and Walter Noddack through Better and Worse: An *Arbeitsgemeinscahft* in Chemistry" in *For Better or For Worse: Collaborative Couples in the Sciences*, ed. Annette Lykknes, Donald L. Opitz, and Brigitte Van Tiggelen (Basel: Birkhäuser Verlag/Springer, Science Networks. Historical Studies 44, 2012), 103–147.
73. See the chapter on Sweden by Lundgren, this volume.
74. Lykknes and Gusland, *Akademi og industri* (note 11).
75. I owe the use of "master" and "servant" as terms for the science of chemistry to Bernadette Bensaude-Vincent and Isabelle Stengers, *A History of Chemistry* (Cambridge, Mass.: Harvard University Press, 1996), 4, 264.

PART V

Response in the Southern European Periphery

CHAPTER 10

cらの

Chemical Classifications, Textbooks, and the Periodic System in Nineteenth-Century Spain

JOSÉ RAMÓN BERTOMEU-SÁNCHEZ
AND ROSA MUÑOZ-BELLO

1. INTRODUCTION

The periodic system is closely linked to chemical pedagogy by many different ways. It is commonly accepted that Mendeleev discovered the periodic law while he was attempting to organize the chapters of a general chemistry textbook for his students at St. Petersburg University.[1] The omnipresence of periodic tables in classrooms and textbooks throughout the twentieth century seems to confirm the decisive impact of Mendeleev's work in chemistry teaching. Thus, one might assume that the advent of the periodic classification was followed by a revolution in late nineteenth-century chemistry classrooms. However, the papers included in this volume have found scarce evidence for a profound transformation of this kind in chemistry education. Our main aim here is to suggest some explanations for this apparent paradox by exploring the rather peripheral context of nineteenth-century Spain.[2]

Our approach is based on new historiographical trends in two interrelated areas: the history of science teaching and the circulation of knowledge. Teaching is no longer regarded by historians as a second-rate activity for scientists, but as a creative context in which new knowledge is produced thanks to the complex interaction of many historical forces and agents.[3] Historians who subscribe to this trend also challenge the common view of textbook writing as repetitive, uninspiring work.[4] Mendeleev was certainly not the first teacher to address the problem of finding an accurate classification for chemistry textbooks.

In fact, when he prepared his *Principles of Chemistry* in 1868, there was already a long tradition of chemistry textbooks dating back to the seventeenth century, and many arrangements had been adopted and discussed by Mendeleev's recent predecessors.[5] Many mid-nineteenth-century textbooks devoted entire chapters to chemical classifications, in which the author presented the debates on artificial and natural classifications and added their own suggestions. One of these books was written in 1855 by Auguste Cahours (1813–1891), a professor of chemistry in Paris, and was translated into Russian with the aid of Mendeleev, just a few years before his work on the periodic system. Cahours's book included a full chapter on chemical classifications, which not only reviewed earlier attempts at ordering the elements but also remarked that several properties (volatility, metallic state, density, affinity for hydrogen and oxygen, and so on) underwent gradual changes when elements in each family were ranged in increasing atomic (or equivalent) weights.[6] We do not mean to suggest, however, that authors such as Cahours were forerunners of Mendeleev, nor do we intend to add more fuel to uninteresting priority controversies. Nineteenth-century chemical classifications were largely a result of the collective creativity associated with chemistry teaching. Cahours's book was just one example of the large group of contemporary chemistry textbooks that addressed the problem of classification. In the following pages we will describe many similar textbooks written in (or translated into) Spanish by relatively little-known authors during the nineteenth century.[7]

Mendeleev's translation of Cahours's textbook highlights the other important historiographical issue dealt with in this paper: the circulation of knowledge.[8] Chemical classifications crossed many national borders during the nineteenth century and were adapted to different local audiences, institutional frameworks, and educational contexts. The following pages will show that this process cannot be grasped in diffusionist terms. Even in a relatively peripheral context such as Spain, chemistry textbook writers introduced changes into previous arrangements, creatively mixed two or more chemical classifications, and sometimes suggested new ones based on their own pedagogical and scientific views.[9] The critical stance of these authors, together with the educational tradition mentioned earlier, broadly shaped the way in which Mendeleev's works were received in Spain during the late nineteenth century.

To explore these issues, the chapter is organized in two parts. In the first section, we describe the controversy surrounding chemical classifications in the Spanish textbooks of the early nineteenth century. We focus on the incorporation of the "artificial" classification to Spanish textbooks during the first third of the century and the subsequent debates on "natural" classifications, paying particular attention to the classification proposed by the pharmacist Josep Antoni Balcells i Camps (1777–1857). The section ends with an analysis of a Spanish textbook published at the time when Mendeleev was preparing his own textbook and periodic system. In the second part, we offer the results of a survey of approximately one hundred Spanish textbooks published between 1870 and 1920, which suggest that the periodic law in fact played a very limited role in the teaching of chemistry during this period. We argue that this was due to the pedagogical tradition described in the previous section. We also discuss two issues, which have been highlighted in other papers on the circulation of the periodic system: the role of the discovery of the predicted elements, and the atomic debates. The last point draws attention to the diverse meanings associated with Mendeleev's work during the late nineteenth century. Rather than as a teaching resource, many Spanish authors saw the periodic system as a means to introduce theoretical topics such as the existence of atoms, the "protyle" hypothesis, the formation of elements, the evolution of inorganic matter, and the formation of the universe.

For the sake of conciseness, in this chapter we present only a brief summary of these last points and will discuss them in more depth in a forthcoming paper.[10] Thus, the following pages cover only one aspect of the circulation of the periodic system in Spain, namely the context of chemistry teaching. In other environments, particularly in popular journals and lectures aimed at a broader audience, the periodic system was appropriated in a very different way, playing diverse and, arguably, more important roles. In order to complete the picture, further research should be conducted on the twentieth century, particularly the 1930s and 1940s, by which time the situation had changed and many of the conclusions of this chapter are no longer applicable.

2. BEFORE MENDELEEV: CHEMICAL CLASSIFICATIONS IN NINETEENTH-CENTURY SPAIN

When Mendeleev's periodic law reached Spain, it was by no means the first chemical classification that Spanish chemists had discussed. Since the late eighteenth century, Spanish textbook writers had argued over the best way to organize their works, which were mainly long lists of descriptive chapters dealing with properties of the substances. For the most part, their intended audience (mainly students of medicine and pharmacy) was interested in the medical properties and technological uses of substances rather than in theoretical issues such as the constitution of matter and the nature of chemical forces. Organizing all the information in an order suitable for teaching involved the formation of classifications in which chemical substances with similar properties could be arranged in the same group. Many different options were available, and the dramatic increase in the number of chemical elements and compounds during the first half of the nineteenth century fueled the controversies on chemical classifications.[11]

These issues are explored in the following four sections. First, we study the introduction of "artificial" classifications (mostly those suggested by the French chemist Louis-Jacques Thénard (1777–1857)). Then, we discuss the mutable character of these arrangements and the changes introduced by the authors and translators of Spanish textbooks. The third section deals with the advent of the "natural" classifications, focusing on the original proposal suggested by the Catalan pharmacist Josep Antoni Balcells (1777–1857). Finally, we provide a description of a Spanish textbook written at almost the same time as Mendeleev's *Principles of Chemistry*.

2.1 Artificial Classifications

During the first half of the nineteenth century, two main classifications were used to organize the expanding section of mineral chemistry in chemistry textbooks: Jacques Thénard's classification, based on the reactions with oxygen, and Jons Jacob Berzelius's arrangement founded on the recently acquired knowledge on electrochemical series. Broadly speaking, Jacques Thénard's *Traité* became an influential model for French textbooks, while Jacob Berzelius's *Lärbok* played a similar role in Sweden and, in a different way, in the German-speaking world.[12] Both textbooks were translated into Spanish but Thénard's artificial classifications had a deeper influence in Spain, because many other French textbooks were translated into Spanish during the first half of the nineteenth century.[13]

Thénard's model textbook was introduced into Spain very early, thanks to the lectures given at the Madrid Royal Laboratory after the end of the Peninsular Wars by a Swiss-born professor, Juan Mieg (1779–1859). Mieg used Thénard's textbook, and the translation was made by his assistant, J. Acosta, who added a small number of notes (some written by Mieg) and omitted many paragraphs, including a whole section on chemical classifications. Mieg's audience was probably less interested in these theoretical points than in dramatic chemical demonstrations.[14] In fact, in 1816 he published a small book on chemistry in which he adopted Fourcroy's old classifications on metals. He mentioned Thénard's classification but claimed that it offered "some details that would distract the reader's attention" in an elementary course.[15] However, in the additions to the textbook he published in 1822, Juan Mieg described Thénard's classification of metals, adding some new ones, which had been recently discovered.[16] In these years, a volume on Thénard's terminology written by Joseph Bienaimé Caventou (a pupil of Thénard) was translated into Spanish in 1818 by the physician Higinio Antonio Lorente, and included a brief section on classifications.[17]

The Madrid Conservatory of Arts and Crafts (the *Conservatorio de Artes y Oficios*) was also involved in the introduction of Thénard's classifications. The chair of chemistry was held by José Luis Casaseca Silván (1800–1869), who had studied in Paris with Thénard,[18] and who translated the book by Eugène Desmarest (1787–1842) into Spanish for his lectures. The book offered a modified version of Thénard's classifications on metals, which was used as the main organizing principle in the inorganic section.[19] Thénard's classifications reached their highest point of influence in Spain with the new translations of the *Traité* in the 1830s. Two complete Spanish versions of the *Traité* were published in France (Nantes, 1830s and Paris, 1836) and another in Spain (Cádiz and Valencia, 1839–40).[20]

Thénard's classifications were also mentioned in translations of other French textbooks influenced by the *Traité*, the most important of which was Mateu Josep Bonaventura Orfila's (1787–1853) *Elémens de chimie médicale*. The first Spanish edition of this book appeared in 1818 and was translated by its author just a year after the first French edition. A new translation based on the second French edition was published in 1822 and an abridged version appeared later.[21] The *Elémens* was recommended as a textbook for medical students in the new syllabus published in 1824. Moreover, textbooks written by Spanish authors like the physician Francisco Alvarez Alcalá (1810–1862) were largely based on Orfila's textbooks. As other translations for medical and pharmaceutical students also drew heavily on Thénard,[22] his classification of metals can be found in many nineteenth-century Spanish textbooks and became part of the chemistry syllabus.[23] For example, in Barcelona in 1846, in a chemistry examination, a student was required to describe the general properties of metals and "(to classify) them according to the system recently established by Mr. Thénard." The student "demonstrated experimentally the characteristics of each group of metals" and ended his dissertation with a complete study of copper.[24]

2.2 Creative Appropriations of Changing Classifications

Many other Spanish students at that time were required to learn Thénard's classification, in which metals were grouped according to their reactivity with water and oxygen. It would be misleading, however, to say that chemistry textbooks merely reproduced Thénard's classifications. On the one hand, authors sometimes disagreed with Thénard on important aspects. The Catalan-born physician Mateu Orfila, for example, argued against Thénard's views on combustion and refused to group chemical substances into "combustible" and "burned" substances, a crucial division in Thénard's table of contents. Even though he adopted Thénard's criteria on the classification of metals, he established different groups based on his own interpretations of the available experimental data.[25] Many Spanish scientists were aware of the differences between the two authors: the physician Alvarez Alcalá, for instance, adopted the classification by Orfila and mentioned its differences with regard to Thénard's.[26]

Other authors introduced minor changes into the classification. Miguel Piñol i Pedret, a pharmacist who probably attended the public lectures at the Madrid Royal Laboratory, used a slightly modified version of Thénard's first classification of metals in his textbook. Piñol praised Thénard's classification, but affirmed that "imperfections" would be discovered and "modifications" introduced when the "attraction degree" between oxygen and metals was better known.[27] The criticism of Thénard's classifications is noted in the new textbooks for secondary students published in Spain during the 1840s. For instance, Fernando Santos de Castro (1809–1890) followed Thénard in his first edition (1842), but, after three years of teaching experience, he stated that he was not "entirely satisfied with the method adopted by this author, or with the sparse style I was forced to adopt in my lectures, taking into account their intended purpose [secondary school]."[28]

Thénard's classifications were by no means set in stone. The substances and arrangements changed from one edition to the next, sometimes with minor amendments and additions, but sometimes with radical variations. The most important change was the group of "still unobtained metals" (Group I), which disappeared in the last edition. Moreover, in the mid-1830s, Henri-Victor Regnault (1810–1878) performed a series of experiments with metals, oxygen, and water and wrote a new version of Thénard's classification, which endured for many years in French text-books.[29] This new version was introduced in Spain by Francesc de Paula Montells i Nadal (1813–1893), a professor in Granada, who wrote a short book on chemical nomenclature in 1837.[30] Some years later, he wrote a very successful textbook in which he also mentioned Regnault's research on metal classifications.[31]

New, updated versions of the Thénard-Regnault classification (including the most recently discovered metals) were used in many other Spanish chemistry text-books between 1840 and 1870.[32] The Spanish authors recognized the risk of using a single organizing principle ("attraction to oxygen"), due to the open-ended charac-ter of experimental research. The emergence of new data compelled them to intro-duce minor (or occasionally substantial) changes in the classification of elements and, therefore, in the arrangement of all the subsequent compounds (oxides, salts, and so on). Moreover, the significance of the experimental data was not always the same for all the chemistry textbook writers, as mentioned earlier. New data could either be accommodated in the old framework or could seriously undermine it. Furthermore, following a trend prevalent in natural history, during the mid-1830s and 1840s it became commonplace for chemistry textbook authors to claim that artificial classifications were deficient and should be replaced by "natural" classifi-cations, which were believed to represent the true arrangements of substances in nature.[33]

2.3 Natural Classifications

During the 1830s and 1840s, Thénard's and Berzelius's classifications were criti-cized by the followers of André-Marie Ampère (1775–1836), who, in a paper pub-lished in 1816, urged chemists to follow the example of naturalists and to design "natural classifications," classifications based on all the characters of the substances to be classified as opposed to the "artificial classifications," which were based on one single character. A natural classification in chemistry would take into account the most numerous and most essential analogies.[34] Ferdinand Hoefer (1811–1878), one of the leading supporters of natural classifications, strongly criticized Thénard's classification for its "exaggeration of the role of oxygen." Hoefer regarded natural classifications as both heuristic and pedagogical tools, which indicated "what still remains to be discovered" and facilitated the learning of chemistry because a group of substances could be studied by paying attention to the "type of a family," that is, the substance whose properties were characteristic of the family.[35]

Many natural classifications were proposed between the mid-1830s and 1840s. In the hands of textbook writers, natural classifications produced excellent results when dealing with the group of metalloids, but problems arose when the method was applied to the larger group of metals. As a result, many French textbooks at the end of the 1840s presented mixtures of artificial classifications (for metals) and natural classifications (for non-metals). These "hybrid" classifications became the cornerstone of the nineteenth-century textbooks and in fact remained in place long after the introduction of Mendeleev's periodic system.[36]

"Natural classifications" rarely feature in Spanish textbooks during the first half of the nineteenth century, but some examples deserve mention. A surprising reference is included in a translation of a textbook written by Jean-Sébastien-Eugène Julia de Fontenelle (1790–1842), a French author who followed Thénard's artificial classification very strictly; the translators, however, included the natural classification of elements suggested by Ampère.[37] Another translation of a French textbook on mineralogy also described Ampère's classification,[38] while a textbook intended for secondary school students (1847) used the classification proposed in France by Adolphe Dupasquier (1793–1848) and Rodolphe Kaeppelin (1810–1871).[39] In 1853 the book on classification and terminology by Ferdinand Hoefer was translated into Spanish, and so the most important natural classifications were presented by one of their leading supporters.[40]

The most original contribution to the discussion on natural classifications was made by Josep Antoni Balcells i Camps, a pharmacist and professor at the School of Pharmacy in Barcelona.[41] In a paper published in 1840, he reviewed the classifications proposed since Lavoisier and included critical remarks on Thénard's and Berzelius's classifications. He described Ampère's natural classification and adopted his general approach. However, he was disappointed by Ampère's disregard of electrochemical properties and pointed out the problems posed by the introduction of some of the recently discovered elements in Ampère's classification. Consequently, he suggested a division of the classification into two classes (electronegative and electropositive elements), in which he included fifteen groups, which were similar to Ampère's natural groups. Like other authors of natural classifications, Balcells did not take into account the old division between metals and non-metals. However, most of the metals were placed among the electropositive bodies, and the groups were not markedly different from the Thénard-Regnault artificial classification. In fact, Balcells used reactions with oxygen and water as the main organizing criteria (although never as the only ones) in many parts of his classification. In his table he included the atomic weights of elements, but he does not seem to have attempted to establish relationships between these values and the chemical properties of elements.[42] (See Table 10.1.)

Balcells used this classification in his lectures at the Faculty of Pharmacy in Barcelona, and later at the Faculty of Medicine. He did not publish a textbook but appears to have continued to work on his classification. Some years later, as an appendix to the translation of Apollinaire Bouchardat's (1806–1886) textbook, the publisher offered a "table with the division of simple bodies by Dr. José Balcells" (by then vicedirector of the Faculty of Medicine in Barcelona). The new classification was almost identical to the first one but contained some modifications, indicating that Balcells was continuing to work on his classification during the period. The main changes were the new names for the classes ("acidifiables" and "basifiables" instead of "electronegative" and "electropositive") and the orders (almost all were renamed). Moreover, some additional suborders (families) were established, two elements (manganese and osmium) were moved from one group to another, and a new element (lanthanum) was incorporated.[43] It is worth noting that Bouchardat's textbook, which included the table, mainly used artificial classifications (mostly

---

Table 10.1 JOSEP ANTONI BALCELLS CLASSIFICATION OF ELEMENTS

First class: Electronegative elements

"Principales" (Permanent gases): O, N, H

"Halógenos": F, Cl, Br, I

"Tiónidos": S, Se, Te

"Arsénicos": P, As

"Bóridos": B, C, Si

"Crómidos": Cr, V, Mo, W, "Colombium", Sb, Ti

Second class: Electropositive elements

"Crísidos": Au, Ir, Pt, Rh

"Argíridos": Os, Pd, Ir, Ag

"Jálkidos": Cu, U, Bi, Pb

"Sidéricos": "Casitéricos" (Sn, Cd, Zn); "Nicólidos" (Co, Ni); "Céridos" (Fe, Mn, Ce)

"Zircónidos": Th, Zr, Al, Y, Be

"Alcalíjenos": "Asbéstidos" (Mg, Ca, Sr, Ba); "Tefrálidos" (Li, Na, K)

Source: see note 42

---

affinity for oxygen). For instance, when dealing with metals, a full discussion of Thénard-Regnault's classifications (based on oxygen) was offered and the translators added a third one "used by Orfila in his 1837 lectures."[44]

These examples show that the translators had no qualms about adding new data (for instance, on recently discovered new elements and compounds), correcting the authors with critical remarks in footnotes, removing entire paragraphs they regarded as uninteresting for their readers, or adapting terminology and weights and measures to the needs of their local audiences.[45] They rarely altered the arrangement of the textbook, but by means of new tables, footnotes, and additions, they could add substantial amounts of information. In the case under analysis here, Bouchardat's textbook was adapted for use by the students attending Balcells's lectures at the Faculty of Medicine and became a hybrid of artificial and natural classifications. By the middle of the nineteenth century, most chemistry textbooks contained hybrid classifications and many options were open to authors and translators.

2.4 A Spanish Textbook Published in 1875

In 1875, just a few years after Mendeleev had published his periodic classifications, the pharmacist and professor of chemistry Rafael Sáez Palacios (1808–1883) wrote a two-volume treatise on chemistry intended for students of pharmacy. Sáez had already published several translations of leading chemistry textbooks, including Berzelius's treatise and a number of French books that had adopted natural classifications. Like many other textbook authors, he devoted a full section to classifications before the chapters on metals, describing Thénard-Regnault's and Berzelius's artificial classifications as well as many examples of natural classifications. He mentioned the taxonomic works of Ampère, Despretz, Henry, Guibourt, and so on, whose natural classifications he regarded as "rather flawed" ("bastante defectuosas") due to their mistaken choice of a small group of characters.[46] He also described in detail (and criticized) the natural classification suggested by Ampère and claimed that the classifications proposed by Alexandre-Édouard Baudrimont (1806–1880) and Hoefer were "a substantial advancement" but "unsatisfactory nowadays." He also mentioned the "philosophical classification" of Dumas, regretting that it remained unfinished. Finally, he mentioned the recent classifications by "Fremy, Naquet and Odling" and adopted a "slightly modified" version of Fremy's hybrid classification.[47]

Like Sáez Palacios's work, many other Spanish textbooks published during the 1870s and 1880s included chapters on chemical classifications. They generally used revised versions of Jean Baptiste André Dumas's (1800–1884) classification of metalloids and Thénard's classification of metals; however, during the 1880s, "dynamicity" (valence) emerged as an important organizing principle, yielding classifications of the elements that were very similar to the groups of Mendeleev's periodic table, thus making it hard to identify the origin of the classification used in a textbook.[48]

The example of Sáez Palacios illustrates the familiarity of Spanish textbook writers of the late nineteenth century with the earlier controversies about chemical classifications. Textbooks mostly included empirical information about an increasing number of compounds and, lacking an accurate order, even an elementary introduction to chemistry was in danger of degenerating into a random collection of short descriptions of chemicals. Textbook writers had to decide on a sequence of chapters that bore in mind the expectations of the readership and the constraints of educational policy. The earlier analysis has shown that many possible arrangements were available: artificial classifications, natural classifications, and even "hybrid" versions. Moreover, the classifications underwent constant modifications due to the discovery of new substances, new empirical data, and evolving ideas about the value of empirical data in the arrangement of substances. Therefore, changes were often introduced in subsequent editions, including critical remarks on previous classifications and minor amendments. Translators also played an active role by finding room for new substances in old classifications, sometimes mixing artificial and natural classifications.

The previous analysis shows how much chemical classifications were a result of the collective creativity of nineteenth-century chemistry teaching, and involved a large group of professors and chemistry textbook writers about whose careers little is known. It is hard to discern how or why they chose a particular classification or why they decided to introduce changes (minor or major) in the arrangements of their books. Probably, many of these authors found themselves in much the same situation as Mendeleev when he started to write his textbook and to devise his classification. The scarcity of sources makes it difficult to know for sure whether they hit upon their classifications during a dream, by means of "chemical solitaire," or, most probably, on the basis of their teaching practice. Be this as it may, they were willing to introduce changes into earlier chemical classifications and to discuss the new proposals in the light of their own experience and the critical points raised by previous controversies. This was the challenging educational context in which Mendeleev's work on the periodic system was introduced, and in fact this context goes a long way toward explaining many of the surprising issues, which are discussed in the next section.

3. THE PERIODIC LAW REACHES SPANISH TEXTBOOKS (1880–1920)

In order to trace the entry of the periodic law into Spanish textbooks, we undertook a survey of a long list of Spanish textbooks published toward the end of the nineteenth century and the beginning of the twentieth.[49] The first book in this list that mentions the periodic law is the *Tratado elemental de química general* [Treatise on General Chemistry] published in 1880 by Santiago Bonilla Mirat (1844–1899), professor of chemistry at the University of Valladolid.[50] The textbook was very positively reviewed by the academic authorities and was adopted by many other universities and schools. In the 1890s Bonilla was promoted to the chair of chemistry at the Central University of Madrid, and his *Tratado* became highly influential, going through several editions and remaining in print throughout the following decade.[51] Bonilla, who supported atomic weights instead of equivalents, mentioned Mendeleev and Lothar Meyer in his chapter on chemical atomism, highlighting the relationship between atomic weights and periodic properties while describing the recent discovery of scandium as proof of the "great importance" of "Mendeleev's ideas." Further on, in a chapter on chemical classifications, Bonilla described the natural classification of metalloids by Dumas and the classification of metals by Thénard (according to their affinity to oxygen), but he decided to use "dynamicity" (valence) as the main criterion for organizing his textbook. He then briefly introduced the classification of Mendeleev and the periodic law.[52] The first edition did not include a table representing the periodic law, but Bonilla added it in 1884,[53] the year in which another textbook including Mendeleev's classification was published by Eugenio Mascareñas Hernández (1853–1934), professor of chemistry at the University of Barcelona.[54]

As Table 10.2 shows, during the 1880s increasing numbers of Spanish textbooks began to include brief descriptions of the periodic system. In the following decade, most of the books analyzed mentioned the periodic classification and, more and more frequently, included a table.[55] The situation, then, was not substantially different from that in France, Britain, the United States, or the other countries discussed in this book, such as Denmark and Sweden. One of the first French textbooks to mention the periodic law was published by Schutzenberger in 1880 and, as in Spain, the references to Mendeleev were not widespread in French textbooks until the 1890s. In a similar survey, Stephen G. Brush (1996) and Gisela Boeck (see her chapter in this book) have shown that German textbooks included references to Mendeleev during the late 1870s. The first British and American textbooks to mention Mendeleev's classification were published in approximately 1877, that is, almost at the same time as the first Czech textbook, and three years before the ones published in France and Spain.[56]

3.1 Critical Points

These quantitative data provide useful information, but they may also mislead us slightly in our attempts to reach comparative conclusions. To begin with, it is worth noting that including the periodic law in a textbook did not automatically suggest that the author accepted it fully and unconditionally. In fact, many Spanish authors expressed doubts about the periodic classification. One of the most critical was José Rodríguez Carracido (1856–1928), who affirmed that Mendeleev's classification was among "the worst" from the point of view of the chemical analogies it highlighted.[57] Other textbook writers expressed reservations about the periodic classification but were not so disparaging. The physician Vicente Martín de Argenta i Teixidor (1829–ca.1896), who published several books on natural history, claimed that many groups of the periodic classifications were "deceptive" ("ilusorios"). The existence of many "gaps" was regarded as another problem, even though some of them had been filled with new elements. Martín de Argenta remarked that the discovery of gallium was not due to the periodic law but to an accurate analysis of spectra by Paul-Emile Lecoq de Boisbaudran (1838–1912).[58] Even more critical was Juan Manuel Bellido Carballo, a priest and professor of physics and chemistry at a Catholic institution in Salamanca, who regarded the periodic law as a form of speculation unsuited to an empirical science like chemistry.[59] At the beginning of the twentieth century, Federico Relimpio Ortega (d. 1919), professor at the University of Seville, wrote a textbook in which he summarized the criticisms of José Muñoz del Castillo (1850–1926) (see below), who regarded the periodic law as an incomplete theory and argued that many other properties apart from atomic weight should be taken into account.[60] Inside this atmosphere of hostility, it is surprising that the puzzling problem of the atomic weight of pairs such as tellurium/iodine and other inconsistencies of the periodic law were hardly mentioned at all by the Spanish authors.[61]

3.2 Successful Predictions

The spread of the periodic table in Spanish textbooks appears to have been associated with the first successful predictions of new elements (gallium and scandium) and the publication of Mendeleev's papers in French journals (*Comptes Rendus de l'Académie des Sciences* and *Moniteur Scientifique*), in which he took credit for the successful prediction of the new elements.[62] The discovery of gallium was mentioned in Spanish newspapers during the fall of 1875, just a few months after Lecoq de Boisbaudran's announcement at the Academy of Science in Paris.[63] After 1880 most of the Spanish textbooks that included references to Mendeleev's classification also mentioned the predictions and the related discoveries as proof of the scientific value of the periodic law.[64] Santiago Bonilla mentioned Nilson's discovery of scandium (quoting the description offered by Marcellin Berthelot (1827–1907) at the Paris Academy of Science in July 1880) as "proof of the importance of Mendeleev's ideas on chemical classification." He added that "the discovery of scandium and gallium" "confirm[ed] the speculations of the Russian chemist, who predicted the existence of these elements and their principal properties."[65] Many other Spanish textbooks mentioned the predictions of gallium, germanium, and scandium.[66] Several years later, in a lecture on chemical classifications at the Madrid Academy of Science, José Muñoz del Castillo, whose work we will analyze briefly in the last section of this chapter, compared the predictions of the periodic law with the mathematical calculations of Urbain Le Verrier (1811–1877), who had forecast the discovery of a new planet (Neptune).[67] Even though he remained critical, Rodríguez Carracido acknowledged in 1887 that the prediction (and later confirmation) of new elements (notably, gallium and scandium) conferred "authority" on a classification that could be regarded as a starting point for a "more precise law."[68] In his view, the other important contribution of the Mendeleev classification was the revival of William Prout's hypothesis concerning the unity of matter.[69]

3.3 Atoms and the Periodic Law

Many other Spanish authors related the periodic law to ongoing debates on the nature of matter.[70] Like Carracido, Santiago Bonilla affirmed in 1893 that the periodic law produced a revival of Prout's hypothesis, even though Prout's ideas had been seriously challenged some years previously by the use of more accurate chemical analyses, which had shown that the atomic weights of elements were not multiples of the atomic weight of hydrogen.[71] Leoncio Más y Zaldúa (1853–1910), professor at the Central Military School, affirmed that Mendeleev's work was largely based on Prout's hypothesis[72] and José Alapont Ibáñez (m. 1922), a secondary school teacher in Valencia, regarded Prout's ideas as "the seed" of Lothar Meyer's and Dmitri Mendeleev's works.[73] Thus, it seems that theoretical issues on the nature of matter, such as Prout's hypothesis, shaped the circulation of periodic law during the late nineteenth century.

This issue has already been discussed by several historians of the periodic system. In 1934 the French chemist Georges Urbain (1872–1938) affirmed that late nineteenth-century supporters of atomism (like himself) were the most enthusiastic "propagandists" for the periodic system, whereas "equivalentists" "attempted to undermine the system of atomic weights by questioning the value of its most prominent expression: the Periodic Classification."[74] This perceptive analysis is supported by the criticisms of the periodic law that the leading anti-atomist Marcellin Berthelot (1827–1907) included in his famous book *Les origines de l'alchimie*.[75] These criticisms became well known in Spain thanks to the work of José Rodríguez Mourelo (1857–1932) and José Rodríguez Carracido, who published many papers and books based on Berthelot's ideas. However, the analysis of French textbooks has already shown that the relation between the periodic system and atomism is not as clear as Urbain claimed: many books using atomic weights did not mention the periodic system at all, while the use of equivalent weights did not prevent some authors from including long descriptions of the periodic classifications in their works: Edmont Fremy's (1814–1894) *Encyclopédie chimique*, for example.[76]

At first sight, Urbain's views seem to apply to the situation in Spain as well. Santiago Bonilla, the author of one of the first Spanish textbooks to include periodic classifications, was a firm supporter of atomism, whereas José Rodríguez Carracido, the author of the critical review of Mendeleev's classification mentioned earlier, regarded atomism as a useless, not to say harmful, hypothesis. Although he accepted that the periodic law could be of some interest in chemistry, he claimed that it could not be used to support the atomic hypothesis: "All conclusions deduced from Mendeleev's classification are claimed by atomists as their own merit because they argue that the classification has been developed by using atomic weights, so had equivalents been employed, the relationships would never have been detected. However, the merit is not so great as they claim, because, as said before, the analogies are very restricted, and the discovery of gallium and the greater part of the new metals were not instigated by this classification, but they were found by other ways."[77]

This text confirms that certain nineteenth-century Spanish chemists regarded the periodic law as a relevant argument in favor of atomic weights. One would thus expect to find reticent equivalentists on the one side, and convinced atomists on the other. However, the historical record suggests a less clear-cut situation. Even while criticizing inconsistencies and noting exceptions, anti-atomist authors such as Rodríguez Carracido and Rodríguez Mourelo played an important role in the appropriation of the periodic law in Spain.[78] Moreover, supporters of atomism were not as enthusiastic about the periodic law as one would have expected if Urbain's dichotomized picture were correct. It is true that some supporters of atomism, like Bonilla, were the most pioneering propagandists of the periodic classifications, but other atomist authors in late nineteenth-century Spain seemed to pay little attention to Mendeleev's work. Gabriel de la Puerta Ródenas y Magaña (1839–1908), for instance, in a paper discussing in atomic terms the issue of the unity of matter in 1882, did not include any reference to the periodic law.[79] José Ramón Luanco y Riego (1825–1905), professor at the University of Barcelona and a leading supporter of atomic weights, only mentioned the periodic law in the third edition of his textbook (1893), and then as an appendix at the end of the book,[80] despite the fact that some years before he had spoken in positive terms about the work of Newlands, Mendeleev, and others in a lecture at the Barcelona Academy of Science in 1888.[81] Another example of the complex situation is Ramón Torres Muñoz de Luna (1822–1890), professor of chemistry at the Central University of Madrid, who had studied in Paris with Charles-Adolphe Wurtz (1817–1884) and in Giessen with Justus Liebig (1803–1873), and attended the famous Karlsruhe meeting in 1860. He used both equivalent and atomic weights in his textbooks but never mentioned the periodic law, even in the last editions of his works published during the 1880s.[82] For instance, in a textbook published in 1886, he devoted two full chapters to classifications, in which he reviewed the different classifications of metalloids and metals suggested by Lavoisier, Berzelius, Dumas, Thénard, and Will, but made no reference at all to the periodic law or to its discoverers.[83]

3.4 Chemical Pedagogy

Torres Muñoz was by no means alone. Even authors who had spoken very positively of the periodic law hardly ever used Mendeleev's classifications as the organizing principle in their textbooks (see figure 10.1 and Table 10.2). The situation was much the same in France, Germany, the United States, Britain, and Denmark, as well as in many other countries discussed in this book.[84] An early twentieth-century Spanish chemist remarked that even Mendeleev maintained nineteenth-century classifications in his own textbooks, and did not use the periodic classification to reorganize the chapters: "It is worth noting that Mendeleev's main goal was not to make a classification of bodies in groups but to establish what he calls the *periodic law or similarities of the elements*, and the proof is that his outstanding work *Principles of Chemistry* does not follow this [his own] classification. The periodic law [. . .] allows us to determine the atomic weight and valence of certain little-known elements, as is the case of iridium, uranium, cerium, ytrium, etc."[85]

Probably, Gabriel de la Puerta's critical remarks were not shared by all the Spanish authors. However, he highlighted several contemporary perceptions about the relevance of Mendeleev's periodic system as a physical law, which could be refined and used for heuristic purposes, but also, as we will discuss later, for supporting broader-ranging views concerning atomism, inorganic Darwinism, and

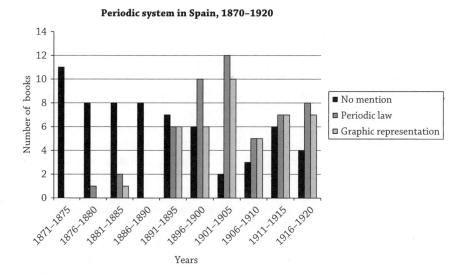

Figure 10.1 Periodic system in Spain, 1870–1920. Number of textbooks mentioning periodic law and including a graphic representation.
Source: See Table 10.2.

Table 10.2 STATISTICAL DATA ON TEXTBOOKS AND PERIODIC TABLE IN SPAIN (1870–1920)

Years	Total number of textbooks	References to periodic law			Classifications employed in textbooks				Predictions and Prout hypothesis	
		0	M	T	Valence / dynamicity	Dumas/ Thénard	Mende-leev	Other	P	Prout
1871–1875	11	11	0	0					0	0
1876–1880	9	8	1	0	1				1	0
1881–1885	10	8	2	· 1	2				2	1
1886–1890	8	8	0	0					0	0
1891–1895	13	7	6	6	6				6	2
1896–1900	16	6	10	6	5	3		2	4	3
1901–1905	14	2	12	10	8	1		3	9	7
1906–1910	8	3	5	5	2			3	4	1
1911–1915	13	6	7	7	3		3	1	6	3
1916–1920	12	4	8	7	3		2	3	5	2
	114	63	51	42	30	4	5	12	37	19

N = number of textbooks (including new editions and translations). 0 = textbooks without any reference to periodic law; M = textbooks mentioning periodic law; T= textbooks with tables representing the periodic law; Classifications = Classifications used as organizing principles of the textbook; P = textbooks including a reference to the prediction of new elements; Prout = textbooks relating the periodic law to Prout's hypothesis in some way.
Source: Bibliography of Spanish chemistry textbooks, 1870–1920. See note 49.

cosmology.[86] Moreover, by alleging that Mendeleev's own textbook neglected the periodic classification, de la Puerta clearly expressed the prevailing idea that the arrival of the periodic law did not greatly alter late nineteenth-century chemical pedagogy. Textbook writers still used the nineteenth-century chemical classifications mentioned previously, especially Dumas's natural classification of metalloids and the Thenard-Regnault artificial classification of metals. In Spain, the most widely used organizing principle during the 1880s and 1890s was "dynamicity" (valence), both by authors who spoke highly of Mendeleev's classification, like Bonilla, and by authors who made no reference to the periodic law, like Torres Muñoz. It was not unusual to find long reviews of various classifications, sometimes including Mendeleev's. Chapters of this kind were more common in textbooks addressed to university students than those written for elementary and secondary schools. In fact, two-thirds of university textbooks included references to the periodic classification, but only a third of secondary school textbooks and none of the elementary textbooks did so (see figure 10.2. and Table 10.3).

The interest aroused by Mendeleev's periodic classification was also limited by its similarities with the main groups of elements, which have been established during the first half of the nineteenth century. The natural families of metalloids were widely used in textbooks, without any great controversy regarding their

Periodic system in Spanish textbooks 1870-1920

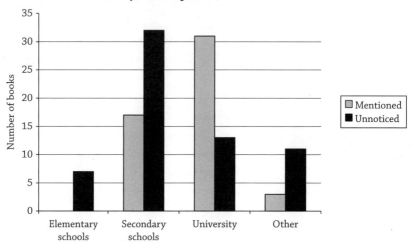

Figure 10.2 Periodic system in Spanish textbooks, 1870–1920. Number of textbooks mentioning periodic law according to educational level.
Source: See Bibliography of Spanish chemistry textbooks, 1871–1920. See note 49.

Table 10.3 NUMBER OF TEXTBOOKS ACCORDING TO
EDUCATIONAL LEVEL, 1871–1920

Mendeleev periodic system	Elementary schools	Secondary schools	University	Other	Total
Mentioned	0	17	31	3	51
Unnoticed	7	32	13	11	63
TOTAL	7	49	44	14	114

Source: Bibliography of Spanish chemistry textbooks, 1871–1920. See note 49.

arrangement. Neither Mendeleev nor the other authors presented any substantial novelty for these groups during the 1870s and 1880s. The main problem facing the introduction of periodic classification was the group of metals, as many Spanish chemistry textbooks writers recognized.[87] Most of the textbooks were organized according to slightly modified versions of Thénard's artificial classification of metals, which was mainly based on their reactions with oxygen and water. These classifications proved to be convenient for both educational and practical reasons because the resulting groups included metals with similar technological uses. Therefore, Thénard's artificial classification of metals was adopted even by authors who supported natural classifications of metalloids, including those who incorporated a discussion of Mendeleev's works. As a result, many hybrid classifications, including some partially based on the new periodic system, sometimes mixed with modified versions of old ones, were common in Spanish textbooks during the early twentieth century.[88] As in the early nineteenth century, many Spanish authors decided to use their own classifications. One of these textbook writers, Edmundo Lozano (ca. 1854–1919), remarked that "a large number of classifications of metals" existed at the beginning of the twentieth century because "not even the humblest authors desist from proposing their own classification."[89]

The new classifications were largely based on previous ones, with slight differences, but there were many exceptions to this general rule. The most original classification was published by José Muñoz del Castillo, a professor of chemistry at the Madrid Faculty of Science. He regarded the periodic law as a sort of geological scale of the evolution of the universe, showing different phases of the nebular condensation. His cyclic classification of chemical elements into three independent groups was broadly molded by Laplacian cosmology and Crookes's inorganic Darwinism. Two groups, noble gases and rare earths, were regarded as two non-periodic series, placed apart from the periodic classification of elements.[90] The remaining third group, the elements whose properties changed according to the periodic law, included several "gaps," meaning elements to be discovered. Muñoz del Castillo thought that he was improving the periodic law, which he regarded as an incomplete and imperfect version of a more comprehensive mathematical function describing the general properties of elements. Though he claimed to have used his classification in his lectures, the impact on Spanish chemistry teaching was scarce and few textbooks included references to his "cyclic classification."[91] Instead, his ideas circulated mostly in popular journals and public lectures addressed to non-specialist audiences who were mainly interested in bold theoretical issues related to cosmology and evolution. In this context, Mendeleev's work was regarded as an imperfect scientific law, which could be improved and used to expand the realm of experimental sciences, to show their useful applications, and to gain more social and economic support. This agenda differed substantially from the issues analyzed in this chapter, namely the controversies about the best organization for teaching purposes, which shaped the circulation of the periodic classifications in late nineteenth-century Spanish chemistry textbooks.[92]

4. CONCLUSIONS

Mendeleev's classification was introduced in Spain at almost the same time as in France, Britain, and the United States—that is, after the discoveries of gallium and scandium. Like other European chemists, Spanish authors regarded the predicted discoveries as crucial to the success of the periodic law.[93] Even so, Mendeleev's classification played a minor role as an organizing principle in Spanish textbooks before the 1930s and the advent of the quantum interpretation of the periodic law.

Chemical classifications were discussed in Spain long before the arrival of Mendeleev's works. In the first part of this chapter, we saw how artificial and natural classifications were discussed by Spanish textbook authors and translators of the early nineteenth century, who tended to introduce small modifications and improvements into the available classifications. By the middle of the century, many textbooks included a chapter on this issue, discussing various classifications and occasionally suggesting new arrangements. The collective creativity of these authors produced not only well-established groups of elements but also an irreverent culture of controversy about chemical classifications and their value in education. This goes some way toward explaining the scarce attention paid to Mendeleev's classification and its original appropriation in Spanish chemistry teaching. Like earlier natural classifications, when Mendeleev's periodic system entered chemistry textbooks, it was generally coupled with previous arrangements, producing new hybrid orders of substances, which revealed the personal preferences of the textbook writers. In other words, during the first decades of its long existence, Mendeleev's classification was accommodated in the collective creativity of nineteenth-century chemical pedagogy. In fact, Mendeleev's natural families were indebted to the earlier debates on chemical classifications reviewed in the first part of this chapter. Thus, when Mendeleev's classification entered Spanish textbooks, it was received as critically as previous classifications with which it shared many similarities: it was never regarded as a dramatic revolution in chemistry teaching.

Even so, many Spanish authors—supporters of atomic and equivalent weights alike—spoke positively of the periodic law. Many of them associated Mendeleev's work and Prout's hypothesis on the unity of matter. These remarks, and the case of Muñoz del Castillo mentioned previously, show that contrasting views concerning the value and meaning of Mendeleev's works existed in Spain, as elsewhere. It was regarded both as a natural classification with potential (though limited) applications in classrooms and as a scientific law with predictive powers that, in addition, could be a starting point for discussing a wide range of topics such as the evolution of matter and the concept of the chemical element. The ambiguous nature of Mendeleev's periodic system offered many opportunities for creative appropriations of his work to suit the interests of the popularizers and their audiences. In this chapter, we have seen how nineteenth-century chemical pedagogy constrained the introduction of Mendeleev's classification in Spanish classrooms. In another paper, we will discuss how Mendeleev's works were capitalized upon in popular magazines and lectures by authors who had very different political and scientific agendas and who encouraged other possible readings of Mendeleev's works in the context of inorganic Darwinist and evolutionary cosmology.

NOTES

1. See Johannes W. Van Spronsen, *The Periodic System of Chemical Elements, A History of the First Hundred Years* (Amsterdam, London, New York: Elsevier, 1969), 125 and 133. Bernadette Bensaude-Vincent, "Mendeleev's periodic system of chemical elements," *British Journal for the History of Science* **19** (1986): 3–17. Nathan M. Brooks, "Dimitrii L. Mendeleev's *Principles of Chemistry* and the Periodic Law of the Elements," in *Communicating Chemistry: Textbooks and Their Audiences*, ed. Anders Lundgren and Bernadette Bensaude-Vincent (Canton: Science History Publications, 2000), 295–311, and Michael D. Gordin, *A Well-Ordered Thing: Dmitrii Mendeleev and the Shadow of the Periodic Table* (New York: Basic Books, 2004).

2. This chapter is part of the project "Science in the classrooms in nineteenth-century Spain" (HUM2006_07206_C03_02_BERTOMEU). We are very grateful to Pedro

Ruiz-Castell, Pieter Thyssen, Antonio García Belmar, and Masanori Kaji for their comments and suggestions.

3. See David Kaiser, ed., *Pedagogy and the Practice of Science: Historical and Contemporary Perspectives* (Boston: MIT, 2005); Kathryn M. Olesko "Science Pedagogy as a Category of Historical Analysis: Past, Present and Future," *Science & Education* 15 (2006): 863–880. For a review, see John L. Rudolph, "Historical Writing on Science Education: a View of the Landscape," *Studies in Science Education* **44** (2008): 63–82.

4. See Anders Lundgren and Bernadette Bensaude-Vincent, eds., *Communicating Chemistry. Textbooks and Their Audiences, 1789–1939* (Canton: Science History Publications, 2000); Bernadette Bensaude-Vincent, Antonio García, and José R. Bertomeu, *L'émergence d'une science des manuels. Les livres de chimie en France (1789–1852)* (Paris: Editions des Archives Contemporaines, 2003); Josep Simon, *Communicating Physics: The Production, Circulation and Appropriation of Ganot's Textbooks in France and England, 1851–1887* (London: Pickering & Chatto, 2011). For a review, see Bernadette Bensaude-Vincent, "Textbooks on the Map of Science Studies," *Science & Education* **15** (2006): 667–670.

5. See José R. Bertomeu, Antonio García, Bernadette Bensaude-Vincent, "Looking for an Order of Things: Textbooks and Chemical Classifications in Nineteenth-Century France," *Ambix* 49 (2002): 227–251. On eighteenth-century chemical pedagogy, see John Perkins, ed., "Chemistry Courses and the Construction of Chemistry, 1750–1830," *Ambix* **57** (2010): 1–103.

6. Auguste Cahours, *Leçons de chimie générales élémentaire, . . .* vol. I (Paris: Mallet-Bachelier, 1855–56), 374–387. On p. 376, Cahours offers a list of properties: "decrease of volatility, convergence to metallic state, increase in density, higher melting point for higher equivalent weights." The Russian translation was published in St. Petersburg, 1859–1862. See Michael D. Gordin, "The Organic Roots of Mendeleev's Periodic Law," *Historical Studies in the Physical and Biological Sciences* **32** (2002): 263–290, on 275. Gordin, however, regards Cahours's book as "pedagogically handicapped by its lack of a coherent structure to introduce inorganic chemistry," a criticism that has also been leveled at other nineteenth-century chemistry textbooks before Mendeleev, thus supporting somehow the idea of a "pedagogical revolution" after the introduction of the periodic system. For a contrasting view, see Bertomeu, García, and Bensaude, "Looking for an Order" (note 5).

7. For a general overview of Spanish nineteenth-century chemistry textbooks, see J. R. Bertomeu; A. García Belmar's chapter in Lundgren and Bensaude, "Communicating Chemistry" (note 4).

8. For a review of the problem, see James E. Secord, "Knowledge in Transit," *Isis* **95** (2004): 654–672. For further discussion, see Raj Kapil, *Relocating Modern Science. Circulation and the Constitution of Knowledge in South Asia and Europe, 1650–1900* (New York: Palgrave, 2007); Simon Schaffer et al., eds., *The Brokered World. Go-Betweens and Global Intelligence, 1770–1820* (Sagamore Beach: Science History Publications, 2009). Concerning the spread of teaching tools, see David Kaiser, Kenji Ito, and Karl Hall "Spreading the Tools of Theory: Feynman Diagrams in the USA, Japan, and the Soviet Union," *Social Studies of Science* **34** (2004): 879–922. For scientific textbooks as go-betweens, see Simon, "Communicating Physics" (note 4).

9. On centers and peripheries, see Kostas Gavroglu et al., "Science and Technology in the European Periphery: Some Historiographical Reflections," *History of Science* **46** (2008): 153–175. See also José R. Bertomeu et al., eds. "Special Issue: Textbooks in the Scientific Periphery," *Science & Education* **15** (2006): 657–680; and Faidra Papanelopoulou, Agustín Nieto-Galán and Enrique Perdiguero, eds., *Popularizing*

Science and Technology in the European Periphery, 1800–2000 (Aldershot: Ashgate, 2009).

10. José Ramón Bertomeu-Sánchez, "Inorganic Darwinism, Chemical Order and Popular Science: The Periodic Law in late Nineteenth-Century Spain" (forthcoming).

11. For an overview, see Bertomeu, García, and Bensaude,"Looking for an order" (note 5).

12. Louis J. Thenard, *Traité de chimie élémentaire*, 4 vols. (Paris: Crochard, 1813–1816). See Bensaude, García, and Bertomeu, "L'émergence d'une science" (note 4), 65–110. On Thenard, see Antonio García, José R. Bertomeu, "Louis Jacques Thenard's Chemistry Courses at the Collège de France, 1804–1835," *Ambix* **57** (2010): 48–64. On Germany, see Bettina Haupt, *Deutschsprachige Chemielehrbucher (1775–1850)* (Stuttgart: Deutscher Apotheker, 1987): 177–192. For Sweden, see Anders Lundgren, "'Theory and Practice in Swedish Chemical Textbooks during the nineteenth century" in *Communicating Chemistry*, Anders Lundgren and Bernadette Bensaude-Vincent, eds., (Canton: Science History Publications, 2000), 91–118. For additional information, see the chapters focused on these countries in this book.

13. Jacob Berzelius's *Lärbok i kemien*, Stockholm, 1808–1830, 6 vols., was translated into Spanish many years after the first edition, when the book had become more a collective than an individual work thanks to the participation of German and French chemists-translators. Cf. Jacob Berzelius, *Tratado de química por . . .* (Madrid: J. Boix and J. M. Alonso, 1845–1852). See Marika Blondel-Megrelis, "Berzelius's Textbook: In Translation and Multiple Editions, as Seen Through His Correspondence" in *Communicating Chemistry*, Anders Lundgren and Bernadette Bensaude-Vincent, eds., (Canton: Science History Publications, 2000), 233–255. As in the case of Thenard, a separate volume dealing with Berzelius's terminology and classifications was published in Spain. Cf. Jacob Berzelius, *Nomenclatura química . . .* (Barcelona: José Torner, 1832).

14. The French edition was published in Paris between 1813 and 1816 and the Spanish translation appeared in Madrid between 1816 and 1819. *Lecciones elementales de Química teórica y práctica para servir de base al curso de las ciencias físico-químicas establecido en el Real Palacio bajo la dirección de los Infantes Don Antonio y Don Carlos* (Madrid, 1816–1819). For instance, it does not include the discussion on chemical classifications offered by Louis J. Thenard in his *Traité élémentaire de chimie*, vol. I (Paris: Crochard, 1813), 118–119.

15. Cf. [MIEG], [JUAN], *Lecciones elementales de Química para uso de los principiantes por M. y A.* (Madrid: Catalina Piñuela, 1816), 82–83. Fourcroy distinguished five different classes of metals according to their brittleness, ductility, and the properties of their oxides (formation, stability, and acid character). On Fourcroy's and Thenard's classifications of metals, see Raoul Jagnaux, *Histoire de la chimie*, vol. II (Paris: Baudry et Cie, 1891), 4 and Bertomeu, García, and Bensaude, "Looking for an order" (note 5), 30–34.

16. Juan Mieg, *[Suplemento a la obrita titulada lecciones elementales de quimica para uso de los principiantes publicada en 1816]* (Madrid: Imprenta de D. M. de Burgos, 1822). He added "thorinium," "lithium," "selenium," and "cadmium."

17. Joseph B. Caventou, *Nueva nomenclatura química, según la clasificación adoptada por Mr. Thenard . . .* (Madrid: Imprenta de la Calle de la Greda, 1818). This work included a brief description of Thenard's classifications on pp. xvi–xvii. On this translation, see José R. Bertomeu and Rosa Muñoz, "Resistencias, novedades y negociaciones: la terminología química durante la primera mitad del siglo XIX en España," *Dynamis* **30** (2010): 213–238.

THE PERIODIC SYSTEM IN SPAIN *(233)*

18. See Rolando E. Misas, "Un químico español del reinado de Fernando VII: José Luís Casaseca y Silván," *Llull* **19** (1996): 131–160.

19. Eugène Desmarest, *Química. Compendio de esta ciencia y sus aplicaciones a las artes, por...* (Madrid: L. Amarita, 1828), 256–257 and 292.

20. Louis J. Thenard, *Tratado completo de química teórica y práctica, por...* (Nantes: Busseil y Compañía, 1830); Louis J. Thenard, *Tratado completo de química teórica y práctica, ...* (Paris: Librería de Lecointe, 1836); Louis J. Thenard, *Tratado elemental teórico-práctico de química ...* (Cádiz y Valencia: Bosch y Jimeno, 1839–1840). A separate volume of Thenard's Traité (the volume on chemical analysis) was also translated into Spanish in 1828, just one year after the new French version.

21. Mateu Orfila, *Elementos de Química Médica, con aplicación a la Farmacia y a las Artes ...* (Madrid: Francisco de la Parte, 1818); Mateu Orfila, *Elementos de Química aplicados a la medicina, farmacia y artes*, 2th ed. (Madrid: Cosme Martínez, 1822); Mateu Orfila, *Lecciones de química de ...* (Barcelona: Manuel Sauri, 1840). On this book, see José R. Bertomeu and Antonio García, "Mateu Orfila's Elémens de chimie médicale and the debate about chemistry applied to medicine during the early XIXth century in France," *Ambix* **47** (2000): 1–28.

22. Examples are the following books: Jean E. S.Julia de Fontenelle, *Compendio elemental de química aplicada a la medicina* (Valladolid: M. Santander, 1829); Jean Lassaigne, *Compendio de Química, mirada como ciencia accesoria al Estudio de la medicina, Farmacia e Historia Natural* (La Habana: Imp. del Comercio, 1837); Jean Lassaigne, *Compendio de Química, mirada como ciencia accesoria al Estudio de la medicina, Farmacia e Historia Natural* (Madrid: A. Gómez Fuentenebro, 1844), and J. B. François Etienne Ajasson de Grandsagne, *Nuevo manual completo de química general aplicada a la medicina...* (Sevilla: Imp. De El Sevillano, 1841).

23. And in other disciplines as well. In mineralogy, for instance, see Santiago Alvarado, *El reino mineral ó sea mineralogía en general y en particular de España ...* (Madrid: Villamil, 1832), 2–19. See Miguel A. Puche, "Difusión de tecnicismos en la lengua de la minería del siglo XIX: La aportación de Sebastián de Alvarado y de la Peña," *Revista de Investigación Lingüística* VII (2004): 199–216; on 204, it seems quite clear that the author used Thénard's classification of metals.

24. *Revista barcelonesa*, t. I (1846/47), 22.

25. See José R. Bertomeu and Antonio García, "Mateu Orfila (1787–1853) y las clasificaciones químicas. Un estudio sobre los libros de texto de química durante la primera mitad del siglo XIX en Francia," *Cronos* **2** (1999): 130–152, on the different interpretations of experiments with lead and silver.

26. See Francisco Alvárez, *Nuevos elementos de química aplicada a la medicina y a las artes...* vol. I (Madrid: Fuentenebro, 1838–1839), forewords.

27. Miguel Piñol, *Elementos de física-química* (Madrid: Villalpando, 1820), 159–160.

28. Fernando Santos, *Nociones elementales de química ...* (Sevilla: Francisco Alvarez, 1845), quoted on pp. i–ii. The general structure of his textbook was still largely based on the groups of metalloids and metals used in Thénard's classifications.

29. Henri V. Regnault, "Recherches relatives à l'action de la vapeur d'eau à une haute température sur les métaux et sur les sulfures métalliques. Essai d'une nouvelle classification des métaux d'après leur degré d'oxidabilité," *Annales de Chimie et de physique* **62** (1836): 337–383.

30. The changes were very quickly introduced by Montells i Nadal, in his short book on chemical nomenclature, in which he mentioned Berzelius and Thenard's classification. See Francesc Montells, *Nomenclatura química, arreglada a los conocimientos modernos* (Granada: V. de Moreno, 1837), 6–9.

31. Francesc Montells, *Curso elemental de química aplicada a las artes*, vol. II (Granada: Benavides, 1845), 38: "metals have been classified in different ways but I shall adopt the broadly accepted organization in six sections according to their affinity for oxygen and their capacity for decomposing water." He described Regnault's classification in an appendix. See V. Regnault, "Recherches relatives" (note 29). On this paper, see Bertomeu, García, and Bensaude, "Looking for an order," (note 5), 250–251.

32. Some examples are Manuel Ramos, *Curso de física y química en definiciones* (Madrid: J. Peña, 1858), 131–133 and Francesc Montells, *Curso de física experimental y nociones de química*, 2nd ed. (Granada: Astudillo y Garrido, 1854), 468–469. See also the very successful textbook by Venancio González and Juan Chavarri, *Programa de un curso elemental de física y nociones de química*, 2nd ed. (Madrid: Imp. del Colegio de Sordo-Mudos, 1851), 379–380. Idem in 8th ed. (Madrid: Imprenta del Colegio de Sordo-mudos, 1866), 405–406.

33. On classifications in natural history, see Peter F. Stevens, *The Development of Biological Systematics: Antoine-Laurent de Jussieu, Nature, and the Natural System* (New York: Columbia University Press, 1994). For a broader analysis, see Ursula Klein, ed., *Spaces of Classification* (Berlin: Max-Planck Institut, 2002).

34. André-Marie Ampère, "Essai d'une classification naturelle pour les corps simples," *Annales de chimie et de physique* **1** (1816): 295–308 and 373–394, quoted on 296. On earlier drafts of his classification, see Catherine L. Dowland-Pillinguer, "A chemist full of bold and ingenious ideas: The Chemical Philosophy of A.M. Ampère" (1775–1836) (PhD diss., University of Cambridge, 1988), 478. And Bertomeu, García, and Bensaude, "Looking for an order" (note 5).

35. Ferdinand Hoefer, *Nomenclature et classifications chimiques* (Paris: Baillière, 1845). The text was translated into Spanish in 1853. See infra.

36. Bertomeu, García, and Bensaude, "Looking for an Order" (note 5).

37. Julia de Fontenelle, "Compendio elemental" (note 22), forewords.

38. François S. Beudant, *Manual de Mineralogia* (Madrid: Razola, 1831), 60. But metals were arranged according to Thenard's classification. Cf. Ibid. on 64–66.

39. Florencio Martín, *Nociones elementales de química* . . . (Cáceres: Concha, 1847).

40. Ferdinand Hoefer, *Nomenclatura y clasificaciones químicas* . . . (Madrid: Manuel Gil, 1853), 151.

41. See Arxiu General Històric Universitat Barcelona, ES CAT-AGHUB 01 EP, José Antonio Balcells. On Balcells, see José L. Gómez, *Historia del Real Colegio de Farmacia de San Victoriano* (Barcelona: 1958), 165–178. See also Carles Puig-Pla, *Física, tècnica i il.lustració a Catalunya* (PhD diss., University of Barcelona, 2006), 570 and ff. (on his son Joaquim Balcells). On Balcells's political ideology, see also Carles Puig-Pla and Jesús Sánchez "Josep Antoni Balcells: la ambigüedad política de un catedrático de física-química," *Actes d'Història de la Ciència i de la Tècnica* **1** (2008): 443–453.

42. Josep A. Balcells, *Memoria sobre los progresos de la física y química*. . . . (Barcelona: Imprenta Brusi, 1838), 5–6. The same classification was published in *Boletín de la Academia de Ciencias y Artes de Barcelona* in 1840. See Gómez, "Historia del Real Colegio" (note 41), 168–174. Another textbook written by Agustí Yañez, and also intended for the students of the Barcelona School of Pharmacy, included a discussion about "artificial" and "natural" mineralogical classifications. See Raimon Sucarrats, *L'ensenyament de la història natural a la Barcelona de la primera meitat del segle XIX. Els llibres de text i la docència d'Agustí Yañez i Girona*, vol. I (Bellaterra: UAB, 2006), 154–169. This work also includes a great deal of interesting data on Balcells. See, for instance, 259–260 and 266–267.

43. Apollinaire Bouchardat, *Elementos de química aplicada a las artes, a la industria y a la Medicina* . . . (Barcelona: A. Gaspar, 1843–1844). The editor offered the subscribers a "table with the division of simple bodies by Dr. José Balcells," Cf. "Cuadro de la división de los cuerpos simples, por el Dr. José Balcells, vicedirector de la Facultad de Ciencias médicas de Barcelona, y adaptado a su cátedra de Farmacia Químico-Operatoria." The publisher recommended that the table should be included just after page 14, that is, after the list of elements and before the notions of chemical terminology. The volume is preserved at the University Library, University of Barcelona, (130/8/19), last page. The classification was also reproduced in Pierre H. Nysten, *Diccionario de medicina*, vol. I (Barcelona: Roger, 1848), 353–354.

44. On metals, see Bouchardat, "Elementos de química" (note 43), 200–206 and on metalloids, see p. 52. Orfila's classification was not included in the other two Spanish translations of Bouchardat's book that were published. See Apollinaire Bouchardat, *Elementos de química* . . . (Valencia: Gimeno, 1843), 180–181 and Apollinaire Bouchardat, *Elementos de química* . . . (Madrid: José Redondo Calleja, 1845), 170–172.

45. On translations in the periphery, see Georgia Petrou, "Translation Studies and the History of Science: The Greek Textbooks of the 18th Century," *Science & Education* **15** (2006): 823–840.

46. Rafael Sáez, *Tratado de Química Inorgánica teórico y práctico, aplicada a la medicina y especialmente a la Farmacia. Segunda edición* (Madrid: Carlos Bailly Baillière, 1875), 912.

47. Ibid., 916–922.

48. Other examples of the broad use of "atomicity" to classify the elements are provided by Anders Lundgren in his paper on Mendeleev in Sweden in this book, which discusses the classification of Christian W. Blomstrand. More evidence can be found in the contemporary discussion of chemical classifications: see Raoul Jagneaux, *Histoire de la chimie*, t. II (Paris: Baudry, 1891), 3–29, in which Mendeleev's periodic system is included. See also Alfred Ditte, "Les Classifications Chimiques," *Revue Scientifique* **46** (1890): 609–619, and William Oechsner de Coninck, *Notes et documents de Chimie générale* (Montpellier: Masson, 1903).

49. One hundred and fourteen books published between 1871 and 1920 were analyzed (see Table 10.2). The statistical data include new editions (just a small group of successful textbooks went through more than five editions and even fewer through ten) and translations (approximately 8 percent of the total). Previous studies on the periodic law in Spain have focused specifically on its introduction in textbooks but have covered only a limited group of works: see the valuable data provided by Santiago Alvarez, Joaquim Sales, and Miquel Seco, "On books and chemical elements," *Foundations of Chemistry* **10** (2008): 79–100. Using an old-fashioned diffusionist approach, some additional data are provided by Francisco Aragón, "Evolución histórica de la clasificación de los elementos" in *Historia de la química* (Madrid: RACFN, 1981), 285–316. A paper on the restoration of a very interesting periodic table from the 1930s at the University of Barcelona was published by Claudi Mans, "La taula periòdica de l'edifici històric de la Universitat de Barcelona", *Notícies Per a Químics* 446 (2009): 4–10.

50. Santiago Bonilla, *Tratado elemental de Química general y descriptiva* (Valladolid: Hijos de Rodríguez, 1880).

51. The eighth edition was published posthumously in 1911. On Bonilla, see José Muñoz Del Castillo, *Química de los cuerpos simples* (Madrid: Imprenta L. Aguado 1901), 69–73. For a contemporary review of his textbook, see Teodoro Moya, *La enseñanza de*

las ciencias. Los orígenes de las Facultades de Ciencias en la Universidad Española (PhD diss., University of Valencia, 1991), 566–567.

52. Bonilla, "Tratado elemental" (note 50), 45–46 (quotation) and 74–77.

53. In his preface, Bonilla explicitly mentioned that this section had been expanded. Cf. Santiago Bonilla, *Tratado elemental de Química general y descriptiva, con nociones de termoquímica. Obra ilustrada con 142 figuras en el texto. Tercera edición* (Valladolid: Hijos de Rodríguez 1884), VII.

54. Eugenio Mascareñas, *Introducción al estudio de la química: compendio de las lecciones explicadas en la Universidad de Barcelona* (Barcelona: Crónica Científica, 1884).

55. Between 1880 and 1920, forty-two books included a chart of the periodic table, that is, 82 percent of the fifty-one books in which periodic classification is mentioned. See Table 10.2.

56. See Stephen G. Brush, "The Reception of Mendeleev's Periodic Law in America and Britain," *Isis* 87 (1996): 595–628, and Ludmilla Nekoval-Chikhaoui, *Diffusion de la classification périodique de Mendeleiev en France entre 1869 et 1934* (PhD diss., University of Paris, 1994). See also the chapter by García Belmar and Bensaude-Vincent in this book. On the earlier circulation in Russia, see the chapter by Masanori Kaji and Nathan Brooks in this volume and his previous paper, "Mendeleev's Discovery of the Periodic Law: The Origin and the Reception," *Foundations of Chemistry* 5 (2003): 189–214. Many chapters in this book offer similar conclusions for other countries.

57. José Rodriguez Carracido, *La nueva Química. Introducción al estudio de la Química según el concepto mecánico* (Madrid: Nicolás Moya, 1887), 162 "la clasificación más imperfecta de cuantas se han fundado."

58. Vicente Martin de Argenta, *Nuevo tratado de física y química ...* vol. 2 (Madrid: Sucesores de Rivadeneyra, 1893), 371. On Argenta, see Joaquim Olmedilla, "Noticia biográfica del Dr. D. Vicente Martín de Argenta," *Anales de la Real Academia Nacional de Medicina*, XVI (1896): 141–154.

59. Juan M. Bellido, *Tratado de Química Inorgánica en armonía con los adelantos modernos de la ciencia* (Madrid: Imprenta del Asilo, 1899), 76.

60. Federico Relimpio, *Compendio de las lecciones de Química General, 2nd ed.*, vol. 2 (Sevilla: Librería Imprenta de Izquierdo, 1902), 28–29. On Muñoz del Castillo, see infra.

61. Most Spanish authors used values between 125 and 128 for the atomic weight of tellurium until early in the twentieth century. Some later authors mentioned the issue; Cf. Simón Vila, *Tratado teórico-experimental de química general y descriptiva* (Barcelona: Agustín Bosch, 1915), 512.

62. On the influence of predictions vs. accommodations, see Brush "The Reception of Mendeleev's Periodic Law" (note 56). Though they offered significant support for the periodic law, the recently discovered elements were hardly described at all in textbooks.

63. A brief report was published in *El Globo (Madrid)*, 211 (October 28, 1875): 3 and reprinted in *La Iberia* **23** (5855), (November 18, 1875): 4. In November 1875, a secondary school teacher, Dionisio Roca Subirana, indicated the place of the new metal in the classification of elements during a session held at the local Academy of Science of Málaga. Lecture on gallium, November 12, 1875: "El nuevo metal llamado Galio" by D. Dionisio Roca Subirana.

Cf. http://www.academiamalaguenadeciencias.com/1872---1899.html. (December 17, 2012). A report of Roca's lecture was published in *La Época* (Madrid) 8491 (January 29, 1876): 1.

64. From the total of textbooks mentioning the periodic system (51), 37 (73 percent) also mentioned the successful predictions.

65. Bonilla, "Tratado elemental" (note 50), 45–46. There is another reference to scandium and gallium on 77.

66. Only a few included references to the corrections of atomic weights (particularly, the case of indium). See Santiago Bonilla, *Tratado elemental de química general y descriptiva. . .* 8th ed. (Madrid: Sanz Calleja, 1911), and Eduardo Lozano, *Elementos de física y química*, 4th ed. (Barcelona: Jaime Jepús, 1895). Also in the translation of Charles-Adolphe Wurtz, *Lecciones elementales de Química moderna . . .* (Barcelona: Librería Religiosa, 1903).

67. José Muñoz del Castillo, *Química de los cuerpos simples . . .* (Madrid: Imprenta L. Aguado, 1901), 27.

68. Rodriguez Carracido, "La nueva Química" (note 58), 164: "There is no doubt—affirmed Rodríguez Carracido—that these confirmations confer authority on this classification."

69. Ibid., 164. He developed his argument using spectroscopic data, the recent discovery of helium, and Crookes's ideas on the evolution of elements in his address to the Madrid Academy of Science in 1888. See José Rodriguez Carracido, *Discurso leído ante la RACEFN en la recepción pública por . . . el 19 de febrero de 1888* (Madrid: Aguado, 1888).

70. See William H. Brock, *From Protyle to Proton. William Prout and the Nature of Matter* (New York: Adam Hilger, 1985).

71. Santiago Bonilla, *Tratado elemental de Química general y descriptiva, con nociones de termoquímica*, 5th ed. (Valladolid: Hijos de Rodríguez, 1893), 73. See also Santiago Bonilla, *Tratado elemental de química general y descriptiva. . .* 7th ed. (Madrid: Gómez Fuentenebro, 1902), 77–78.

72. Leoncio Mas, *Lecciones de Química e Industria militar, explicadas en la Escuela Superior de Guerra*, vol 1 (Madrid: Imprenta del Cuerpo de Artillería, 1895), 183.

73. José Alapont, *Nociones de química*, 2nd ed. (Valencia: Manuel Alufre, 1902), 30.

74. Georges Urbain, "Comment les idées de Mendéléïev ont été accueillies en France," *Revue Scientifique* **72** (1934): 657–661: "s'efforcèrent alors de miner, en critiquant la valeur, le système des poids atomiques, c'est-à-dire, son expression la plus saillante: la Classification Périodique." On nineteenth-century chemical atomism, see Alan J. Rocke, *Chemical Atomism in 19th Century: From Dalton to Cannizzaro* (Columbus: Ohio State University Press, 1984). On French nineteenth-century textbooks and chemical atomism, see José R. Bertomeu and Antonio García, "Atoms in French Chemistry Textbooks during the First Half of the Nineteenth Century: The Elémens de Chimie Médicale by Mateu Orfila (1787–1853)," *Nuncius* **19** (2004): 78–119.

75. Marcellin Berthelot, *Les origines de l'alquimie* (Paris: Steinheil, 1885), 302–315. On this issue, see the chapter by Bernadette Bensaude-Vincent and Antonio García-Belmar in this book.

76. See Nekoval-Chikhaoui "Diffusion de la classification" (note 56), 93–94 and 129–130. Although Wurtz accepted the periodic system as support for the new atomic weights, he included some critical remarks concerning the prediction of new elements in his famous book *La théorie atomique*. See Charles-Adolphe Wurtz, *La théorie atomique*, 4th ed. (Paris: F. Alcan 1886), 108–134, critical remarks in pp. 117–118. See the chapters by Marco Ciardi and Antonio García-Belmar and Bernadette Bensaude-Vincent for more data about the points of view of Wurtz on the periodic table and atomic theory.

77. Rodriguez Carracido, "La nueva Química" (note 57), 169.

78. See Bertomeu, "Inorganic Darwinism, Chemical Order and Popular Science" (note 10).

79. Gabriel de la Puerta, "La unidad de la materia," *El Globo*, **8** (2159) (1882): 1.

80. José R. Luanco, *Compendio de las lecciones de química general . . .* 3rd ed. (Barcelona: La Academia Barcelona, 1893), 879–881. It was an addition at the end of the book, including a figure with Mendeleev's table on 880.

81. See the report of his conference in *La Dinastía* (Barcelona) (January 19, 1888): 2. See Claudi Mans, "José Ramón de Luanco: químico y química en transición'" NPQ 434 (2007): 9–20. On atomic theory in nineteenth-century Spain, see Inés Pellon, *La recepción de la Teoría Atómica Química en la España del siglo XIX* (Leioa: UPV-EHU, 1998).

82. See Pellón, "La recepción de la Teoría Atómica" (note 81), 235–283.

83. Ramón Torres, *Tratado de química general y descriptiva . . .* 5th ed. (Madrid: Ricardo Fe, 1885), 93–97 and 347–348.

84. See Brush, "The Reception of Mendeleev's Periodic Law" (note 56) and Nekoval-Chikhaoui, "Diffusion de la classification" (note 56).

85. Gabriel de la Puerta, *Suplemento al Tratado de Química Inorgánica* (Madrid: Perlado, 1904), 15. (The emphasis is his own). On Mendeleev's changes in the contents and arrangement of his textbook, see Masanori Kaji, "D.I. Mendeleev's Concept of Chemical Elements and the Principles of Chemistry," *Bulletin of History of Chemistry* **27** (2002): 4–17, especially pp. 7–13. Nathan M. Brooks, "Dimitrii L. Mendeleev's Principles of Chemistry and the Periodic Law of the Elements" in: *Communicating Chemistry: Textbooks and Their Audiences* eds. Anders Lundgren and Bernadette Bensaude-Vincent (Canton: Science History Publications, 2000), 295–311.

86. See the last section of this chapter.

87. The problem was widely recognized by textbook writers. See, for instance, Eugenio Mascareñas, *Elementos de Química General y Descriptiva* (Barcelona: Pedro Ortega, 1903), 110.

88. For a full discussion, see Eugenio Piñerúa, *Principios de Química Mineral y Orgánica* (Valladolid: Impr. y Librería Nacional y Extranjera de Andrés Martín, 1898). Other examples are Enrique Roca, *Tratado de Química inorgánica aplicada a la Farmacia . . .* (Santiago: Imprenta de la Gaceta, 1897) and Conrado Granell, *Tratado elemental de química moderna* (Madrid: Bailly-Baillière, 1906), who employed a classification of metals based on chemical analysis. See also Luis Bermejo, *Elementos de quimica general y descriptiva* (Valencia: Pubul y Morales, 1909), who discussed the difficulties concerning the classification of metals. Other authors, like Gabriel de la Puerta, remarked that chemical classifications were very deficient due to the practical character of chemistry, so he adopted a classification adapted to teaching purposes. Cf. Gabriel de la Puerta, *Tratado de Química Inorgánica . . .* vol. I (Madrid: Vda. Hernando, 1896–97), 124.

89. Cf. Edmundo Lozano, *Ciencias Físico-Químicas* (Madrid: Ediciones de La Lectura, 1916), 264. See, for instance, Piñerúa "Principios de Química Mineral" (note 88), 66–67. Eduardo Lozano wrote a large number of innovative textbooks and polemical writings dealing with new trends on science education.

90. On the accommodation of rare earths and noble gases, see Pieter Thyssen and Koen Binnemans, "Accommodation of the rare earths in the periodic table: a historical analysis" in Handbook on the Physics and Chemistry of Rare Earths **41**, eds. Karl A. Gschneidner Jr., Jean-Claude Bünzli, and Vitalij K. Pecharsky (Amsterdam:

Elsevier, 2011), 1–95. See also the chapter by Helge Kragh, who reviews the work of Julius Thomsen on inorganic Darwinism.

91. Two books including references to Muñoz del Castillo's classifications are Gabriel de la Puerta, "*Suplemento*" (note 85), 15–18 and Relimpio, "Compendio" (note 60), 28–29, which includes some critical remarks on the periodic law based on Muñoz del Castillo's ideas.

92. This context is fully discussed in another paper. See Bertomeu, "Inorganic Darwinism, Chemical Order and Popular Science" (note 10).

93. See, for instance, the papers by H. Kragh on Denmark and Anders Lundgren on Sweden (in which the discovery of scandium played an important role).

CHAPTER 11

ᐁ

Echoes from the Reception of Periodic Classification in Portugal

ISABEL MALAQUIAS

1. INTRODUCTORY OVERVIEW

This essay attempts to evidence the remaining echoes of the reception of Mendeleev's periodic classification in Portugal during the last quarter of the nineteenth century. The research involved the identification of remaining traces at different higher-level teaching institutions as well as with books, textbooks, and programs from beginner's level to advanced level that appeared in the period between 1876 and 1904.

Following the institutionalization of chemistry as an independent scientific discipline in Portugal, after the 1772 reform of Coimbra University, the 1844 reform of the university serves as a breath of fresh air in terms of the study of chemistry as being split into different chemistry specialties, an action much related to developments abroad. By 1851 the Coimbra professor J. A. Simões de Carvalho (1822–1902) was opposing chemistry being taught at the university with French textbooks. Instead, he wrote a modernized text, which included recent research for chemical philosophy lessons and advocated a much greater "attention to the day-to-day communications in scientific journals and newsletters than to more complete and extensive manuals."[1]

In the two decades immediately before the period we are interested in, there was a resurgence of the country's economy and some developments also affected the still unique university. In this context, reform of curricula took place and a positivist wind blew through the faculties, starting in the Law Faculty and then spreading to the other faculties with a symptomatic decline in the influence of the Canonical Faculty. The Law Faculty had the biggest number of students and several of its members took up higher administrative or governmental positions. The

intellectual atmosphere of the Faculty of Natural Philosophy was deeply impreg-
nated with positivism, whereby it was intended to regenerate the sciences body.[2]

The program of sending some of its members abroad was reactivated, to keep
them up to date with the modern experimental sciences,[3] particularly chemistry,
while some foreign staff came to work in the university.[4]

During the same period, a few doctorate theses were being defended, namely on
the relations among different disciplines, including chemistry and the other sciences. Although the subjects were up to date, they were developed academically
in the sense that they were not emerging directly from research carried out by the
candidates. This came to be perceived as a frailty.[5]

As Amorim da Costa wrote, the teaching of chemistry was not old fashioned, but
a gap was emerging when relating experimental teaching and research. This was
connected with both insufficient governmental financial support and insignificant
links to a weak industrial milieu.[6]

Although it had only a single university at the time, nineteenth-century Portugal
tried a particular emphasis to teaching, and the scientific and technical preparation
of younger middle-class students was thought to be what society wanted from the
university. As had already happened in France, Germany, and England, and been
planned by the Pombalian educational reform program,[7] courses, schools, and
institutions were created with new profiles.[8] This orientation was also imposed at
the secondary level, with the introduction of several disciplines in which the first
scientific and technical elements of mathematics, physics, chemistry, natural history, political economy, public administration, and commerce were taught.[9] Later
in the century, technical teaching was improved, leading to a deep restructuring
of the agricultural and industrial teaching, both of which became more practical.

Two institutions turn out to be prominent by the middle century: the Polytechnic
School in Lisbon and the Porto Polytechnic Academy. The Lisbon school gained,
by 1859, a revitalized momentum with the amplification of its curricular program
and the transfer of its jurisdiction to the Kingdom Ministry, as the Instruction
Ministry did not yet exist.[10] It was also influential in the spreading of positivism,
not only in the articulation of the disciplines in a general educational system, but
also in the hesitant way it dealt with the epistemological demands of inductive support for these, and the School's involvement in the social and economic endeavor
that underpinned the training it was intended to produce.[11] The Polytechnic School
was created to intervene in the modernization of the country, in harmony with the
interests of society's forces of production, and it was open to middle-class students.
The formation of Porto Polytechnic Academy, together with its Lisbon counterpart,
was intended "to implement in the country the industrial sciences, quite different
from classical and purely scientific studies." Both should prepare students for the
practice of agriculture, industry and commerce and chemistry was considered of
utmost importance, also for the students coming from the Porto Medical Surgical
School. A diploma was produced some time later for chemists, who were licensed to
manufacture and handle chemical products.[12] Notwithstanding its eminently practical character, laboratory chemical experimentation was nonexistent; there was
only the occasional demonstration. These were performed by the professor and an

auxiliary technician. In a certain sense, the understanding of the experiment as a pedagogical means of demonstration of the concepts emerging from the theoretical models lost sight of the inventive creativity pure science enables. In this sense, it can be understood why the work produced by authors connected with positivism was more erudite than innovative. They had a greater propensity to popularize foreign research results and present primers containing the ideas of Comte, Émile Littré, Darwin, Haeckel, and Spencer.[13]

The earliest diffusion, in Portugal, of Comte's ideas took place through the teaching of mathematics in the polytechnic schools.[14] As already mentioned, in the first years of the 1870s, the university was also full of these positivist ideals, albeit ones following different testimonies.[15] Positivism entered the Portuguese culture not so much for its scientific or philosophical value, but mainly to serve a historical political movement. The episode known as the "Casino Conferences" illustrates the intention to "Europeanize" the country. The creation in 1873 of the Republican Party and some of its leading figures marked the way. Positivism increasingly turned the philosophy of liberals and republicans.[16] The reasons for a generalized interest in Comte's ideas amongst a range of professors in higher-education institutions is related to the way Comte conceived the true nature of this discipline in his *Course of Positive Philosophy*. After 1872 the positivist philosophical system was generally accepted as being the path to take.[17] Subsequent to the setting up of the Porto Municipal Chemical Laboratory (1884/87) with António Joaquim Ferreira da Silva (1853–1923), practical chemistry teaching stood out. Theoretical chemistry teaching was updated.

A similar situation occurred at the Lisbon Polytechnic School, where, during its first years, the teaching of chemistry was mainly expository and speculative, with some demonstrations. The organization and main features of this did not differ much from the chemistry teaching at Coimbra University in the same period. However, the Polytechnic had "younger and much more motivated and possibly better-prepared professors," including Agostinho Lourenço (1822–93), António Augusto de Aguiar (1838–87), José Júlio Rodrigues (1843–93), Roberto Duarte Silva (1837–89), and Achilles Machado (1896–1932). These few teachers did their best, either publishing textbooks or organizing and obtaining good equipment for their laboratories.[18]

This introduction offers a general framework within which we can appreciate a number of surviving traces concerned with the reception of Mendeleev's periodic classification in Portugal during the last quarter of the nineteenth century.

For three decades until 1872, a course in Chemistry and the Chemical Arts was taught at Porto Polytechnic Academy,[19] Joaquim de Santa Clara Sousa Pinto (1802–76) the professor in charge. Between 1872 and 1876, a new chemistry professor joined the Porto Polytechnic, António Luis Carneiro de Vasconcelos Ferreira Girão (1823–76). A former student at Coimbra University, Ferreira Girão took degrees in philosophy and mathematics. He was well known in the cultural and academic media not only for his scientific skills in chemistry, mineralogy, and metallurgy, but also for his writings on science as well as his poems and humoristic writing.[20]

2. TRACES

2.1 1876—The Theory of Atoms and The Limits of Science

In 1879 Ferreira Girão published a book entitled *The Theory of Atoms and The Limits of Science* in Porto.[21]

The publication date does not correspond to the year in which it was written, 1876, the year the author passed away, as we can see from the preamble. The book is structured into three chapters followed by an appendix, and the subtitle presents it as "three chapters of general physics." The preamble refers to this, arguing that the subject is atomic theory, which seems to be more chemistry than physics. Nevertheless, between chemistry and physics there are several contact points, so that they look more like one science than two.

Ferreira Girão begins citing Maxwell on the question of the final atom and the creation of matter. He produces an abridgment of the historical background since the Greek time to the development of atomic theories, Morveau, Wenzel, Richter, and Gay-Lussac's contributions. He also presents Dalton's theory and the need for this theory to explain the chemistry facts. At the same time, he states that the atomic theory, as put forward by chemists, does not fulfill the philosophical spirit, and next offers a number of questions: "What is the nature of these corpuscles and what are their essential properties? Will the atoms of the simple corps be the last level of divisible matter?"[22] Following this, in another chapter, he differentiates chemical atoms and prime atoms, considering that the law of definite proportions shows that those bodies are made by the juxtaposition of materially invariable units. As this reason is not reconcilable with indefinite division, one must conclude that chemical atoms exist. He also mentions the materials that constitute the earth's crust and the huge number of different minerals, in order to ask: "Is it credible, to form such a large number of compounds, that nature possesses just sixty-five elementary or simple bodies?"[23] "Will the atoms of sulfur, lead, and iron be the last plots of matter, or should we consider these elements as different condensed states of a unique and sole substance?"[24] He pursues his historical examples, mentioning that the idea of considering the simple bodies as compounds was very old, but when alchemy stepped down, it took with it the unity of matter`. More recent discoveries, like those of cyanogen and ammonium hypothesis, together with some isomeric facts, brought again the unity of matter to the fore. But these theoretical visions were not experimentally confirmed, namely Prout's after Stas's accurate determinations. Thus, two directions could be followed: to conclude that the ultimate particles of hydrogen, iron, and copper are the unique prime atoms, or to follow another order and see if the direct analysis can leave it further. In this last case, everywhere and in all circumstances, the last expression of decomposition is always a simple body. He discusses proofs for this hypothesis on the basis of evidence relating to the body's free fall, optical phenomena, the dissociation of gases at high temperatures, and the spectral analysis of celestial bodies.

"It seems to us that certain luminous and calorific phenomena cannot be understood if we do not admit that the chemical atom far from being the unique active center of ponderable matter is, on the contrary, the last arrangement of several previous and inferior groups of prime atoms."[25]

Newton, Herschel, Ritter, Wundt, and Secchi are mentioned as well as Pedro Norberto, a former professor of chemistry at the Coimbra University[26] who in 1836 "presented the idea that the chemical atoms decompose in inferior groups to give place to the luminous, calorific and electrical phenomena," considering that "atoms vibrations produce electricity while the sub atoms vibrations produce the calorific and luminous waves."[27] Ferreira Girão, however, concluded that "such a conception neither explains the color variety, nor the origin of the different waves. It was to vainly multiply the hypothesis without solving the difficulties."[28] Considering the spectral analysis presents some authors like Secchi,[29] who consider that "*maybe* the electrical energy dissociates the gases in its elements, dissociation that is maintained only if the temperature is very high, time during which atomic vibrations analogue to allotropic changes operate."[30]

Comparing the spectrum of a platinum wire crossed by an electrical current, of crescent variable temperature, and those of dissociating gases, Ferreira Girão considers that instead of being opposite to his interpretation, they confirm it. It is just a matter of considering the gases' dissociation complete—the vibrations were strong enough to break the links that united the elements, while in the case of platinum the movements were not able to break the links in the prime atoms, so that the different systems subsist and this seems to be in complete accordance with facts of physics. His considerations pursue, also referring to the nebulae spectral analysis and Laplace's theory recuperation about the origin of Earth after Huggins's observations on more than sixty nebulae. "Shouldn't we conclude . . . that there exists a more elementary matter than that of the known bodies, matter that the direct analysis has not been able to discover?"[31] Secchi is also in favor of such interpretation. The examples of different authors' known experiments are presented, after which Ferreira Girão systematizes and discusses a certain number of objections to considering the unity of matter and the existence of prime atoms. Ferreira Girão then deduces that "apart the enormous authority of the illustrious professor of the College of France,"[32] Berthelot's objections are not sufficiently strong to destroy the hypothesis of the simplicity of matter. The only justifiable consequence is that the chemist's atoms are already the result of inferior arrangements of prime atoms. Thus, the existence of prime atoms, and the unity of matter, seems to be beyond doubt.

In the last chapter he refers to the dynamist school that opposes the atomic defenders, namely in the consideration that the chemical atoms are really extensible and actually indivisible. This would imply that prime atoms could also be indivisible and extensible. He deduces that, in this field, the position of the atomists is no safer, because if, on the one hand, the inextensible monad is incomprehensible, the philosophical atom, indivisible by definition, is counterintuitive, as indivisibility cannot coexist with extension. He also quotes du Bois Raymond's position on these subjects, namely "if inside certain limits, the theory of atoms rends excellent services to the physico-mathematical analysis of phenomena and even until a certain point turns it indispensable, since we surpass these limits we exaggerate its range, and are dragged to thousand insoluble contradictions, that have been in all the times the terrible hindrance of corpuscular philosophy."[33] Ferreira Girão then concludes that, "if reason and experiment lead us to infer that chemical atoms cannot be the last expression of the divisibility of matter, it does not follow that we know what the nature and properties of the first principles are."[34]

In the appendix, he dedicates a few pages to the separation between organic and inorganic chemistry, the difference between organic and organized matter, vital force and spontaneous generations. He also leaves a question regarding the future great problem of science, that is, to explain what life is. Will it be the result of the work of a higher energy, previous and exterior to matter, or something else? While mentioning several authors, Ferreira Girão cites neither Mendeleev nor the periodic classification, ending with some extracts from the *British Medical Journal* on the production and propagation of the contagious principle (May 1876).

2.2 1880—A Student's Perspective

In 1880 a booklet entitled *La loi périodique—de M. Mendéléjeff en ce qui concerne le problème de l'unité de la matière et la théorie de l'atomicité* was published.[35] The author was a certain D. Agostinho de Sousa, presented as a student from Porto Polytechnic Academy.

This unknown student wrote a fifty-two page booklet in French that, curiously, was published in Porto by Ernest Chardron,[36] bookseller and publisher. It was also sold in Lisbon at the José António Rodrigues National and Foreign Bookseller. One may ask why it is in French and what its context of publication was, but this remains unclear for lack of information. Nevertheless, we know it had its readers, as there are copies in academic libraries and, as being written in French, it would be accessible to non-Portuguese readers. In the booklet, reference is made to several foreign and national authors, including Ferreira Girão, mentioned above, who taught the author at the Polytechnic Academy in Porto, as well as to various recent foreign publications. He introduces the subject by stating that chemistry is, at the time, undergoing a great revolution, and that it is ready to be transformed. This revolution will probably be similar to the one physics suffered as a result of the dynamic theory of heat. He divides his presentation into two chapters. The first deals with "The periodic law and the question of the unity of matter," and the other with "The theory of atomicity and Mr. Mendeleev."

Chapter 1 begins with a statement on the tendency of chemistry, physics, and astronomy to establish the unity of matter, despite the brilliant opposition of MM. Stas and Berthelot. The spectral analysis of nebulae demonstrated the generation of simple bodies from hydrogen. Thus, the author believes that the periodic classification gives an unexpected support to Prout's theory that even Mendeleev would not believe, as he demonstrates later. He goes on to say that Mendeleev is supported by the previous periodical findings of Dumas, Marignac, and Lothar Meyer.

This is briefly the famous periodic law of Mr. Mendelejeff. We are not remarking here on the imperfections found, imperfections in fact inherent to a subject both complicated and difficult. We did not regard it but as a whole and in this respect, we must admit that the periodic law is a broad synthesis, a rational history of simple bodies, and especially a powerful affirmation of the unity of matter. Indeed, Mr. Dumas had established the natural families of simple bodies, but he did not know the link that connects one group to another. That honor belongs without question to Mr. Mendelejeff, who has filled the gap, noting that the difference between the atomic weights of two neighboring bodies does not surpass an average of two or three units, and where this interval is greater, there are gaps to be filled by later discoveries[37]—the case of gallium and scandium would confirm this.

Thus, the famous Russian chemist came to the aid of Prout's thesis, and supported Sousa's opinion on the unity of matter. "So, we believe that the idea that physical and chemical properties of the simple bodies are dependent on their atomic weight, is an appropriate development of the demonstration of the unity of matter, as shown by M. Abbot Moigno in his consideration of the masses, a well known demonstration to be reiterated here."[38]

The author concludes by noting that Mendeleev has shaken the theory of atomicity by recognizing that the hydrogen, the chlorine, and the oxygen cannot serve as a standard for measuring the atomicity of elements, an observation that led him to propose the periodic principle. Yet this principle is not rigorously exact and in some cases one needs to duplicate or split the formulae. The author believes that if the theory of atomicity is still incomplete, this results from a lack of research in determining the cause of atomicity, what its dependence is, and what the role attributed to the atomic movement is in terms of both rotation and translation. This occurred because the problem had not yet been transferred into the field of mechanics, which meant that the theory of atomicity did not yet have a solid basis.

Given this, the author moves on to Berthelot, stating that when chemistry relates its laws to those of pure mechanics and the physical sciences, then it will raise itself to the level of the positive physical sciences and, concurrently, will contribute to reaching the unity of the universal law of movements and natural forces.

2.3 1881

At Porto Polytechnic Academy Ferreira Girão was succeeded by António Joaquim Ferreira da Silva in 1876, who also obtained a philosophy degree at Coimbra University. Although the University had offered him a position teaching chemistry, he preferred to return to his hometown, Porto, where he developed a reputable career[39] and took chemistry to a higher level at the Academy and at the Municipal Laboratory, "an Institute to protect the consumer against fraud" where there was significant teaching of practical chemistry. Among the recommended textbooks was Ferreira da Silva's *Elementary Chemical Treatise*.[40]

From the preface to the second edition, we know that the first edition was published in 1884, although the first part of this book, entitled *Noções Gerais* (General
notions), had already been published in 1881 and quickly sold out. The author states
that in the second edition the structure of mineral chemistry was largely similar
to that in the first edition. Nonetheless, some more recent topics were inserted,
namely those related to Moissan's[41] research that led to the preparation of fluorine
(1886), as well as the atomic notation (1895), for: "being more rational than any of
its equivalents, as already mentioned, and universally adopted. Some doctrines of
general chemistry were enlarged, namely those concerning: the periodic law of elements, etc. while some special theories were suppressed that were thought to be not
necessary on an elementary course (. . .)."[42]

According to the first edition, we know that the textbook was intended to prepare the students taking the chemistry course at the Polytechnic Academy (Porto)
and that it was written in a simple way, so that they could more easily learn the
basics of this wonderful science. He dedicates several pages to the general laws concerned with the composition of bodies.

When referring to the atomic theory, Ferreira da Silva points out that "the
atomic theory in chemistry is independent of the general theories generally admitted on the constitution of matter. But the exposition method followed in the elementary books, even the more popular ones, hides rather than presents this truth,
as admitted by the leader of this doctrine in France, Mr. Wurtz, and even more
clearly expressed by another savant of the same school, P. Schutzenberger,[43] professor at the College of France and in whose last work is there a lot to learn."[44] Ferreira
da Silva believed that this was the reason why the leading chemists in France had
used their influence to put obstacles in the way of the official introduction of the
atomic theory in secondary level programs.

Next, Ferreira da Silva mentions that, in his paperback, he is attempting to
introduce the atomic theory without enslaving it to the litigious hypothesis on
the constitution of matter, as only in this way is it possible to compare it with the
theory of equivalents. Concerning the theory of atomicity, similar reflections can
be made. He thus concludes that, deprived of the amount of hypothesis with which
it is usually covered, atomicity is a valuable notion.

However, Ferreira da Silva recognizes that there is an abyss between the notions
of atomic valence, as it results from comparing the atomic weight with the substitution equivalents, and the notion of atomicity, as accepted by atomists, and that
this is not one of the theoretically smallest difficulties to get around. He therefore
agrees with Schützenberger that "things occur in chemistry as if atoms or particles
would attract and weld to form complex molecules. But it is impossible to go further and admit that a chlorine particle, for instance, would really possess a special
attractive force exerting on a hydrogen particle and be able to precipitate on it."
"Impossible," he adds, "in the field of positive science."[45]

Either in the reprint of the first part of this book or in the rest of it, Ferreira da
Silva tried to maintain the structure of the first publication, while making an effort
to clarify the relations between the system of equivalents and the atomic one, and
to realize and compare the two languages, based in each of the systems. He affirms

that this had never been done in an introductory work, with the extension of the subject importance required.

To base the atomic notion in hypothesis on the constitution of matter, acceptable as this may seem, would be, as Berthelot says, to confuse the general and positive laws of science, considered in its abstract and certain expression, with the representative more-or-less arbitrary hypothesis, by means of which we intend to translate those laws. Neither the equivalent nor the atomic systems exclude those hypotheses; but one thing is to consider them as instruments or support for the exposition, another as a basis of the same science, which in reality they did not constitute.

In the second edition, Mendeleev's periodic classification is presented in six pages, beginning with the "Relations between the elements properties and their atomic weight." Under this title the author presents what to expect when comparing the physical and chemical properties of the elements in a horizontal line in the table, setting hydrogen apart. He adds that, after a certain number, and in regular intervals, one can find a new series of bodies that reproduce more or less completely the properties of the former. When superimposing these series, the bodies that align in the vertical column are very similar, and constitute groups that may form natural families. He quotes Mendeleev's law as "the properties of the simple bodies, the constitution and properties of their combinations are periodic functions of their atomic weights." He then presents the periodic table, with some description of the analogies between elements in each column. The author also reiterates that the periodic law is again expressed in the similitude of the physical properties (not just the chemical ones), namely the specific weight, the atomic volume, fusibility, tenacity, malleability, volatility, specific heat, and heat and electrical conductibility. These facts can be understood considering the graph that follows, which is a summary copy of Lothar Meyer's one in his *Les théories modernes de la chimie*.[46] No more additional commentaries to Meyer are presented.

> As we can see from the joint map, all these properties increase or decrease regularly in each period, so that in the medium terms it acquires a maximum or a minimum value. The metals in group 8 possess very near atomic weights and form the passage between two neighborhood horizontal series.[47]

In the last paragraph of the "General notions" section, we can perceive the author's position concerning the periodic classification. He states:

> Some reservations should be maintained regarding Mendeleev's classification of the place occupied by some metals, like gold, which usually act as trivalent and not as monovalent. // But as it is organized, it represents a great and happy attempt. It enabled this author to foresee the existence and the properties of some new elements, which were missing in the table; three of these bodies were discovered: *gallium* (Mendeleev's ekaluminium), *germanium* (ekasilicium), and *scandium* (ekaboron). It also enabled the atomic weights of many little known elements to

be corrected, weights which were confirmed by new experimental determinations (uranium, cerium, etc.)."[48]

In the third edition of his *Elementary Chemical Treatise* (1903), Ferreira da Silva mentions some continuing difficulties with the periodic classification, namely as regards the noble gases, and concludes: "For this reason, the principle that underpins Mendeleev's classification does not seem to be a natural law."[49]

2.4 1888–1889

2.4.1 Lisbon Polytechnic School

Eduardo Burnay (1853–1924) was a physician, a bachelor in philosophy and substitute professor of chemistry at Lisbon Polytechnic School (from 1893 on, he became a full professor of organic chemistry), and member of the Lisbon Academy of Sciences. In his application to become a corresponding member of the Academy (December 1890), Burnay mentions his *Theoretical Introduction to the Study of Chemistry* (1888),[50] among other works.[51]

In this book, supposedly a detailed summary of lessons, he considers classification, determining that bodies can be divided into simple and compound. Simple bodies had been classified on various bases,—upon the electrochemical character, the atomic weight, and the atomicity and the natural physicochemical characteristics. He then briefly refers, without much discussion, to the above-mentioned classifications. Burnay illustrates his two-page presentation of Mendeleev's classification with a table that includes density as well as atomic weight. Referring to this, he highlights what he calls the typical period—the first one, the small periods (where density attains a maximum that sometimes is the first term, and then decreases), and the great periods (where density increases progressively) with examples and then mentions the vertical sequences as the true natural chemical families. He ends by declaring that

> this taxonomic display of the elements offers a certain plausibility, although it possesses some imperfections and exceptions. The table has gaps, but these probably correspond to as yet unknown bodies. In fact the place of *gallium* was established before it was even discovered by Lecoq de Boisbaudran.[*] This element corresponds to the *ékaluminium* Mendeleev forecast. [* Today, the following metals are suspected to exist: *decipium, philipium, mosandrum, iterbium, scandium, holmium, thulium, samarium, davyum* and *norwegium*.].[52]

Presenting the classification according to atomicity, he is brief but ends by stating that in taking into consideration a "typical" atomicity, this classification is really useful in foreseeing the atomic proportions in which bodies are involved in reactions and in determining their constitution.

As for the natural classification (metalloids and metals), he considers that a rudimental test shows how this can be equivocal. In a footnote, he mentions the modern vision of chemistry as integrating both organic and inorganic aspects.

Considering what he calls "the legitimacy and value of the atomic theory," Burnay states in three points that: (1) its basis and conclusions do not go against the facts and experimental laws of chemistry,—on the contrary, it helps systematization and explanation; (2) the rational predictions derived from the theory have been practically verified and confirmed; and (3) the verification and confirmation of numerous theoretical predictions have been, and will continue to be, the focus of progress in chemistry.

In 1896 Achilles Machado (1862–1940) became full professor of chemistry. The jury included the Porto Polytechnic professor A. J. Ferreira da Silva and the Coimbra University professor Francisco José de Sousa Gomes (1860–1911). There were other candidates for the place, including the Polytechnic Zoology professor Balthazar Ozorio (1855–1926), who was internally mastering the chemistry classes, and two others. Ozorio gave up. Achilles Machado was chosen unanimously for the place.[53] He worked in the profession for thirty-six years (1896–1932), first as a professor at the Lisbon Polytechnic School and later at the new Faculty of Sciences of the Lisbon University (May 1911). He also rose to the rank of general in the Portuguese Army. On October 1934, as vice president of the *Office International de la Chimie* (Paris), he was elected president after the death of Paul Painlevé.[54]

Reading the history of chemistry at the Lisbon Polytechnic School, of which he was one of the authors, we find that, in 1897, a comprehensive program was published that contained a greater focus on modern theoretical notions than had previously been the case. It included thermochemistry, chemical equilibrium, ionization, facts explained by Arrhenius's theory, Mendeleev's classification, and so on. This program was published just as Achilles Machado became a professor.[55]

Machado published several texts for his students dealing with the whole theory, as well as textbooks for secondary schools, among other publications. Reading the book he wrote with his brother on *General Chemistry and Chemical Analysis* (1892),[56] used at the Lisbon Industrial Institute as well as at the Polytechnic, we find, under the item "Elements classification," some reference to "Mendeleeff's classification."[57]

2.4.2 Coimbra University

In a curious publication presented to the Hispano-Portuguese-American Pedagogical Congress, dated 1892, the Coimbra professor Francisco José Sousa Gomes refers to what he called *Note on the teaching of chemistry at Coimbra University*.[58] He was then full professor of chemistry.

Beginning with a contextualization of the origins of chemistry as an independent course in Portugal during the second half of the eighteenth century, he quickly comes, in his *Note*, to recount his own last three years of teaching experience. Taking as a reference frame student secondary education and their insufficient training in experimentation, he says he made some decisions relating to his own teaching. He would avoid long explanations on the fundamental laws and clarify how it is possible to associate one characteristic number to each element and to each compound, illustrating its combination or reaction value, as well as how to adopt a system of constants to the simple bodies, making it possible to represent by symbolic formulae and by chemical equations the composition of bodies and their reactions. This follows the presentation of the atomic weights table, assumed as established, and the system of molecular formulae. Next, he turns to descriptive chemistry, after which the remaining time would be used to teach the philosophy of chemistry more freely. Sousa Gomes clearly mentions his adoption of Mendeleev's periodic classification, and presents his reasons: because it enables the summary presentation of several properties, reactions, and processes of preparation —repeated in each group and so almost always abridgeable in a general schema.

He reveals that the classification had been adopted from the time he entered as professor of mineral chemistry—between 1888 and 1889—and that he has followed a German translation of Mendeleev's textbook, since it appeared in 1891.[59]

In assessing the use of this classification, Sousa Gomes considers that students learn all the related facts more easily when using Mendeleev's classification in contrast to when they were taught descriptive chemistry in a disconnected manner based on the old arbitrary classifications. This clearly contrasts with the findings of many of the other essays in this collection.[60]

3. SECONDARY SCHOOL TEACHING

The Portuguese secondary school system, including science teaching, evolved more affirmatively from the mid-nineteenth century, with an increasing number of schools around the country, and not just in Coimbra, Lisbon, and Porto. Chemistry, along with physics, was taught during the last five years of secondary school. In the two final years, a kind of spiral method was used to delve deeper into the subjects than before. In reading textbooks from that time, one realizes that the methodological advice was to teach practical knowledge and the laws already established, mentioning some notions of atomic theory (1887), while the official secondary program (1895) concerning chemistry recommends the examination of the bodies, experimentation, the use of equipment, and the inspection of appropriate pictures.[61] The language should be clear and simple.

From the official textbooks used in 1893 in all the secondary schools, and after some preliminary considerations on chemistry, affinity, and electropositive and electronegative bodies, attention is given to the states of matter and notions on crystallography, bodies' mutual action, classification and nomenclature, and chemical theories (equivalents, atomic theory, molecular and atomic weight, dissociation, chemical formulae, types theory; comparison of atomic formulae and equivalent formulae; isomeric interpretation). Then, in a second volume, metalloids, metalloid compounds, and metals and theirs compounds are dealt with and a third book is dedicated to organic chemistry and different organic compounds. In the last page the official program is presented together with the correspondent pages in the volumes as the order in these was not exactly the same.[62]

The periodic classification only appears officially as a topic to be taught in the programs of the 1940s in 1948.[63] This does not mean, however, that the subject was unknown at this level in the preceding decades. Frequently, not to say always, in those years, some university/polytechnic professors were involved in the production of textbooks and sometimes also gave classes either in the general secondary or in the technical/industrial schools. Achilles Machado was one such example. His books were either used as the only official textbooks, or with others, for several decades. As far as is known, the first published presentation of the periodic classification at this level was in 1906, in one of Achilles Machado's textbooks (Figure 11.1), as extra reading material. It contained two items discussed in eight pages. They were: (1) Relations between the properties of different elements and their atomic weights, Mendeleev's classification; and (2) Applications of the periodic law.[64]

Quadro A

	1.º grupo	2.º grupo	3.º grupo	4.º grupo	5.º grupo	6.º grupo	7.º grupo	8 º grupo
1.ª série	Li	Be	B	C	N	O	F	
2.ª »		Na	Mg	Al	Si	P	S	Cl	
3.ª »	K	Ca	Sc	Ti	Va	Cr	Mn	Fe, Co, Ni
4.ª »		Cu	Zn	Ga	Ge	As	Se	Br	
5.ª »	Rb	Sr	Yt	Zr	Nb	Mo	—	Ru, Rh, Pd
6.ª »		Ag	Cd	In	Sn	Sb	Te	I	
7.ª »	Cs	Ba	La	Ce	Di	—	—	— — —
8.ª »	—	—	—	—	—	—	—	
9.ª »	—	—	Yb	—	Ta	W	—	Os, Ir, Pt
10.ª »	Au	Hg	Tl	Pb	Bi	—	—	
11.ª »	—	—	—	Th	—	U	—	— — —

A 1.ª e a 2.ª *series* (1.ª e 2.ª linhas horizontaes) teem cada uma 7 elementos, a 3ª *série* tem 10 elementos; as diversas *series* (a partir da 2.ª

Figure 11.1 Most likely the first presentation of the periodic classification in a Portuguese secondary chemistry textbook. Photo by I. Malaquias.

4. POPULARIZING

In terms of popular knowledge, the traces found were not connected with the periodic classification, but with atoms and elements. In a collection entitled *Bibliotheca das Ideias Modernas* (Modern Ideas Library) conveying modern scientific ideas, two titles stand out: Wurtz's *The Theory of Atoms and The World General Conception*[65] and Berthelot's *The Nature of Chemical Elements*.[66] Both were translated and published in 1883, in Porto.

A well-known populist collection intended for both Portuguese and Brazilian learners, the *Bibliotheca do Povo e das Escholas* (People and Schools Library)[67] does not include the classification in any of the various books dealing with chemistry or chemistry subjects. In the book *Inorganic Chemistry* and in the item "The atomic theory and its adversaries," the author points out that, considering the populist aims of the collection, and the public to which it is addressed, he had simplified the presentation, whereas trying not to sacrifice scientific accuracy. While having followed the notation and nomenclature based on the reasoning of the atomic theory, that thesis had not yet been unanimously accepted by all chemists of renown. He thus simply attempted to harmonize his unpretentious and modest work with the very latest studies that had been published in the field of the chemical philosophy.[68]

5. RESEARCHING

5.1 1894

One of the last traces found during this research is connected with the book *Funcção Chimica da Luz* (Chemical Function of Light) (1894), written by Balthazar Ozorio, a member of the Academy of Sciences, and published in Lisbon by the National Press.[69] We have already mentioned this professor from the Lisbon Polytechnic School. In this curious book, published shortly after his withdrawal from the nomination to the chemistry chair, and in three chapters, he deals with the origin of the elements and the influence of light. Subsequently offers a summary on a number of topics.

The first chapter is the more interesting in terms of the periodic classification, as it begins with the ancient idea of the unity of matter and unity of force, before reflecting on the modern tendency to go back to the old principles. In the following Ozorio turns to the metalloids and metals and compares these with the organic series, after which he presents Mendeleev's table and considers the possibility of isomeric metals as well as the pre-elements. He then discusses the ideas of Crookes, Berthelot, and Nordenkiold on the unity of matter. The study of gadolinite is also referred to, and he states that this always contains the three oxides of yttrium, ytterbium, and erbium, all of which have the same atomic weight. Gadolinite behaves as if it was a simple body despite being a compound substance. Crookes's spectroscopic experiments are mentioned, and then the possibility of an evolution from a primordial substance, as in biology. A primitive matter or protyle, as physicists and chemists call it, begins from hydrogen and then successively builds the other bodies with greater atomic weights—a kind of evolution that does not need to come to an end, in the author's opinion.

Ozorio goes on to consider Newton's law of attraction as applied to the formation of chemical compounds as well as the approximation of chemistry and astronomy and the formation of atoms according to A. Duponchel[70] before ending the chapter by presenting Mendeleev's opinions on, for instance, ammoniac with a planetary system, sodium chloride as a double star of sodium and chloride, and so on.

Ozorio was a prolific writer whose production refers mainly to researches on ichthyology from several parts of the Portuguese empire. The above-mentioned book is very singular in all his written work. In that same year (1894) he published seven additional papers. Another curious trace found is a letter, dated February 4, 1904, and addressed to St. Petersburg, from Mendeleev to this same Balthazar Ozorio (Figure 11.2), which reads as follows.

5.2 1904

To Professor Baltazar Ozorio.
St. Petersbourg. Labalcansky. 19.

Dear Sir and Colleague
Having received your kind letter on your discovery of a substance accompanying Iodine and microscopic samples of this substance firstly I must thank you and express my deep regret as, at this moment I'm resting after an operation to the eye (cataract) so I cannot yet see through the telescope clearly. But I hasten to say that I find your discovery very interesting, not only in general terms but also relative to the periodic system, because you know, sir, that the atomic weight of Tellurium represents an anomaly with respect to Iodine and for a long time I think there is now a certainty, precisely in the impurities of Iodine, received by the method of Mr. Stass.[71] You would give me great pleasure, if you have the kindness, to bring me your book on this subject.[72] Accept, dear sir and colleague my sincere compliments.

February 4th 1904
D Mendeleeff
Lissabonne Lisboa
Portugale
A Monsieur le Professeur
M. Baltazar Ozorio
Ecole Politechnique[73]

We tried to find other traces of this relationship with Mendeleev in the Russian Archives, as well as in Portugal, but unfortunately nothing else seems to remain, including neither the original nor the copy of Ozorio's letter. It would be interesting to unearth additional information, but in its absence we can only imagine that

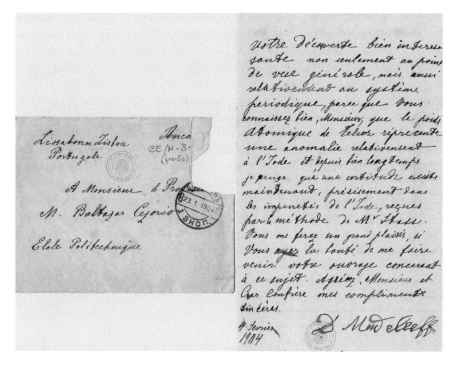

Figure 11.2 Part of Mendeleev's letter to Balthazar Ozorio. Courtesy of Arquivo Histórico do Museu Bocage, University of Lisbon.

some more precise work could have been developed around the atomic weight of iodine, probably when doing research around other subjects. That could have happened, as it also did in the less well-known case of Charles Lepierre (1867–1945), the French chemical engineer that published an article on the atomic weight of thallium (1893), defying the determinations of William Crookes, while he was already appointed as professor in Portugal.[74]

6. CONCLUSION

Based on printed (and handwritten) material, it is possible to arrive at conclusions from three different aspects. The first is the acquaintance with a contemporary chemical topic and some reaction to it, either in support or not; the second pertains to the knowledge to be taught at two different levels: higher education and secondary education; the last one relates to interest in popularizing the topic. We would like to consider another aspect that is based on weaker research.

At first glance it seems that Mendeleev's classification was more or less commonly known in 1880s Portuguese higher education, and by its more visible leaders. Some rational intersection between the periodic behavior of elements and the atomic theory appears to have been stated together with some astronomical analytical findings. In some cases, the supporters consider it as a broad synthesis, a rational history of simple bodies, and especially a powerful affirmation of the unity of matter, while others do not go further in what concerns speculations on this last category, as it does not fit positive science.

This kind of discussion was more or less avoided when presenting chemistry to students, although it was a self-presented student that explicitly writes on the periodic table in one of the traces found. Disagreements from the textbooks do not emerge immediately, if at all. While a pedagogical tool, textbooks or writings about decisions on how to teach reveal a kind of operational interest in teaching the classification in the sense that the introduction of chemistry to students should be synthetic and more or less acquired. In this sense, and by 1888, adoption of the periodic classification was deemed useful, as it efficiently summarizes chemical and physical properties and enables the prediction of other, as yet unknown, elements. This mitigates in its favor, but against other prior developments that might only confuse students.

Populist books did not appear to immediately address Mendeleev's classification, preferring to focus on Wurtz or Berthelot's ideas on atoms and elements.

Mendeleev's letter, and the publication on the chemical function of light, offers some insight into research that may have been performed, not specifically on the development of the periodic classification but on subjects around elements, or at least some elements.

The development of science in the secondary school system had evolved, particularly after the 1850s. The subject of the periodic classification appears explicitly in the official secondary programs only much later, in the 1940s. Nevertheless, during the period prior to its official publication, some university professors were involved in the publication of textbooks for secondary education, as in the case of Achilles Machado, whose books were used either as stand-alone textbooks, or with others, for several decades. As far as we have been able to establish, the first published version of the periodic classification at this level was in 1906 in one of Achilles Machado's textbooks.

In all the books for students, and bearing in mind the topic under analysis, there was an obvious intention to use a clear and simple language—to avoid big theoretical discussions, particularly as regards causes or implications, and particularly the more recent ones, but instead to use them in a pragmatic sense if they could offer a measure of practical return.

In the cases where the intention can be considered more speculative, it seems that there is an acceptance of the periodic classification, but also recognition of some of its transitory weaknesses, while waiting for a more comprehensive and unifying law or even questioning the possible future recognition of limits to science.

ACKNOWLEDGMENTS

I wish to express my gratitude to Maria Estela Guedes for having turned my attention some years ago to some manuscripts existent at Arquivo Histórico do Museu Bocage, University of Lisbon (AHMB-UL), to Prof. Maria Judite Alves for all the facilities she provided me with when consulting that Archive (AHMB-UL), to Nikolai Sobolev for his invaluable help in contacting Mendeleev Archives, and to Vítor Bonifácio for his careful reading.

NOTES

1. António M. Amorim da Costa, "Chemistry and the scientific development of the country: the nineteenth century in Portugal," in *The Making of the Chemist—The Social History of Chemistry in Europe 1789-1914*, ed. David Knight and Helge Kragh (Cambridge: Cambridge University Press, 1998), 265–287, on 268.

2. Adelino Cardoso, "Filosofia e História das Ciências: a inteligibilidade científica no Portugal oitocentista [Philosophy and History of Science: the scientific intelligibility in nineteenth-century Portugal]," in *História do Pensamento Filosófico Português—O século XIX*, ed. Pedro Calafate (Círculo de Leitores, 2004), vol. IV, Tomo 2, 14–41, on 18.

3. These scientific periods abroad were already common in the last quarter of the eighteenth century, but declined severely in the first half of the nineteenth century, mainly due to the Portuguese political situation that included the three tentative Napoleonic invasions, followed by civil war and other political turmoil.

4. Perhaps the most mentioned was Bernhard Tollens (1841–1918), whose short stay (1868–69) induced a great enthusiasm in organic chemistry, while he was in charge of the experimental laboratory work. Leaving for Göttingen in the following year, he induced a Portuguese member of the university staff to study there under his supervision.

5. Amorim da Costa (note 1), 269–270.

6. Amorim da Costa (note 1), 272; Vanda Leitão, Ana Carneiro, and Ana Simões, "Portugal: Tackling a complex chemical equation: the Portuguese Society of Chemistry, 1911–1926," in *Creating Networks in Chemistry—The Founding and Early History of Chemical Societies in Europe*, ed. Anita Kildebæk Nielsen and Soňa Štrbáňová (Cambridge: RSC Publishing, 2008): 257–280, on 259.

7. The 1772 reform created at the Coimbra University the faculties of Natural Philosophy and Mathematics, with annexed cabinets and a chemical laboratory, aimed the introduction of experimental teaching and learning of physical and natural sciences.

8. Isabel Malaquias and Manuel S. Pinto, "Searching for Modernization—Instruments in the Development of Earth Sciences in Portugal (18th Century)," *Centaurus* **53** (2011): 116–134.

9. J. L. Brandão da Luz, "A propagação do positivismo em Portugal," in *O Século XIX*, ed. P. Calafate (Círculo de Leitores, 2004), vol. IV, Tomo I, 239–261, on 246.

10. It was initially created under the Ministry of War. Pedro José da Cunha, *A Escola Politécnica de Lisboa—Breve Notícia Histórica* [The Lisbon Polytechnic School—Short Historical Note] (Lisboa: Faculdade de Ciências de Lisboa, 1937), 33.

11. Ana Luísa Janeira, *Sistemas Epistémicos e Ciências. Do Noviciado da Cotovia à Faculdade de Ciências de Lisboa* [Epistemic Systems and Sciences. From the Cotovia Novitiate to the Lisbon Faculty of Sciences] (Lisboa: Imprensa Nacional-Casa da Moeda, 1987), 108.

12. Amorim da Costa (note 1), 273–274.

13. Brandão da Luz (note 8), 250–251.

14. Fernando J. A. Catroga, "Os inícios do positivismo em Portugal. O seu significado político-social [The beginning of positivism in Portugal. Its sociopolitical meaning]," *Revista de História das Ideias*, I (1977): 287–394, on 314.

15. Joaquim de Carvalho, *Estudos sobre a cultura portuguesa do século XIX* [Studies on Portuguese nineteenth-century culture], vol. I (Coimbra, Por ordem da Universidade, 1955), 163.

16. A. M. Martins, "Positivismo em Portugal," *Enciclopédia Luso-Brasileira de Cultura*, vol. 15, 843-845 (Lisboa: Editorial VERBO, 1973).

17. Brandão da Luz (note 9), 253–259.

18. Amorim da Costa (note 1), 274–275.

19. Following the book by Pedro Mata, *Sinopsis filosófica de la química, obra escrita para facilitar y abreviar el estudio de esta ciência* [Chemistry philosophical synopsis, work written to facilitate and abbreviate the study of this science] (Madrid: Imprenta de Higinio Benesps, 1849).

20. Rodrigo A. Guedes de Carvalho, *História do Ensino da Engenharia Química na Universidade do Porto* (1762–1995) [History of Chemical Engineering Teaching at Porto University] (Porto: FEUP Edições, 1998), 37–42, 50; Girao familia [http://www.giraofamilia.com/historia.html], accessed August 27, 2011.

21. António Luiz Ferreira Girão, *A theoria dos átomos e os limites da sciencia* (Porto: Typographia de Antonio José da Silva Teixeira, 1879).

22. Ibid., 32.

23. Ibid., 35.

24. Ibid., 36.

25. Ibid., 41.

26. Pedro Norberto Correia Pinto de Almeida (1806–1849) defended in 1836 a doctoral thesis on "*Affinitas chemical ab electricitate pendet?* [Does Chemical affinity depend on electricity?]."

27. Ferreira Girão (note 21), 45.

28. Ibid.

29. Quoting Secchi's *Unité des forces physiques*, p. 211.—*Le Soleil*, p. 248.

30. Ferreira Girão (note 21), 46.

31. Ibid., 52.

32. Ibid., 60.

33. Ibid., 70.

34. Ibid., 71.

35. D. Agostinho de Sousa, La Loi périodique de M. Mendéléjeev en ce qui concerne le problème de l'unité de la matière et la théorie de l'atomicité (Porto: Ernest Chardron, Libraire-Éditeur, 1880).

36. Ernesto Chardon was born in France in 1840 and set up in Porto as a bookseller and editor. There he created in 1869 the *Livraria Internacional* (International Bookseller) at Rua dos Clérigos, n°96-98. He was a well-known figure in the city.

37. Sousa (note 35), 16–17.

38. Ibid., 20. He is probably referring to Moigno's translation of John Tyndall's lectures, given at Dundee, September 5, 1867, and at the Royal Institution of Great Britain, London, June 6, 1862, which was published as *La Matière et la Force* (Paris: Gauthier-Villars, 1867).

39. Ferreira da Silva was the first director of both the Municipal and the Polytechnic School Chemical Laboratories. His lectures on chemistry and chemical works and experimental studies on the toxicology of wines, oils, vinegars, and water, carried out by him and his team in these laboratories, attracted the attention of an influential audience. His intense activity and his international contacts, namely with Marcelin Berthelot, led to the publication of the first Portuguese chemical periodical, the *Revista de Chimica Pura e Applicada*, and to the creation of the *Portuguese Chemical*

Society, two precious and fundamental milestones in the development of chemistry in Portugal, in the first decades of the twentieth century. *Revista de Chimica Pura e Applicada* was first published in 1905 in Porto (Amorim da Costa, 1998, 277).

40. António Joaquim Ferreira da Silva, *Tratado de Chimica Elementar—Tomo I—Chimica Mineral*. 2nd edition (Porto: Typographia de A. J. da Silva Teixeira, 1895).
41. Ferdinand Frédéric Henri Moissan (1852–1907).
42. Ferreira da Silva (note 40), vii.
43. Paul Schutzenberger (1829–1897).
44. Ferreira da Silva (note 40), ix–x.
45. Ibid., x.
46. Probably a copy of the French translation from the fifth German edition made by Albert Bloch (*Les théories modernes de la chimie et leur application à la mécanique chimique* par Lothar Meyer; ouvrage traduit de l'allemand sur la 5e édition, par M. Albert Bloch, Paris: G. Carré, 1887–1889), still existent in Coimbra.
47. Ferreira da Silva (note 40), 102.
48. Ibid.
49. Ferreira da Silva quotes the following articles: Étard, *Th.ch.*, p. 51; *Act. chim.*, t. I, p.18-31, Wyrobouf's art.
50. Eduardo Burnay, *Introducção Theorica ao Estudo da Chimica* (Lisboa: Livraria A. Ferin, 1888).
51. Academia das Ciências de Lisboa, Processos Académicos, ff. 19-20-21.
52. Burnay (note 50), 93.
53. Achilles Machado and António Pereira Forjaz, *Escola Politécnica de Lisboa—As cadeiras de Química e os seus Professores* [Lisbon Polytechnic School—the Chemistry courses and their professors] (Lisboa: Faculdade de Ciências de Lisboa, 1937), 34.
54. Announced in the Portuguese newspaper *O Século*, October 18, 1934, as recent news from Paris, October 17—"O sábio químico Aquiles Machado foi eleito presidente do 'Office International de la Chimie,' para a vaga deixada por Painlevé."
55. Achilles Machado (coord.), *Apontamentos de chimica: 6ª cadeira (Chimica Mineral)*. [da Escola Politecnica] ([Lisboa]: s.n., 1896–1897); Achilles Machado, *Programa da 6ª cadeira [da Escola Politécnica]: chimica mineral* (Lisboa: Imprensa Nacional, 1898); C. P. C. Mello and E. M. Fernandes, *Apontamentos [do] curso de Chimica Mineral da Escola Polytechnica: segundo as liçoes e o programa do Professor Conselheiro Achilles Machado* (Lisboa: Imprensa Lucas, 1907).
56. Virgilio Machado and Achilles Machado, *Chimica Geral e Analyse Chimica*, Volume I *Metalloides* (Lisboa: Por ordem e na Typographia da Academia Real das Sciencias, 1892).
57. Ibid., 66–67.
58. Francisco José Sousa Gomes, *Nota sobre o ensino da Chimica na Universidade de Coimbra* (Coimbra: Imprensa da Universidade, 1892).
59. D. Mendelejeff, *Grundlagen der Chemie* (Petersburg: Verlag von Carl Ricker, 1891).
60. See other chapters of this book, for example on France, Germany, Norway, Sweden, and Spain.
61. Ministerio dos Negocios do Reino—Direcção Geral de Instrucção Publica, *Ensino Secundario—Decreto Organico—Regulamentos e Programas—Edição Official* (Lisboa: Imprensa Nacional, 1895), 745.
62. F. R. Nobre, *Tratado de Chimica Elementar contendo as Materias dos Programmas Officiaes para o ensino d'esta sciencia em todos os estabelecimentos de instrucção secundaria* [Elementary Chemistry Treatise containing the subjects of the Official Programs to teach this science in all the secondary instruction institutions]. (Porto: Typographia de José da Silva Mendonça, 1893).

63. *Diário do Governo*, I Série, Nº 247 (22 Outubro, 1948), 1159.
64. Achilles Machado, *Elementos de Chimica –IV e V Partes*, Ensino Secundário Official (Lisboa: Imprensa de Libanio da Silva, 1906), 95–98.
65. Charles Adolphe Wurtz, transl. A. Correia Barreto, *A Theoria Atomica na Concepção Geral do Mundo* (Lisboa: Nova Livraria Internacional, 1883).
66. Marcelin Berthelot, transl. A. Correia Barreto, *A Natureza dos Elementos Chimicos* (Lisboa: Nova Livraria Internacional, 1883).
67. Vitor Bonifácio, Isabel Malaquias, and João Fernandes,"The earliest astrophotography popularisation book? Ernesto de Vasconcellos's 1886 Astronomia Photographica," *Journal of Astronomical History and Heritage* **11** (2008): 116–123.
68. J. M. G. de Mello, *Química Inorgânica*. 5ªed. (Lisboa: A Editora, 1907); Isabel Malaquias, "Stirring towards a chemical modernization—a curious popularising collection," Paper presented at the 9th International Conference for the History of Chemistry (Rostock, Germany, September 14–16 2011).
69. Balthazar Ozorio, Funcção Chimica da Luz (Lisboa: Imprensa Nacional, 1894).
70. *Comptes Rendus*. . . 1893 (T.117) p. 336. Here it is recorded that Duponchel demanded the opening of his "pli cachete," submitted to the Academy of Sciences on March 27, 1893 and entitled "Principes de Cosmogonie générale" after having been published as *Principes de cosmogonie rationnelle* (Paris, 1895).
71. Jean Servais Stas (1813–1891).
72. Not found.
73. A M.ʳ le Professeur Baltazar Ozorio.
    St. Petersbourg. Labalcansky. 19.
    Monsieur et cher Confrère

    Ayant reçu votre aimable lettre relative à un [sic] *votre découverte de substance acompagnant l'Iode* et les échantillons microscopiques de ce substance Je dois premierement *d'abord* [sic] Vous remercier sincèrement et d'exprimer mes regrets profonds, qu'en ce moment je m'établi seulement apres une opération de l'oeil (cataracte) c'est pourquoi je ne puis pas encore voir clairement ni à la lunette. Cependant je m'empresse de vous dire que je *trouve votre découverte bien interessante* non seulement *au point de vue générale, mais aussi relativement au système periodique*, parce que vous connaissez bien, Monsieur, que le poids atomique de Telur réprésente une anomalie relativement à l'Iode et depuis bien longtemps je pense qu'une certitude existe maintenant, précisement dans les impuretés de l'Iode, reçues par la méthode de Mʳ. Stass. Vous me ferez un grand plaisir, si vous ayez la bonté de me faire venir votre ouvrage concernant à ce sujet. Agréez, Monsieur et Cher Confrère mes compliments sincères.

    4 fevrier 1904
    D Mendeleeff
    Lissabonne Lisboa
    Portugale
    A Monsieur le Professeur
    M. Baltazar Ozorio
    Ecole Politechnique

    The postmark indicates a delivery date of 23.1.1904. *Underscoring and italics is ours.* Arquivo Histórico do Museu Bocage, Universidade de Lisboa.
74. In 1888 Paul Charles Lepierre was appointed by the Portuguese government to the Lisbon Polytechnic School as director of the chemistry laboratory through the Portuguese professor at the *École de Chimie et Physique Industrielles* in Paris, Roberto Duarte Silva, who was contacted by José Júlio Rodrigues, professor in Lisbon. Lepierre stayed in

the Lisbon Polytechnic for just a few months. In 1889 he settled in Coimbra, where he stayed for the next twenty years as professor at the University and at the Industrial School Brotero. He was a remarkable chemist, having established a renowned school of Chemical Analysis. In 1911 he moved to the new Technical University in Lisbon, where he stayed for the next twenty-six years. On Lepierre see Mário N. Berberan Santos and Miguel A. R. B "Os laboratórios de Química do Instituto Superior Técnico (1911–1955) [The Chemistry laboratories of the Technic Superior Institute (1911–1955)]," in *Divórcio entre Cabeça e Mãos? Laboratórios de química em Portugal (1772–1955)* [Divorce between Hands and Head? Chemistry laboratories in Portugal (1772–1955)], ed. Ana Luísa Janeira, Maria Estela Guedes and Raquel Gonçalves, 159–179 (Lisboa: Livraria Escolar Editora, 1998). See also Crookes, "Editorial: Researches on thallium. Redetermination of its atomic weight." *Chemical News* 69 (April 14, 1893): 171; W. H. Brock, *William Crookes (1832–1919) and the Commercialization of Science* (Burlington, Vt.: Ashgate Publishing Company, 2008), 336.

CHAPTER 12

ᐂ

Popular Science, Textbooks, and Scientists

The Periodic Law in Italy

MARCO CIARDI AND MARCO TADDIA

1. INTRODUCTION

This essay deals with an issue that has never before been the focus of attention in the field of research on the history of chemistry in Italy: the diffusion of Mendeleev's periodic system in our nation.* In the following text we will analyze the situation in the period preceding the arrival of Mendeleev's theory in Italy with regard to the matter of classifying elements. By doing so, it will be possible to demonstrate that—despite the superficiality and lack of accuracy of certain studies—Italian chemistry was already very willing to consider new proposals relating to the classification of elements. We will then attempt to illustrate how Mendeleev's work not only attracted the attention of the most renowned Italian chemists, such as Augusto Piccini and Giacomo Ciamician, but also became widely used in university texts and secondary school textbooks.

2. BEFORE MENDELEEV'S LAW

In order to understand the classification criteria for elements adopted by Italian chemists before Mendeleev and therefore the cultural terrain the law of periodicity was to take root in, it would be better to refer to a number of texts used widely for teaching in universities. We will examine four of these, published between 1819 and 1867. In all these texts, the term "simple bodies" appears, with the expression "simple substances" used less frequently, while Antoine-Laurent Lavoisier (1743–94), in his 1789 *Traité élémentaire de chimie* (*Traité* thereafter), uses the same term "simple substances" or "simple substances . . . which may be considered as the elements of bodies." It is interesting to note that Vincenzo Dandolo's Italian translation (first edition 1792) uses the expression *"sostanze semplici,"* interpreting quite literally the Frenchman's choice of term.[1] Thirty years after publication of the *Traité*, Antonio Santagata (1774–1858), professor of general chemistry at the Pontificia Università di Bologna, published his *Lezioni di chimica elementare* [Lessons in elementary chemistry],[2] derived from *Lezioni di chimica elementare: applicata alla medicina e alle arti* [Lessons in Elementary Chemistry: Applied to Medicine and the Arts] (Bologna, 1804), written by his predecessor in the university chair, Pellegrino Salvigni (1777–1841). Santagata dedicates the third chapter of his book to the "general division of bodies," in other words to an attempt to classify elements. This work is particularly interesting because the initial classification was then developed by looking for confirmation from experience. The first group, with thirteen components, includes caloric, light, and electric and magnetic fluids, followed by nine non-metals. The second, denominated "metallic substances," contains forty-one components. Simple substances are in turn divided into "ponderable" (i.e., having their own mass) and "imponderable." Caloric, light, and electric fluids belong to the latter group, while the ponderable elements are further subdivided into metals and non-metals. The author also mentions another form of classification, evidently again derived from Lavoisier. The criterion adopted is the division of elements into those that maintain combustion (e.g., oxygen, chlorine, and iodine) and those that are combustible themselves (e.g., hydrogen, sulfur, phosphorous, etc.). Domenico Mamone Capria (1807–88) also refers to this last characteristic in his *Elementi di chimica filosofica-sperimentale* [Elements of Philosophical-Experimental Chemistry] aimed at medicine and pharmacy students, the fifth edition of which was published in 1846.[3] In this work, Domenico Mamone Capria, a native of the Calabria region and lecturer in pharmaceutical chemistry at the University of Naples, maintains the distinction between "ponderable" and "imponderable" bodies, including the usual four fluids (caloric, light, electricity, and magnetism) in the latter category but going much further. He organizes the simple bodies into groups according to the various compounds obtained when combined with oxygen and the electric charge the molecules that constitute all the simple and compound bodies are born from. It is clear that Mamone Capria had taken on board Berzelius's theory of duality, according to which all molecules (including organic ones) are derived from the union of two parts carrying an opposite electrical charge. His classification based on combustion, a reaction upon which every chemical combination is based, divides the bodies into combustibles, incombustibles, and oxidants (or combustion aids). Even bodies that are not capable of burning to produce light and heat, but can simply be combined with oxygen, chlorine, or iodine, are included under the combustibles heading. Electrochemistry came to his aid here because every body with a positive charge in relation to a battery can be called combustible due to its relationship with another negatively charged body.

The last two texts examined (both published shortly before Mendeleev's first paper on the periodic law) are those written by Paolo Tassinari (1829–1909), professor of general chemistry in Pisa, and Raffaele Napoli (1813–1866), professor of organic chemistry in Naples. Paolo Tassinari, a pupil of Raffaele Piria's and able experimenter, was among the first Italian authors that included atomic theory in university textbooks and programs. In his *Manuale di chimica—chimica inorganica* [Chemistry Textbook—Inorganic Chemistry],[4] he deals first with hydrogen and then with all the other elements, subdividing them simply into metalloids and metals. Regarding the compounds made from each metal, he identifies the so-called "minimum" and "maximum" salts, basically a reference to valence, or in other words to what we today call oxidation number. He also introduces the concept of "atomicity" as the comparison admitted between the quantities proportional to the relative atomic weights, and he applies this both to metalloids and metals. This leads him to divide the elements into columns in a way that anticipates, albeit in a much simpler manner, Mendeleev's groupings. Therefore, alkalis and halogens are placed in the first column, alkaline earth metals in the second, carbon and silicon in the third, antimony, arsenic, and nitrogen in the fourth, and so on.

Moving on to the last of the texts analyzed—Napoli's[5]—the most striking point is the "general comments on metals." Preceded by the usual distinction between metalloids and metals, as well as a mention of atomicity where it is defined as the "saturation capacity with which elements come to combine with others," by which definition oxygen is a "bi-atomic" body and hydrogen "mono-atomic," these comments are rather curious. Napoli lists eight possible criteria for classifying bodies that range from density to electrical conductivity to behavior in water and so on. After this, he asserts that the scholar, when assigning importance to these, should bear in mind the current state of science. Given that any number of classifications was at that time possible and their purpose was to coordinate ideas without hindering the mind, he encouraged young students to classify metals as they saw fit, in such a way as to provide them with immediate practical advantages. He cites various examples of classification, dedicating particular attention to the method of Louis Jacques Thénard (1777–1857) as modified by Jules Henri Debray (1827–1888), which focused on the possibility of reaction with water and air. To finish, he also mentions the "molecular" classification, considered the "most perfect" because based on atomicity.

From this brief survey it clearly emerges that just before the arrival of the theory of periodicity, the situation was rather confused; the incitement some scientists made to young people to "do it themselves" could only increase the confusion.

3. MENDELEEV IN ITALY

It may be interesting to mention that Dmitrii Ivanovich Mendeleev (1834–1907) had his first encounter with Italy in October 1860, after the Karlsruhe Congress. Together with his friend Aleksandr Porfir' evich Borodin (1833–87), who was also a member of the Russian contingent at the Congress, Mendeleev crossed over the St.

Gotthard Pass and traveled to Genoa, where he spent the night of the fifteenth of October. The following day, they both left for Civitavecchia. Mendeleev apparently liked Italy very much, as he wrote: "We enjoyed Italy immensely after the suffocating and reserved life in Heidelberg." On arriving in Rome, Mendeleev and Borodin visited St. Peter's Basilica and the Sistine Chapel, where they not only admired Michelangelo's frescoes, but also attended a solemn mass celebrated by the Pope.[6]

The Russian chemists stayed in close contact with their Italian counterparts. Borodin lived in Pisa for a time between 1861 and 1862. While there, he had occasion to work in the renowned Pisa University laboratory founded by Raffaele Piria (1814–65). Piria, who at that time was teaching in Turin, had played a determining role in the creation of the national school of chemistry, thanks to his recruiting brilliant young scholars such as Stanislao Cannizzaro (1826–1910) and Cesare Bertagnini (1827–57).[7] In Pisa, Borodin met two of Piria's pupils—Sebastiano de Luca (1820–80) and Paolo Tassinari (cited before)—and he took advantage of the opportunity to send his regards to Cannizzaro. As Tassinari wrote to Cannizzaro in June 1862, "Borodin, the Russian chemist whom you met at the German conference, has been working in Pisa all year and has asked me to send you his regards."[8]

The years passed. Borodin went on to write fabulous pieces of music, while Mendeleev was destined to become one of the most renowned scientists in the world. The most important Italian scientific institutions, the Reale Accademia dei Lincei, based in the capital, nominated him as a Corresponding Member (1893). To this recognition was added that of the Reale Accademia delle Scienze of Turin (1893) and, some years later, the Accademia of Bologna (1901). In the early twentieth century, Mendeleev was extremely popular in Italy and his theory had widespread support among chemists in universities, in secondary schools, and even in the field of popular science.

The first important initiative that helped to explain Mendeleev's new theory and the periodic system to the Italian public was the opening of the Italian section of the *International Scientific Series*, the brainchild of Edward L. Youmans (1821–87), founder and editor of the widely read *Popular Science Monthly*.[9] The series was designed to provide information about the latest scientific discoveries in a style that was both accessible to laymen and suitable for a more informed readership. Scientists themselves were to be the authors of the volumes. The series, which came out in various countries (including the United States, Great Britain, France, and Germany), was published in Italy by the Fratelli Dumolard publishers in Milan, under the name *Biblioteca Scientifica Internazionale* [International Science Library]. The Dumolard were among the most active publishers in Italy in the field of popular science and the *Biblioteca* "constituted the high point of popular science works published in Italy."[10] In their advertising for the publication of the *Biblioteca* in the summer of 1875, the Dumolard brothers stressed that the volumes were aimed equally at scholars in the field and educated persons in general, as they were the work of "distinguished specialists" but written "in such a way as to be accessible to intelligent people in general." This initiative was designed to "facilitate the international exchange of scientific literature" and provide a platform for both the "natural" and the "social" sciences. Indeed, as Michele Nani rightly points out, "although academics—and others besides—in those days had a certain familiarity with languages such as French, German and English, and therefore could consult the original editions of works, we must not underestimate the importance of translations in broadening and transforming cultural horizons."[11]

In 1879 the eighteenth issue of the Italian edition was dedicated to Wurtz's volume *La teoria atomica* [Atomic Theory],[12] which also came out at the same time in the other nations taking part in the project (the volumes did not always turn out together). Wurtz's book was a fabulous historical digest of atomic theory, from the formulation of the first laws on proportions (Richter, Proust, and Dalton) to the presentation of Mendeleev's recently devised periodic system. In it, Wurtz clearly stressed the merits of Italian scientists in the construction of atomic theory, from Avogadro's hypothesis (which the physicist from Turin formulated in 1811),[13] to Cannizzaro's decisive contribution: "This eminent chemist has doubled the atomic weight of a large number of metals in order to bring them into harmony with Dulong and Petit's law and the Avogadro rule."[14]

The sixth chapter of Wurtz's volume was entirely given over to a presentation of the periodic table, which was explained to Italian readers using a simple and effective language that covered all the aspects: "In recent years, Mendeleev's work has brought greater clarity to the relationships that exist between the atomic weights of base elements and their properties. The latter are functions of atomic weights and this function is *periodical*. Such is the proportion stated by Mendeleev; it is not, however, limited to one or other group of elements, but it embraces all the simple bodies in chemistry. It does not content itself with seeking certain analogies but considers the total of physical and chemical properties. It is simple in its conception and fertile in its consequences."[15]

Wurtz underlined the fact that one of the strengths of the new system devised by Mendeleev lay in its incredible ability to organize existing elements and predict the properties of those that had yet to be identified. The discovery of gallium (1875), one of the "gaps" left by Mendeleev in his table, made a big impression on the scientific community: "Rather remarkably, one of these gaps has now been filled. Lecoq de Boisbaudran's gallium had its place marked out in Mendeleev's framework. As well as the number assigned to its atomic weight, very close to its actual value, its density had been predicted exactly right. The Russian chemist's table is a powerful summary and from now on, every time elements are to be classified according to their properties and their reactions—in short, when scholars need to view chemistry matters from above and as a whole—they will have to take it into consideration."[16]

How the periodic table worked, its characteristics, and its structure were explained by Wurtz following the criteria set by the publishers of the International Scientific Series—simply, capably, and with verve. This was in 1879, when the periodic table was barely ten years old. Wurtz set out clearly the potential and the limits of Mendeleev's work: "We have printed a full version of the table in order to allow our readers to appreciate more fully and correctly the work of classification in question, which, for the first time, embraces all the elements known to chemistry. Undoubtedly this work still displays some imperfections, but these are due in part to the state of uncertainty surrounding our current knowledge, especially as regards rare earth elements."[17] Nevertheless, the conclusion was unequivocal: despite the uncertainties, he had no hesitation in stating that "the principle outlined by the eminent Russian chemist has come to form one of the foundation stones of chemical classification." The results obtained by Mendeleev could never have been arrived at "if one had attempted to deduce them from 'equivalents.'"[18] This is why he felt it necessary to stress a particular point: the periodic system was "a solid argument" in favor of atomic theory. As the *grand finale*, he used Mendeleev's own words: "Our concepts about atomic weights have in recent times reached such a level of solidity, above all after the application of the Avogadro-Ampère law, and after the work of Laurent, Gerhardt, Regnault, Rose and Cannizzaro, that we can confidently claim that the idea of atomic weight, that is, the smallest quantity of an element contained in a molecule composed of its combinations, will remain unaltered despite the variations that chemical theories may undergo."[19]

After his death in St. Petersburg on February 2, 1907, Mendeleev was commemorated at the Accademia dei Lincei by Raffaello Nasini (1854–1931),[20] a physical chemist of considerable prestige and one of Cannizzaro's many pupils. According to Nasini,[21] Mendeleev's ideas started to be taken seriously around 1878 in Italy, in large part thanks to Cannizzaro, who was in charge of the Istituto Chimico of Rome. The date corresponds both as far as the initiatives of the Dumolard publishing house are concerned and regarding the role played by Cannizzaro in spreading knowledge about the periodic system. In fact, in 1880, the Tuscan chemist Augusto Piccini (1854–1905) joined the Cannizzaro group as an assistant. It was probably Cannizzaro who assigned the task of studying the periodic system to Piccini and Francesco Mauro (1840–93), another of his pupils. The two chemists together undertook experimental research on certain rare elements, but it was Piccini above all who stood out, introducing an addition to the periodic system necessary to support its mode of classification and bringing the ideas of Mendeleev to Italy through his writings.

4. "SO PROFOUNDLY WELL-VERSED IN THE PRINCIPLES OF THE PERIODIC SYSTEM": AUGUSTO PICCINI

Piccini was born in 1854 in San Miniato in the province of Pisa and studied chemistry at the University of Padua, from which he graduated in 1876.[22] As we will see, at the same time as Piccini, Giacomo Ciamician joined the Cannizzaro group and the two became friends, so much so that on the death of Piccini it was Ciamician who gave the funeral oration.[23] After several years spent at Cannizzaro's school of chemistry, in 1885 he was called to teach general chemistry at the University of Catania. Two years later, he returned to Rome to teach docimastic chemistry at the School of Application for Engineers in Rome. Finally, in 1893, he moved to Florence, where he was invited to teach pharmaceutical and toxicological chemistry at the Istituto di Studi Superiori.

Piccini applied himself to organic chemistry and then to chemical analysis, later starting to cultivate a special commitment to inorganic chemistry. He soon demonstrated his complete familiarity with the system by means of a dissertation on the oxidisation of titanic acid published in the *Gazzetta Chimica Italiana* between 1882 and 1883[24] that was followed by similar articles in the Records of the Reale Accademia dei Lincei. In these works, Piccini makes reference to the periodic system for the first time. "This is the first case," wrote Piccini, "in which scientists have attempted to obtain oxides superior to the form RO_2 from the elements in group IV of Mendeleev's system."[25] To better understand Mendeleev's work, Piccini started to study Russian in 1883. From that moment Piccini became "the foremost propounder and at the same time the foremost critic of the periodic system," as Giulio Provenzal (1872–1954) noted (Provenzal was a student of Piccini's from 1893 to 1895).[26] In 1885 he succeeded in supplementing his translation of Victor von Richter's treatise on inorganic chemistry with a long appendix (fifty-five pages, thirty of which were dedicated to Mendeleev's system)[27] that was published separately, but still with the Loescher publishing house.[28]

Piccini dwelled in particular upon the controversial highest and super-highest combination formulas of some elements, posing the question of whether peroxides (BaO_2 and others) should also be taken into consideration, because in that case it would be necessary to shift the maximum limit. After a general review of the groups, he concluded by suggesting a closer study of the dependence of properties on the form of combination, a dependence that creates "a quantity of partial relations of the same element with many other elements that are different from each other." His objective was to construct a "large framework in which the truths discovered by Mendeleev would be harmonised with those that would be discovered in his time." Giulio Provenzal shared with his master the excitement of the historic time of the discovery of the first noble gases: "a new and enormous problem with regards to the periodic system."[29] The discovery did not affect Piccini's faith in the Mendeleev system, however. In 1901 Piccini published (in the first volume of the *Nuova enciclopedia di chimica* [New Chemical Encyclopaedia],[30] edited by the chemist Icilio Guareschi (1847–1918), a chemist and historian of chemistry) a long article that constituted not only his scientific testament but also an extraordinary work of popular science on Mendeleev's work.

The Tuscan chemist explains: "For more than thirty years the periodic law has been the subject of authoritative confirmations and questionable confutations. No definitively proven fact has been discovered that could invalidate its spirit or its content." Naturally, everything was meant as "healthy criticism," without demanding more from the periodic law than it could deliver. The same Mendeleev, writes Piccini, "although recognizing that the analogy concept of the elements, summarized in the identity of the highest combination formulas, is one which is destined to remain in science," was aware that "the theoretical foundation of the periodic law" was missing. In short, the periodic law was like "Kepler's laws waiting for Newton to explain them."

Piccini's presentation of Mendeleev's work is found principally in the fifth chapter of the article. In this chapter, the Tuscan chemist retraces the story that led to the "idea of establishing the true function that connects all atomic weights with the physical and chemical behavior of all elements." Piccini here supports the view that precursors to Mendeleev's work did not exist and the work of the Russian chemist was an absolutely original creation. After a discussion of William Prout's theory, Piccini then dedicates ample space to an illustration of Mendeleev's articles of 1869 and 1871, which he compares to the articles of Lothar Meyer. At the end of his analysis, Piccini's judgment is unequivocal: "Meyer did not discover the periodic law, either independently of Mendeleev or otherwise." According to Piccini, Mendeleev's work was comparable to that of Lavoisier, also the father of a revolution, because he was able to "interrogate both long-known facts and freshly discovered ones, forcing them to reveal their jealously guarded secrets."

An illustration of the periodic system proper follows the historical part, with the presentation of Mendeleev's most recent table. In the final part of the article, Piccini also deals with the difficult question of "objections to the periodic law." In particular, he again dwells upon the problem of the choice of "combination limits," summarizing his contribution to the debate and his investigations on peroxides, claiming they should become a priority.

Mendeleev gave credit to Piccini for the diffusion of the periodic system. In a letter sent to Piccini on January 29, 1903, Mendeleev wrote: "I am heartily pleased to see a scientist in a far-off country who is so profoundly well-versed in the principles of the Periodic System." [31] (Figure 12.1)

5. CIAMICIAN'S CONTRIBUTION

The name of Giacomo Ciamician (Trieste, 1857–Bologna, 1922), a pioneer in both organic photochemistry and green chemistry,[32] can be mentioned in relation to Mendeleev's periodic system for two reasons: (1) his early research on spectroscopy; (2) his role as professor of chemistry at Bologna (1889–1921).

Ciamician, whose family was of Armenian origin, studied in Trieste and then in Vienna, graduating from Giessen University in 1880. While a student at Vienna Polytechnic, he published his first research paper entitled "Spectra of Chemical Elements and Their Compounds,"[33] which was followed by another paper dealing with the effect of density and temperature on emission spectra.[34] In his first work, Ciamician highlighted the spectral analogies between elements of the same group of the periodic system, reaching the analogy that was defined as the "law of homology" and surpassing the previous research of Huggins and Thalén relative to the emission spectra of metals, as well as Salet's work on non-metals. In 1880 he extended his research on analogies to twenty elements and to a certain number of compounds. Ciamician found that the spectral lines of chemically similar elements (e.g., O, S, Se, Te) could be compared singly or in groups. He concluded that each natural group of elements had its own characteristic spectrum, which differed from the one for single members of the same group, except for the fact that the homologous lines (i.e., lines that fluctuate to the same extent from discharge to discharge) shifted toward one side or the other side of the spectrum. This meant

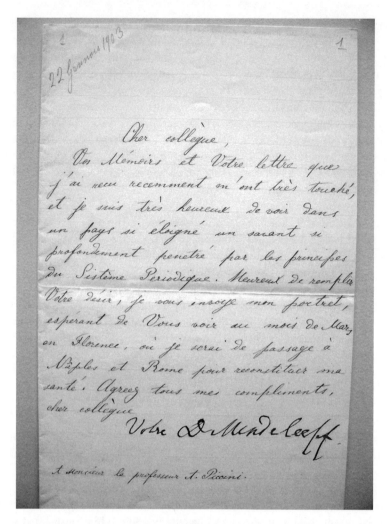

Figure 12.1 Mendeleev's letter to Piccini, dated January 29, 1903. Source: Museo Galileo - Institute and Museum of the History of Science, Florence.

that the wavelengths of the transitions increased or decreased and that, amongst other things, certain lines or groups of lines disappeared. In the second paper, "The Influence of Density and Temperature on Vapor and Gas Spectra,"[35] he analyzed the spectral behavior of chlorine, bromine, and iodine, as well as the vapors of mercury and sodium. He noticed that the spectra of elements belonging to groups of halogens changed notably according to temperature and pressure conditions as far as the number of spectral lines and the intensity of the same were concerned. Taking into examination the variable lines caused by pressure, he formulated the relationship between the various partial spectra of different elements. For example, he saw that the spectrum of diluted bromine vapors came closer to the spectrum of

chlorine the more he diluted the vapors, whilst the spectrum of concentrated bromine vapors seemed more similar to the spectrum of iodine. Mendeleev acknowledged the contribution made by Ciamician to the confirmation of the periodic
system in *The Principles of Chemistry*. In the 1891 German version, two citations of
Ciamician's work are found.[36] Again in the article "The Periodic Law of the Chemical
Elements," published in 1889 by the *Journal of the Chemical Society* and taken from
the Faraday Lecture, Mendeleev acknowledged Ciamician's contribution.[37]

After his degree from Giessen, Ciamician joined Stanislao Cannizzaro's group in
Rome. He became a visiting professor at Padua in 1887 and after only two years was
invited to hold the chair of General Chemistry at Bologna. At Bologna, Ciamician
carried out important research in the field of photochemistry and the chemistry
of natural substances and plants. Ciamician, however, never collated his lessons
into books. Notes of his lessons were taken and kept by students. The *Notes on
General and Inorganic Chemistry* taken by Adolfo Baschieri and dated 1899–1900[38]
are the oldest of those available today and differ considerably from all the others
in terms of the extensiveness of the historical background recorded through quotations and classic experiments. Chemistry is split into general chemistry and special
chemistry. The first, which also comprises physical chemistry, studies the chemical
properties of bodies in general, without any special reference to individual ones,
and is divided into two parts: stoichiometry and the theory of affinity. The second
part is divided into inorganic and organic chemistry. The concept of valence and
the periodic system of elements according to Mendeleev brings the stoichiometry
section to a close. The periodic system is discussed for approximately eleven pages,
out of a total of 228 dedicated to general chemistry as a whole. The presentation
recorded from the notes begins with some reflections on the scale of oxides placed
in order of the valence of each element and continues with observations on their
respective hydrates. The groups derive from regularity observed in the elements,
and reveal that atomic weight is the best criterion to classify elements. Two tables
are presented: one compiled by Mendeleev (Figure 12.2) and the other by Meyer.
Ciamician also dwells on the issues raised by the discovery of noble gases, which
"cause no little disruption to the periodic system, since they are not elements fore-
seen by the theory." In the inorganic chemistry section (246 pages), the elements
are studied in groups, in the same order in which they are presented in the periodic
system, but for each group considerations of a general nature are made. In the notes
by Bruno Maggesi and Andrea Stagni, too, the periodic system occupies less than
eleven pages.[39]

According to these notes, Ciamician's presentation began with a definition of
the periodic system of elements, which he considered to be, "in the field of chemistry, one of the most successful attempts at expressing the qualifying differences
between the various objects of the study using numerical data." After briefly mentioning the theory of triads, Newland's law of octaves, and the studies of Lothar
Meyer, Ciamician passed swiftly to Mendeleev and reminded his listeners of the
guiding concept: "a connection that we are as yet unable to express algebraically
exists between the atomic weight of an element and its properties." He recounted
that Mendeleev, by distributing the elements according to the order of increasing

19 H

Sistema periodico degli elementi secondo Mendelejeff.

Gruppi	Ossidi	Tipi piccoli periodi	Grandi periodi I	II	III	IV	V
1	R_2O	Li 7	K 39	Rb 85	Cs 133	—	—
2	R O	Be 9	Ca 40	Sr 87	Ba 137	—	—
3	R_2O_3	B 11	Sc 44	Y 89	La 138	Yb 173	—
4	$R O_2$	C 12	Ti 48	Zr 90	Ce 140	—	Th 232
5	R_2O_5	N 14	V 51	Nb 94	—	Ta 182	—
6	$R O_3$	O 16	Cr 52	Mo 96	—	W 184	U 240
7	R_2O_7	Fl 19	Mn 55	—	—	—	—
8	R_2O_8		Fe 56	Ru 103	—	Os 192	—
			Co 58½	Rh 104	—.	Ir 193	—
			Ni 59	Pd 106	—	Pt 194	—
1	R_2O	Na 23	Cu 63	Ag 108	—	Au 197	—
2	R O	Mg 24	Zn 65	Cd 112	—	Hg 200	—
3	R_2O_3	Al 27	Ga 70	In 113	—	Tl 204	—
4	$R O_2$	Si 28	Ge 72	Sn 118	—	Pb 206	—
5	R_2O_5	P 31	As 75	Sb 120	—	Bi 208	—
6	$R O_3$	S 32	Se 79	Te 126	—	—	—
7	R_2O_7	Cl 35.5	Br 80	I 127	—	—	—

Figure 12.2 Mendeleev's table from *Notes on General and Inorganic Chemistry (Ciamician Lessons)* taken by Adolfo Baschieri and dated 1899–1900.

atomic weights (H being a unit of measure), had observed an evident regularity corresponding to a distribution in periods, and a repetition of the same kind in the physical properties and chemical behavior. Ciamician affirmed that periodicity was a complex function that was not completely clear. A number of comments followed on the general properties of the groups and the observation was made that from the position of an element its valence could be deduced immediately, and thus, also its highest combination formula. A table was presented in which elements were put in order depending on the valence and oxide limit formulas, highlighting their more or less accentuated acidic or basic character. Moving onto a discussion about compounds other than oxides, Ciamician acknowledged the difficulty in correlating the periodic system with Dalton's law. The position of an element in the periodic system did not therefore depend only on its atomic weight but also on its physical and chemical properties. Ciamician affirmed that Mendeleev's systematic ordering was "up until now the best attempt" and queried whether it would be possible to discover new elements. Maybe one day, "periodicity would become greater and the elements of the system would be subject to a different arrangement." Ciamician also spoke about the recently discovered rare earths, to be positioned between lanthanum and ytterbium. A law so general created many problems, which, according to Ciamician, caused scientists to be transported from the field of experimental chemistry into a "fantasy" world. Ciamician gave space to the interpretation of those who thought that the order of the elements in the Mendeleev classification did not depend on the weight but on the electrical charge of the atoms.

On November 7, 1903, Giacomo Ciamician held the solemn inauguration of the 1903–04 academic year at the University of Bologna. His dissertation was entitled "Chemical Problems of the New Century." Quoting Mendeleev's system, he said that "as our Piccini demonstrated by words and facts, it has been our only guide through the intricate labyrinth of inorganic chemistry for thirty years and numerous researches based upon it have inspired a great quantity of results."[40] And he was right.

6. UNIVERSITY AND SECONDARY SCHOOL TEXTS

We have seen that Italy provided fertile ground for the periodic system, perhaps in part due to the ties between the Italian and Russian schools that were mentioned earlier. This is further confirmed by the progressive spread of Mendeleev's ideas through educational texts produced by Italian hands. One of the first scholars to take an interest in the periodic table from a didactic point of view was Paolo Tassinari, who preferred to dedicate himself more to teaching than to scientific research, despite being so well esteemed as an analytical chemist that he was used as a consultant on some historic occasions.[41] Before the publication of Mendeleev's works, Tassinari had published a *Manuale di chimica* [Chemistry Textbook],[42] which became very popular around Italy. It was arranged in accordance with the classic distinction between metalloids and metals, and contained groupings of metal-based compounds entitled "minimum salts" and "maximum salts," with minimum and maximum referring to their valences. Before the French translation of Mendeleev's work was published, Tassinari, as soon as he heard the news of the discovery of gallium (1875), urged a young Russian-born student in his laboratory named Alessio Alessi (Moscow, 1857–Reggio Emilia, 1934) to translate it into Italian. Alessi asked Mendeleev's permission to publish it; he kindly agreed, but it was not easy to find the means. He entrusted the publication to a small journal called the *Monitore dei Farmacisti* and after numerous hesitations it was decided to publish a page every now and again, unfortunately with various typographical errors. Already discouraged, once the French translation came out Alessi interrupted his publication. This unfortunate occurrence did not, however, slow down the uptake of the system by the authors of textbooks, who were very much aware of what was being published elsewhere. From our research it appears that the first texts to mention the periodic system were written by Felice Marco (Vico Canavese, 1836–?). Marco, a mathematician and physicist, worked as a teacher in Turin. His scientific interests were mainly in the field of physics and he showed quite some zeal in striving toward some fairly ambitious theoretical goals. He expounded his own ethereal theory on electricity in a short volume entitled *Principi della teoria meccanica, dell'elettricità e del magnetismo* [Principles of Theoretical Mechanics, Electricity and Magnetism] (1867).[43] He also wrote a number of successful textbooks that combined physics and chemistry or discussed chemistry alone. The two texts by Marco examined here were entitled, respectively: *Nozioni di chimica secondo il sistema periodico* [Basics of Chemistry According to the Periodic System of Elements][44] and *Elementi di chimica secondo il sistema periodico* [Elements of Chemistry According to the Periodic System of Elements].[45] The first was written for technical institutes and university faculties, while the second was designed for secondary and professional training schools.

Initially, Marco wrote a single textbook, but then, from the seventh edition onward, decided to divide the material according to the type of institution the book was to be used in. One important change made for the seventh edition (1887) was its reworking "according to the periodic system." It should be noted that this already appeared as part of the title of both works, although in his introduction to the Primer, Marco specifies that it was not necessary to include the periodic table in teaching programs, but it would be useful to mention it in order to provide "a clear and precise idea of the current progress of science." Thanks to this system, Marco writes (somewhat emphatically), "Chemistry no longer has to envy Physics its ability to prophesy facts that experiments later prove." The Primer discusses the periodic system in chapter 4 of the first part, the same chapter that deals with thermochemistry. Just under ten pages are reserved for the periodic system. The author specifies in a note that the chapter was taken in large part from the fifth edition of Richter's treatise on inorganic chemistry (1886).[46] He also adds that study of this chapter could be "alternated with the study of individual elements and their compounds, in the manner the teacher deems most appropriate." The section headed "Periodic System" is divided into eight parts: (1) Basis of the Classification; (2) Series and Periods; (3) Analogies between the Elements in Different Periods; (4) General Table of the Periodic System; (5) Relationship between the Physical and Chemical Properties of the Elements and Their Position in the Table; (6) Relationship between Valence and Atomic Weight; (7) Scientific Worth of the Periodic System; (8) Groups in the Periodic System and Normal Elements. As regards the Basic Chemistry textbook, we were able to consult a copy of the tenth edition (1896), which still contained the introduction written for the seventh edition, identical to the one found in the Primer and dated November 1886. The "brief mention" of the periodic system is partnered with thermochemistry in chapter 4 and the contents are identical, as are the notes and the length, which again is just under ten pages.

Another textbook that found room for a presentation of the theoretical system in its pages is a work by Fausto Sestini and Angelo Funaro (in this essay we will discuss the first and fourth editions). The first edition dates back to 1886 and was compiled according to the new government programs for technical institutes;[47] the fourth was designed for use in secondary schools in general.[48] Regarding the authors, the first, an agricultural chemist also involved in research, is undoubt-edly the best known of the two. Fausto Sestini (Campi Bisenzio, Florence, 1839–Lucca, 1904) originally taught in technical schools—first in Forlì, then in Udine, where he directed the local Agriculture Centre founded in 1870. From Udine, he then moved to Rome. Here he founded another Agriculture Centre, after which, in 1873, he was called to the Agriculture Ministry and given the role of inspector of technical schools. He remained in Rome until 1876, when he was awarded the chair in Agricultural Chemistry at the University of Pisa.[49] Angelo Funaro, on the other hand, is known mainly as an author of technical and didactic texts. He was a teacher at the Royal Senior Secondary School in Leghorn, lecturer for the University of Pisa, and also occupied the post of director at the Municipal Chemistry Laboratory in Leghorn. Funaro put his name to numerous technical texts in the field of agricul-ture, as well as a number of chemistry and physics textbooks. The Sestini-Funaro book was received favorably by both students and teachers. It reached four editions between 1886 and 1896, the first three of which sold out quickly. In 1914 and again in 1921 new editions were brought out on the initiative of Quirino Sestini (1872–1942), Fausto's son, who kept the names of the original authors despite the fact that his father had died a decade before. The most renowned of the chemists who cut their scientific teeth on the Sestini-Funaro textbook, and one known to the public at large, is the writer Primo Levi (Turin, 1919–87). Levi studied chemistry at the University of Turin, from which he graduated in 1941. A Jew, he was captured by the Germans and in 1943 was deported to Monowitz, near Auschwitz, to a con-centration camp where the prisoners were forced to work in a rubber factory. On returning to Turin, he found work as a chemist in a paints and inks factory, but he progressively spent more time on literature. In his collection of writings on chem-ical elements entitled *The Periodic System* (1975),[50] Levi recalls his deeds as a young student getting to grips with the directions given in the Sestini-Funaro.[51]

For the purposes of this study, the first and the fourth editions of the Sestini-Faunaro textbook were consulted, dated, respectively, 1886 and 1896. The 1886 edition of the book was aimed specifically at technical schools in accordance with the new government programs released on June 21, 1885, which required the publication of new textbooks. The first edition is 517 pages long and is divided into twenty-one chapters. The first eleven chapters discuss general and inorganic chemistry (310 pages) and the remaining ten chapters deal with organic chem-istry. As for chemical elements, non-metallic substances are discussed first, fol-lowed by metals. The issue of classifying chemical elements is taken up in chapter ten, after an exhaustive description of them in the previous nine chapters, exactly the opposite of how textbooks are organized today. The authors dedicate just five pages to the issue, one of which is entirely occupied by a table showing Mendeleev's classification. The most interesting part of the chapter, however, is the section headed "The Periodic Ordering of the Elements in Brief." It rightly begins with an acknowledgment of the limitations of the classification criteria in use at the time. The distinction between metals and non-metals did not have a clear basis and the groupings based on valences contained a number of exceptions and raised some doubts. Everything seemed to point to the idea "that a law of great importance for the study of chemistry is still unknown to us." This is why "the ingenious and elegant classification of chemical elements" devised by Mendeleev deserved special attention, as he had ordered them systematically in such a way as to "satisfy the intimate exchange that passes between their properties and the sizes of their respective atomic weights." When introducing Mendeleev's classification, the authors recall Dumas, who had previously realized that there was a numerical relationship between atomic weights. They go on to point out that there were elements whose atomic weight was basically an average of those of the closest similar elements and that this relationship could not be the product of chance. Having mentioned this, they move on to a description of Mendeleev's classification with its periods and groups, and to the combinatory power of elements with oxygen, hydrogen, and chlorine, according to the different groups. They then underline that the law of periodicity had assumed particular importance as it had made it possible to predict the existence of as-yet-undiscovered elements. According to the authors, as chemistry progressed, so did the classification of the natural world; Mendeleev's periodic system, although imperfect, should be considered the best because "it has its foundations in experience and it serves to establish the greatest number of comparisons between elements and their chemical combinations." The students at technical schools should also have knowledge of this vital progress for science, albeit only in summary form. The fourth edition of the textbook, published in 1896, displays some significant differences from the first; indeed, it has been "completely reworked." The changes, however, do not concern the classification of elements. The "brief mention" of Mendeleev's system has been lifted whole and inserted into chapter thirteen, where it can be read alongside other topics such as the properties of metals and alloys, the reactions between salts, and the basics of electrochemistry. What is new was that this chapter is no longer found at the end of the section on inorganic chemistry, but in the middle between the section on general chemistry (which included some important non-metals) and a discussion of metallic elements—a hard-fought step forward toward reason.

7. CONCLUSION

As we mentioned in the introduction, our work is practically the first historical study examining the introduction and uptake of the periodic system in Italy, excepting a few essays, now rather dated, dedicated to a comparison of the work of Piccini and Mendeleev. Naturally, this essay only scratches the surfaces of the subject, which is one that will require further investigation and confirmation in order to understand the reasons behind the swift and whole-hearted acceptance of the periodic system in Italy. We sincerely hope that our work will provide a stimulus for further research in this field.

NOTES
* Sections 2, 5, 6 are attributed to Marco Taddia, sections 3, 4 to Marco Ciardi.

1. On Dandolo's translation, see Ferdinando Abbri, "Lavoisier e Dandolo. Le edizioni italiane del «Traité élémentaire de chimie»" [Lavoisier and Dandolo. The Italian editions of the "Traité élémentaire de chimie"], *Annali dell'Istituto di filosofia dell'Università di Firenze*, 6 (1984): 163–182; Marco Beretta, "Italian translations of the *"Méthode de nomenclature chimique"* and the *"Traité élémentaire de chimie*: The case of Vincenzo Dandolo" in *Lavoisier in the European Context*, ed. Bernadette Bensaude-Vincent and Ferdinando Abbri (Canton, Mass.: History of Science Publications, 1995), 225–248.

2. Antonio Santagata, *Lezioni di chimica elementare* (Bologna: Annesio Nobili, 1819), I, 51–79.

3. Domenico Mamone Capria, *Elementi di chimica filosofica-sperimentale* (Napoli: Andrea Festa, 1846).

4. Paolo Tassinari, *Manuale di chimica—chimica inorganica* (Pisa: Nistri, 1866).

5. Raffaele Napoli, *Prontuario di chimica elementare moderna* [Handbook of modern elementary chemistry] (Napoli: Vincenzo Pasquale, 1867), I, 44, 137–144.

6. Aleksandr Porfir'evič Borodin, *Biografia. Tutti gli scritti musicali, le lettere e i saggi scientifici del compositore* [Biography. Musical Writings, Letters and Scientific Papers of the Composer], trans. and ed. Valerij Voskobojnikov (Napoli: Edizioni Scientifiche Italiane, 1994), 357–366; Marco Ciardi, "San Pietroburgo, Karlsruhe, Pisa: Aleksandr Porfirievič Borodin e la chimica italiana nell'età del Risorgimento [St. Petersburg, Karlsruhe, Pisa: Porfirievich Aleksandr Borodin and Italian Chemistry in the Age of the Risorgimento]," in *Atti del X Convegno Nazionale di Storia e Fondamenti della Chimica*, ed. Marco Ciardi and Franco Giudice (Roma: Accademia Nazionale delle Scienze, 2003): 305–322.

7. Marco Ciardi, *Reazioni tricolori. Aspetti della chimica italiana nell'età del Risorgimento* [Tricolor Reactions. Aspects of the Italian Chemistry in the Age of the Risorgimento] (Milano: Franco Angeli, 2010).

8. Stanislao Cannizzaro, *Scritti e carteggi (1857–1862)* [Writings and Correspondence, 1857–1862], ed. Leonello Paoloni (Palermo: Università di Palermo, 1992), 252.

9. R. M. MacLeod, "Evolution, internationalism and commercial enterprise in science: The International Scientific Series, 1871–1910," in *Development of Science Publishing in Europe*, ed. A. J. Meadows (Amsterdam-New York-Oxford: Elsevier, 1980); Id., *The 'Creed of science' in Victorian England* (Aldershot: Ashgate-Variorum, 2000); L. Howsam, "An experiment with science for the nineteenth-century book trade: The International Scientific Series," *British Journal for the History of Science* **33** (2000): 187–207.

10. Paola Govoni, *Un pubblico per la scienza. La divulgazione scientifica nell'Italia in formazione* [An Audience for Science. The Popularization of Science in Italy](Roma: Carocci, 2002), 132.

11. Michele Nani, "Editoria e culture scientifiche nell'Italia postunitaria. Appunti sulle edizioni Dumolard [Publishing and scientific cultures in post-unity. Notes on Dumolard publisher]," *Ricerche storiche 2* (1999): 267.

12. Charles-Adolphe Wurtz, *La teoria atomica* (Milano: Dumolard, 1879).

13. Marco Ciardi, "Avogadro's concept of the atom: some new remarks," *Ambix* **48** (2001): 17–24; Id., "Theory and Technology. The Avogadro Manuscripts at the Turin Academy of Sciences," *Nuncius* **13** (1998): 625–656. Id., *Amedeo Avogadro. Una politica per la scienza* [Amedeo Avogadro. A Policy for Science] (Roma: Carocci, 2006).

14. Wurtz (note 12), 86.

15. Ibid., 113.

16. Ibid., 113–114.

17. Ibid., 148.

18. Ibid., 161.

19. Ibid., 161–162.

20. On Nasini, see "Luigi Cerruti, Scienza, industria, estetica. Raffaello Nasini e i soffioni boraciferi [Science, Industry, Aesthetics. Raphael Nasini and the Italian Geysers]," in *Il calore della Terra. Contributo alla storia della geotermia in Italia* [The Heat of the Earth. Contribution to the history of geothermal energy in Italy], ed. Marco Ciardi and Raffaele Cataldi (Pisa: ETS, 2005), 276–292.

21. Raffaello Nasini, "D. I. Mendeleev," *Rendiconti dell'Accademia dei Lincei* **16** (1907): 823–832.

22. Giulio Provenzal, "Augusto Piccini (1854-1905)," *Rivista di storia delle scienze mediche e naturali*, 21 (1930): 189–208; Id., *Profili bio-bibliografici di chimici italiani. Sec. XV–Sec. XIX* [Bio-bibliographical Profiles of Italian Chemist] (Roma: Istituto Nazionale Medico Farmacologico 'Serono,' 1938), 249–254.

23. Giacomo Ciamician, *Discorso per la inaugurazione del Busto nella ricorrenza del 2° anniversario della morte di Augusto Piccini* [Address for the inauguration of the bust on the occasion of the 2nd anniversary of the death of Augustus Piccini] (Firenze: Ramella, 1907).

24. Augusto Piccini, "Ossidazione dell'acido titanico [Oxidation of Titanic Acid]," *Gazzetta chimica italiana* **12** (1882): 151; **13** (1883): 57.

25. Provenzal, "Augusto Piccini" (note 21), 195.

26. Ibid., 196.

27. Augusto Piccini, "Appendice del traduttore [Appendix of the Translator]," in Victor von Richter, *Trattato di chimica inorganica* [Treatise of Inorganic Chemistry] (Torino: Loescher, 1885), 404–459.

28. Augusto Piccini, *Sul limite delle combinazioni e sul sistema periodico degli elementi* [On the Limit of the Combinations and the Periodic System of Elements] (Torino: Loescher, 1885).

29. Provenzal, "Augusto Piccini" (note 22), 195.

30. Augusto Piccini, "Correlazioni numeriche fra i pesi atomici e Classificazione degli elementi [Correlation between atomic weights and Classification of Elements]," in *Nuova enciclopedia di chimica scientifica, tecnologica e industriale: con le applicazioni a tutte le industrie chimiche e manifatturiere, della medicina, farmacia, fisica*, ed. Icilio Guareschi (Torino: UTET, 1901), I, 322–364.

31. Museo Galileo. Istituto di Storia della Scienza di Firenze, *Carte Piccini (1901–1905)*, raccolte mss. 06 (Letter Idiographic, with Signature of Mendeleev). On the correspondence between Piccini and Mendeleev, see also Bonifatij Mikhailovich Kedrov, *Mendeleev e gli scienziati italiani*, in *Actes du VIIIe Congrès International d'Histoire des Sciences*, Florence, Milan, Septembre 3–9, 1956 (Vinci: Gruppo Italiano di Storia delle Scienze; Paris: Hermann, 1958), 577.

32. Marco Taddia, "Lo sguardo oltre il confine [A View Beyond the Border]," *Sapere* **73** (2007): 44–49.

33. Giacomo Ciamician, "Über die Spectren der chemischen Elemente und ihrer Verbindungen," *Wiener Sitzungsberichte,* 76 Bd II, Abth. October-Heft (1877): 499–517.

34. Giacomo Ciamician, "Über den Einfluss der Dichte und der Temperatur auf die Spectren von Dämpfen und Gasen [The Effect of Density and Temperature on the Spectra of Vapors and Gases]," *Wiener Sitzungsberichte* **78** Bd II, Abth, October-Heft (1878): 181.

35. Giacomo Ciamician, "Über den Einfluss der Dichte und der Temperatur auf die Spectren von Dämpfen und Gasen," *Wiener Sitzungsberichte*, 1878, **78**, Bd. II. Abth, October-Heft 1878.

36. D. Mendelejeff, *Grundlagen der Chemie*, aus dem russischen übersetzt [The Principles of Chemistry, translated from the Russian by] von L. Jawein und A. Thillot (St. Petersburg: Ricker, 1891), p. 607.

37. D. Mendeléeff, "The Periodic Law of the Chemical Elements," *Journal of the Chemical Society* **55** (1889): 634–656.

38. *Chimica generale ed inorganica; appunti presi alle lezioni del Prof. Giacomo Ciamician dal dott. Adolfo Baschieri* [General and inorganic chemistry/notes taken during the lectures of professor Giacomo Ciamician/by Dr. Adolfo Baschieri] (Bologna: Universitas, 1899–1900).

39. *Appunti di chimica generale ed inorganica: presi dalle lezioni del Prof. Giacomo Ciamician dagli studenti Bruno Maggesi e Andrea Stagni, II Ed. riveduta corretta e aumentata* [Notes of general and inorganic chemistry: taken during the lectures of professor Giacomo Ciamician/by the students Bruno Maggesi, Andrea Stagni.—2. Reviewed corrected and supplemented edition] (Bologna: Tip. Minarelli, 1900).

40. Giacomo Ciamician, "I problemi chimici del nuovo secolo," in *Chimica, filosofia, energia. Conferenze e discorsi* [Chemistry, Philosophy, and Energy. Conferences and Talks], ed. Marco Ciardi and Sandra Linguerri (Bologna: Bononia University Press 2007), 89–90.

41. Marco Taddia, "Il chimico di Garibaldi. Paolo Tassinari e l'arte dell'analisi [Giuseppe Garibaldi's Chemist. Paolo Tassinari and the Art of Chemical Analysis]," *La Chimica e l'Industria* **91**, (10) (2009): 124–125.

42. Tassinari, *Manuale di chimica—chimica inorganica* (note 4).

43. Felice Marco, *Principi della teoria meccanica, dell'elettricità e del magnetismo* (Torino: Paravia, 1867).

44. Felice Marco, *Nozioni di chimica secondo il sistema periodico* [Basics of Chemistry According to the Periodic System of Elements], VII ed. (Torino: Paravia, 1887).

45. Felice Marco, *Elementi di chimica secondo il sistema periodico* [Elements of Chemistry According to the Periodic System of Elements], X ed. (Torino: Paravia, 1896).

46. On Richter's textbook of inorganic chemistry, see the chapter on Russia (editor).

47. Fausto Sestini, Angiolo Funaro, *Elementi di Chimica* [Elements of Chemistry] (Livorno: Giusti, 1886).

48. Fausto Sestini, Angiolo Funaro, *Corso di Chimica* [Chemistry Course], IV ed. (Livorno: Giusti, 1896).

49. Italo Giglioli, "Commemorazione del Prof. Fausto Sestini" [In commemoration of Prof. Fausto Sestini], *Rendiconti della Società Chimica di Roma* (1904), pp. 165–176. http://w3.uniroma1.it/nicolini/Sestini.html, accessed November 23, 2013.

50. Primo Levi, *Il sistema periodico* (Torino: Einaudi 1975); *The Periodic Table*, translated from the Italian by Raymond Rosenthal (London: Campbell, 1995).

51. Marco Taddia, "Primo Levi e le insidie del gas esilarante [Primo Levi and the threats of laughing gas]," *La Chimica e l'industria* **89** (2007): 175–176.

PART VI

Response Beyond Europe

CHAPTER 13

☙

Chemical Classification and the Response to the Periodic Law of Elements in Japan in the Nineteenth and Early Twentieth Centuries

MASANORI KAJI

1. CHEMISTRY BEFORE THE MEIJI RESTORATION

The year 1868 is usually considered to be the beginning of modern Japan. In that year the Tokugawa government, a feudal samurai government in Edo (today's Tokyo), was replaced by a modern imperial government (initially based in Kyoto, the old imperial capital) at a time of internal crisis and the fear of colonization by European imperial powers. This revolutionary political change is named *the Meiji Restoration* because the ancient imperial system was nominally restored under Emperor Meiji. The new government began as a mixture of ancient Japanese and modern Western imperial systems, but it soon became a completely Westernized government, which adopted a policy of full-fledged modernization.

However, the introduction of Western science had already started long before the Meiji Restoration. During the Edo Period, the Tokugawa Shogunate (1603–1867) strictly controlled overseas trade and the Netherlands was the only European country with which Japan had diplomatic relationship from the middle of the seventeenth century until 1853. In the second half of the eighteenth century some books in Dutch on science, technology, and medicine were imported into Japan. For the introduction of Western medicine, physicians played an important role. During the Edo Period there was a class system: the samurai class (warrior) controlled the common people in villages and towns. All the professions were considered to be hereditary. However, physicians could move rather freely along the social ladder (hierarchy). If physicians were employed by feudal lords, they became accepted as members of the samurai.

There was a reform movement among physicians during the eighteenth century. In 1754 Yamawaki Toyo[1] (1706–62), a physician in Kyoto, received official permission to inspect the anatomy of a human body, using a cadaver of a condemned criminal, after he had inspected otters (a small animal with four webbed feet), the structure of which was quite different from Chinese medicine's teaching. After him physicians were allowed to inspect condemned criminals' bodies. Sugita Genpaku (1733–1817) and Maeno Ryotaku (1723–1803) inspected such a body in 1771 and compared it to the pictures in an imported Dutch anatomy book (a textbook for military physicians). They were amazed by the exactness of the anatomy atlas and decided to translate this book for Japanese medicine. A group of Japanese medical doctors headed by Sugita and Maeno began to learn Dutch and translated one Dutch anatomy textbook into Japanese.[2] This was the beginning of the so-called "Dutch learning (*Rangaku*)," which was a process of studying Western culture, generally through Dutch language.

An anatomy book was a good starting point for translations from Dutch, because an anatomical atlas helped the Japanese physicians to guess the meanings of words. After this introduction, these physicians started to extend their knowledge of medicine from surgery to other areas such as internal medicine. The latter needed pharmaceutical knowledge: one must find Chinese or Japanese counterparts of Western drugs, described in medical books in Dutch. Some who studied Western pharmacy found a completely new area of knowledge behind it: the natural sciences, including chemistry. The Udagawa family, a house of physicians, followed this trajectory. Udagawa Genzui (1756–98) started to learn Dutch after Sugita Genpaku and Maeno Ryotaku translated a book on internal medicine from Dutch into Japanese in 1793. His adopted son, Udagawa Genshin (1769–1834), revised and enlarged his father's book and studied pharmacy. Between the 1830s and 1840s, Udagawa Yoan (1798–1846), Udagawa Genshin's adopted son, was the first Japanese to write a chemistry textbook, *Seimi Kaiso* [An Introduction to Chemistry], which was grounded on Lavoisier's chemistry (Fig. 13.1).[3] The textbook was based on a Dutch version of a German translation of William Henry's *Elements of Experimental Chemistry*.[4] However, Udagawa expanded the Japanese translation substantially by adding various relevant information and arguments, which included his own opinions and comments with some original analysis of hot spring waters in Japan. In the process of writing *Seimi Kaiso*, Udagawa read a number of contemporary Dutch chemistry textbooks, imported into Japan, including the Dutch translation of Lavoisier's *Traité Elémentaire de Chimie*. In the *Seimi Kaiso* Udagawa listed fifty-eight elements, five of which turned out to be wrong later, including light and caloric. He employed a traditional classification of the elements, dividing them between non-metals and metals.

In 1854, after the feudal government abandoned its policy of tight control over foreign trade, the political situation became destabilized. During the political

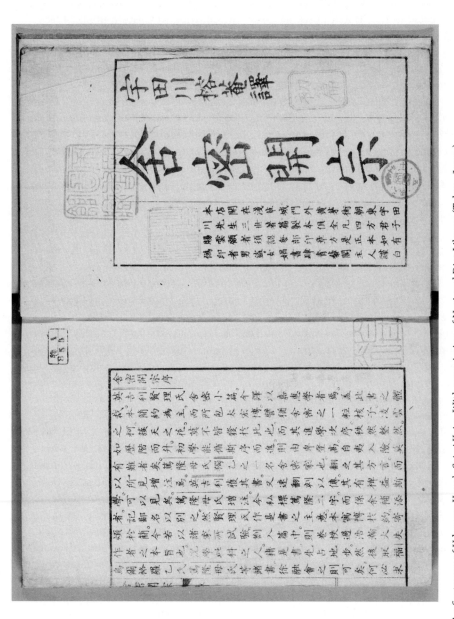

Figure 13.1 The first page of Udagawa Yōan's *Seimi Kaiso*. With permission of National Diet Library (Tokyo, Japan).

turmoil and colonization crisis, some intellectuals who belonged to the samurai began to study Western science and technology, including chemistry and chemical technology, as militarily useful subjects.

In the 1840s and 1850s samurai intellectuals other than physicians became interested in the "Dutch Learning" as an introduction to Western military technology due to a colonization crisis after the Opium War in China (1840–42) and the opening of the country in 1854, first to the United States and then to major European countries. During this period, scholars studied Western knowledge not only through Dutch, but also other European languages, such as English, French, and German. They were called yogaku-sha, scholars of "Western Learning." Even the Tokugawa government established various institutions of "Western Learning" at this time. Under this favorable atmosphere, some scholars of Western Learning developed Udagawa's chemistry. One of these scholars, Kawamoto Komin (1810–71),[5] a teacher of chemistry at the Bansho Shirabesho [School of Western Learning] of the Tokugawa government in Edo, established in 1856, translated a number of chemistry textbooks, such as Kagaku shinsho [A New Book of Chemistry] and wrote a textbook based on Dalton's atomic theory.[6]

Nagasaki Kaigun Denshu-jo [Nagasaki Naval Training Center] was established in 1855 by the Tokugawa government in Nagasaki, a port in Kyushu, the most southern main island in Japan, where the Dutch representatives were stationed. When Pompe (Johannes Lydius Catherinus Pompe van Meerdervoort, 1829–1908), a Dutch military doctor, came to the Naval Training Center as one of the teachers, he established *Igaku Denshu-jo* [Medical Training Center] attached to the Naval Training Center and started a systematic training course of medicine, including chemistry classes. Uyeno Hikoma (1838–1904), the son of a watchmaker from Nagasaki with broad intellectual interests, studied chemistry under Pompe. In 1862 Ueno published a chemistry textbook, *Seimi-kyoku Hikkei* [A Handbook of Chemistry Department][7] based on various chemistry textbooks in Dutch translation, including Söckehardt's *Die schule der Chemie* and Fresenius's *Anleitung zur qualitativen chemischen Analyse*. It is said that Ueno wrote both inorganic and organic chemistry, but only the first part, which contained non-metals in inorganic chemistry, was published. This included descriptions of chemical technology (the manufacture of sulfuric acid, chloric acid, matches, and glass) and photography. Ueno listed sixty-four elements with Japanese names and chemical equivalents in the traditional Japanese alphabetical order of the names. His well-organized and well-written textbook, along with Udagawa's *Seimi Kaiso*, mentioned earlier, served as good introductions to chemistry before the Meiji Restoration. Ueno became a famous commercial photographer and was considered to be one of the founders of Japanese photography.[8]

Some Chinese translations of European elementary chemistry textbooks were also good sources of chemical knowledge in those days for Japanese intellectuals who were well-versed in the Chinese literary language. *Huaxue Chujie*, a translation of Wells's *Principles and Applications of Chemistry* by John Kerr and Ho Liao-jan [He Liaoran], was imported, reprinted, and even translated into Japanese.[9]

2. INSTITUTIONALIZATION OF CHEMISTRY IN JAPAN AND EARLY CHEMISTRY TEXTBOOKS

After the Meiji Restoration in 1868 even more foreign textbooks of chemistry, mostly at the elementary level, were translated into Japanese by scholars of the Western Learning, who had been working from the 1850s.[10] These scholars helped to promote Western science in this transitional period. For example, *Kagaku-shinron Mondo* [Dialog on a New Chemistry] was based on interviews on chemistry by an official of the Ministry of Education with Utsunomiya Saburo (1834–1902) and Katsuragawa Hosaku (1839–90), both famous scholars of the Western Learning with a good knowledge of chemistry.[11] These interviews were all based on chemical atomism and mentioned atomic weights or chemical equivalents. They listed more than sixty elements and divided them into non-metals and metals.

In the hope of institutionalizing science and technology and introducing the modern Western education system, the new government employed many foreign teachers in higher education. Foreign teachers' lecture notes were translated into Japanese and published by the Ministry of Education. Herman R. Ritter (1828–74), a German chemist, came to Japan in 1870 and taught chemistry and physics in Osaka and Tokyo in English. His lecture notes on physics and chemistry were translated into Japanese and published as *Rika Nikki* [A Diary of Physics and Chemistry] in 1870, and its part on chemistry was published separately as *Kagaku Nikki* [A Diary of Chemistry] in 1876.[12] *Kagaku Nikki* contains information on elements, atomic weights, and valence, and an elementary explanation of the structural theory of organic compounds, but with no particular mention of the classification of elements.

During this early period of modernization, foreign elementary chemistry textbooks were translated into Japanese in a more systematic way. While scholars of Western Learning translated books at hand by chance in the earlier period, popular textbooks in Europe and the United States (mostly in English) now were translated, such as those of the English chemist Henry E. Roscoe and the American chemist Ira Remsen[13] after the 1870s. Roscoe's introductory textbook, *A Primer of Chemistry*, from the series "Science Primers for Elementary Schools" was translated into Japanese several times during the 1870s and 1880s.[14] Since it was an elementary textbook, it mentioned sixty-three elements and only classified them into non-metals (fifteen elements) and metals (forty-eight elements). Ira Remsen's *The Elements of Chemistry: A Textbook for Beginners*[15] was another popular book. The original was published in 1887, which included a classification of elements based on valency, but made no mention of the periodic law.

In the process of the institutionalization of science and technology, each ministry of the government had its own higher education school before all of these higher education schools were unified into the Imperial University under the Ministry of Education in 1886. Some of the first Japanese graduates of the Imperial University, who majored in subjects other than chemistry but gained some knowledge of chemistry during an earlier stage of their education, started publishing textbooks of chemistry as well. For example, Shiga Taizan (1854–1934), who studied chemistry under Herman Ritter and later became a forestry specialist, published *Kagaku Saishin* [The Newest of Chemistry] in 1877.[16] Shimoyama Jun-ichiro (1853–1912), who studied pharmacy at the Tokyo Medical School under the Ministry of Education and became the first professor of pharmacy at the Imperial University, published his lecture notes on chemistry as *Kagaku Shinri* [The Truth of Chemistry] in 1880.[17] These textbooks were not for chemists, but rather for those who majored in other fields and needed a basic knowledge of chemistry in Japanese. They listed approximately sixty elements, classified them into non-metal and metal, and employed a classification based on groups of elements or atomic valency.

Even though most children did not have the opportunity to study beyond the level of elementary education, some middle and upper class children, including those belonging to the former samurai class, studied western languages from an early age and continued their studies at higher educational institutions in which westerners taught. The Meiji government sent the most successful students to Europe and the United States to study further. When they returned, they replaced the foreign teachers, thus becoming the first generation of Japanese professors, with the responsibility of educating future generations of Japanese students.

Sakurai Joji (1858–1939)[18] was a chemist who belonged to the first generation of Japanese scientists. He was born in 1858 into a samurai family in Kanazawa, the capital of one of the largest and most powerful feudal lords. He started to learn English at an early age and entered a School of Western Learning[19] in 1871. He studied the three-year preliminary course and the two-year specialist chemistry course in the Chemistry Department under the English chemist Robert William Atkinson (1850–1929).[20] He was then sent to England and studied at University College London between 1876 and 1881 under Atkinson's teacher, the famous chemist Alexander William Williamson (1824–1904).

In 1877, while Sakurai was in London, Tokyo University was created out of the School of Western Learning at which Sakurai had studied, as well as the School of Medicine. The new university comprised the faculties of law, science, literature, and medicine. In the same year Kuhara Mitsuru (1856–1919), Takasu Rokuro, and Miyazaki Michimasa (1852–1909), three students who had finished a chemistry course at the School of Western Learning that year, were considered to be the first graduates of the Department of Chemistry at the newly established Tokyo University.

When Sakurai returned to Japan in 1881 at the age of twenty-three, he obtained a teaching position as a lecturer in the Faculty of Science at Tokyo University as the successor to his mentor, Atkinson, and was promoted to professor the next year. He was the second Japanese professor of chemistry after the American-trained chemist Matsui Naokichi (1857–1911), who had been appointed a year earlier.

With the foundation of the first Imperial University in Tokyo in 1886, the education system in Japan was fully established. Sakurai became the head of the Department of Chemistry at the College of Science at the Imperial University, teaching organic chemistry as well as physical and theoretical chemistry. His lectures, especially those on organic chemistry, were highly praised by students, even though Sakurai's own research was in physical chemistry rather than organic chemistry.

3. THE FOUNDATION OF THE TOKYO CHEMICAL SOCIETY AND ITS INFLUENCE

The first three graduates of the Department of Chemistry of Tokyo University, Kuhara Mitsuru, Takasu Rokuro, and Miyazaki Michimasa, formed the Tokyo Chemical Society in 1878, the year following their graduation. The Society played an important role in the dissemination of the periodic law in Japan.[21]

Atkinson, the professor of chemistry at Tokyo University, read a lecture on the history of the theory of elements at the first annual meeting of the Society on April 19, 1879. This was one of the earliest mentions of the periodic law in a Japanese publication. Atkinson referred to Lothar Meyer's paper on the relationship between atomic weights and atomic volumes, but did not mention Mendeleev.[22]

Matsui Naokichi, the first Japanese professor of chemistry in the Department of Chemistry at Tokyo University, mentioned Mendeleev's Periodic Law for the first time in a paper on the recent history of atomic theory in the nineteenth century, which appeared in *the Journal of the Tokyo Chemical Society* in 1882.[23] He explained the law as a recent discovery based on Cannizzaro's new atomic weights.

4. CHEMISTRY TEXTBOOKS BY THE FIRST GENERATION OF CHEMISTRY PROFESSORS

In the 1890s the first generation of Japanese professors of chemistry and those of applied chemistry, who also played an active role in the Chemical Society as its earliest members, started to write chemistry textbooks in Japanese for secondary schools and for introductory courses of chemistry in higher education. They all mentioned Mendeleev's Periodic Law. The basic structure of their textbooks on inorganic chemistry was based on the periodic law.

One of the first chemistry textbooks to mention Mendeleev's Periodic Law was *Chemistry Textbook, First Part, Inorganic Chemistry*, edited by Takamatsu Toyokichi (1852–1931)[24] and published between 1890 and 1891 (Fig. 13.2). In the last chapter Takamatsu states that Mendeleev's Periodic Law had been proven by the discovery of the three elements scandium, gallium, and germanium. Takamatsu was one of the earliest graduates of the Department of Chemistry at Tokyo University. After a period of study in England and Germany, he became a professor in the Department of Chemistry. When the Imperial University was established, he became the first professor of applied chemistry at the College of Engineering of the Imperial University.

Yoshida Hikorokuro (1859–1929)[25] published a two-volume chemistry textbook for secondary schools, normal schools (teachers colleges), and liberal arts colleges in 1893,[26] soon after the publication of Takamatsu's textbook. Yoshida studied chemistry under Atkinson in the Department of Chemistry at Tokyo University and graduated in 1880. After working in the Geological Survey of the Ministry of Agriculture and Commerce as an analytical chemist until 1886, he became associate professor of the College of Science at Tokyo Imperial University in 1886, professor

週期律

上表ニ一括セル元素中、殊ニ第二ニ第一ニ存在スルモノニハ、原子量ノ詳密ナラザル為ニ、其ノ位置ノ確定シ難キモノアリ、又上表中尚新元素ヲ以テ充スベキ数多ノ空位アリテ、未ダ之ヲ完全スルニ至ラザレバ、メンデレーフ氏ハ仮ニ週期律ヲ応用シテ原子量ノ判然セザルモノヲ改定シ、又未知元素ノ存在及ビ性質ヲ頗ル概略豫言スルヲ得タリ。

今其例ヲ挙グルニ、インヂウム(In)ノ原子量ハ從前75.6ヲ以テ表ヘシモ、之ヲ週期律ニ照スニ、其位置 As(=75)ト Se(=79)ノ間ニ在リテ若モ該元素ノ性質ト符合セズ、依テ其原子量ヲ 1.5×75.6 =113.4ト改定シ、之ヲ Cd(=112)ト Sn(=118)トノ間ニ位セシナルニ、爾後ニ其後ニインヂウムノ比熱ヲ測定シ、以テ之ガ原子量ヲ計算シ得ルニ、恰モ113.4ニ等シキ数ヲ得タリ。(問題71ヲ參照セヨ)。

又メンデレーフ氏ガ該表ヲ創設セシ時、スカンヂウム(Sc)、ガリウム(Ga)及ビセルマニウム(Ge)ノ三元素ハ尚未ダ之ヲ發見スルニ至ラザリシガ、氏ハ其大ニ乏シキ位置ノ關係ヨリ推シテ、各未知元素ニ有スベキ原子量及ビ主要ナル性質ヲ豫言シ、爾ルニ其後上ノ三元素

舎密化學

	第一属 R₂O	第二属 RO	第三属 R₂O₃	第四属 RH₄ RO₂	第五属 RH₃ R₂O₅	第六属 RH₂ RO₃	第七属 RH R₂O₇	第八属 (R₂O₄)
第一列	H=1							
第二列	Li=7	Be=9	B=11	C=12	N=14	O=16	F=19	
第三列	Na=23	Mg=24	Al=27	Si=28	P=31	S=32	Cl=35.5	
第四列	K=39	Ca=40	—=44	Ti=48	V=51	Cr=52	Mn=55	Fe=56 Ni=58.5 Co=59
第五列	Cu=63	Zn=65	—=70	—=72	As=75	Se=79	Br=80	
第六列	Rb=85	Sr=87.5	Yt=89	Zr=90	Nb=94	Mo=96	—	Ru=104 Rh=104 Pd=106
第七列	Ag=108	Cd=112	In=113	Sn=118	Sb=120	Te=125	I=127	
第八列	Cs=133	Ba=137	La=138	Ce=140	Di=142	—	—	
第九列	—	—	Yt=173	—	—	—	—	
第十列	—	—	Er=166	La=180	Ta=182	W=184	—	Os=198 Ir=198 Pt=198
第十一列	Au=196.5	Hg=200	Tl=204	Pb=207	Bi=208	—	—	
第十二列	—	—	—	Th=231	—	U=240	—	

Figure 13.2 The periodic table in Takamatsu Toyokichi's *A Textbook of Chemistry* (Tokyo, 1891). With permission of National Diet Library (Tokyo, Japan).

of *Gakushuin*, a school for nobility, in 1892 and Third High School, one of privileged three-year liberal arts colleges in Kyoto in 1896, where he was sent to Germany in 1898 to study further for two years. When he returned, he was appointed one of the first professors of chemistry at Kyoto Imperial University, the second imperial university in Japan. He was known for his pioneering work on the oxidation enzyme in Japanese lacquer.

Yoshida's book was one of the first chemistry textbooks thoroughly based on the periodic law. In the preface Yoshida mentions Mendeleev, Richter, Ramsay, Ostwald, Remsen, Bloxam,[27] and Takamatsu.[28] After fourteen introductory chapters on basic chemical concepts, such as elements, energy, atomic theory, and water, oxygen, hydrogen, chlorine, and their compounds, chapter fifteen explains the periodic law. Further on, he describes the families of elements in the order of the seventh family, sixth, fifth, fourth, first, second, third, and, lastly, the eighth group of elements. This structure is reminiscent of that of Mendeleev's *Principles of Chemistry*.

Yoshida's chemistry textbook seemed to be read not only by students, but also by various intellectuals who were interested in science. I found its second edition, published in 1895, in the book collection of Mori Ogai (Rintaro) (1862–1922)[29], a Japanese Army Surgeon general officer, one of the most famous novelists of the Meiji period and a graduate of the Faculty of Medicine at Tokyo University. There were handwritten notes in the margin, which suggests that Mori read the textbook with great attention.

5. THE INFLUENCE OF CHEMISTRY TEXTBOOKS ON HIGHER EDUCATION: LOTHAR MEYER, MENDELEEV, IRA REMSEN

As the preface of Yoshida's textbook shows, Japanese chemists of the Meiji period read various chemistry textbooks published in Europe and the United States. One can find both Lothar Meyer's famous *Die modernen Theorien der Chemie und ihre Bedeutung für die chemische Statik*[30] and Mendeleev's *Principles of Chemistry* in university libraries in Japan.[31] Lothar Meyer's *Outlines of Theoretical Chemistry* was translated into Japanese between 1894 and 1895.[32] One can thus surmise that there must be direct influences on Japanese chemistry from the textbooks by discovers of the periodic law.

Unlike earlier times, in the 1890s more advanced textbooks of chemistry were translated into Japanese. For example, an introductory textbook of chemistry for college students, *Introduction to the Study of Chemistry* (1886) by Ira Remsen (1846–1927), chemistry professor at John Hopkins University, was translated into Japanese between 1893 and 1894 under the supervision of Kuhara Mitsuru and Shimomura Jun-ichiro.[33] Kuhara studied organic chemistry at Johns Hopkins University under Remsen in the 1880s. This was not a direct translation, but rather the translation of the German version of Remsen's textbook (*Einleitung in das Studium Chemie*) by Karl Seubert (1851–1942) into Japanese. The German version was expanded in the chapters on carbon (chapter 9), atomic theory (chapter 12), and the natural system of elements (chapter 26). The last concluding chapter dealt with the Periodic System of Elements as "Natural Groups of Elements." The German translation of this part was supplemented by a more detailed discussion of groups of elements, along with a mention of Prout's hypothesis. Seubert, the German translator, studied chemistry under Lothar Meyer in Tübingen and worked with Meyer until Meyer's death in 1895. This means that a more detailed discussion on the natural system in the line of Lothar Meyer was conveyed to Japanese readers through this German edition.

6. THE SYSTEMATIZATION OF EDUCATION: THE INTRODUCTION OF PERIODIC LAW INTO SECONDARY EDUCATION

After the Meiji Restoration the systematization of education was also promoted. After several changes by the 1890s, the principal form of the secondary educational system was finally settled: ordinary secondary schools, secondary schools for girls, and normal schools (teachers' training schools). The secondary school (for boys) system was established by an ordinance of the secondary schools in 1886 and substantially revised by another ordinance in 1899. The establishment of the so-called "outline program of instruction" (official curricula) for the secondary schools in 1902 was especially important.[34] It contained detailed contents of instruction for each subject and each year. Thanks to this outline, education in secondary schools in Japan became highly systematized.[35]

According to this outline program of instruction for the secondary schools, chemistry was taught in the fourth year of the five-year education program in secondary schools three to four hours every week for a whole year. Subjects one was required to teach were as follows: ordinary gas; oxygen and its compounds; halogen and its compounds; sulfur and its compounds; solutions nitrogen, phosphorus, arsenic and their compounds; activity (thermodynamics); carbon, silicon, boron and their compounds; metals and their compounds; the periodic law; organic compounds; aliphatic compounds; aromatic compounds; fermentation; and decomposition. Under the periodic law, "the comparison of elements arranged according to the order of atomic weights" and "the comparison of atomic weights and physical and chemical properties" were indicated.

Almost all of the chemistry textbooks for secondary schools around the end of 1890s and all of the textbooks for secondary schools published after 1902 mentioned the periodic law.[36]

7. AN IN-DEPTH APPROACH TO THE PERIODIC LAW: THE CASE OF IKEDA KIKUNAE

Even though almost all of the chemistry textbooks for secondary schools accepted the periodic law as the classification of chemical elements, one can see a subtle difference among them in their definitions of an element. Takamatsu Toyokichi defined an element as the same as a simple body in his textbook in 1890.[37] Yoshida Hikorokuro, whose textbook was one of the first chemistry textbooks thoroughly based on the periodic law in Japan, defined elements as the same as simple bodies in 1893, as did Takamatsu.[38] However, Ikeda Kikunae (1864–1936)[39] showed a somewhat different approach to those important chemical concepts.

Ikeda Kikunae was born in Kyoto as a son of a samurai family stationed in the Kyoto residence for the Satsuma feudal lord, a very powerful lord in Kyushu, the most southern main island.[40] In 1882, at the age of eighteen, Ikeda moved to Tokyo to enter a privileged national liberal arts college attached to the University of Tokyo. He graduated from the Department of Chemistry in the College of Science of the Imperial University in 1889.[41] After graduation Ikeda first became a professor at the Higher Normal School in Tokyo and later, in 1896, associate professor at his alma mater. He was sent to Europe for further training in 1899. During a two-year stay in Europe, he studied under Wilhelm Ostwald (1853–1932). When he returned, he was promoted to full professor in the Department of Chemistry at Tokyo Imperial University. During his professorship at the Normal School and the early period of his professorship at the Imperial University (1894–1906), he wrote many chemistry textbooks for secondary schools and higher education, including a translation of Ostwald's Inorganic Chemistry as well as elementary textbooks. His prolific authorship showed an in-depth understanding of important chemical concepts.

Ikeda clearly distinguished elements from simple bodies in his first textbook of chemistry for beginners, *Kagaku* [Chemistry] in 1895,[42] explaining as follows:

> Material existence common among various substances, which turns out to be a simple body when it appears alone, is called element . . . Material existence common among carbonic acid, carbon oxide and charcoal is called carbon element. Charcoal is only one of simple appearance.[43]

This part almost reminds one of Mendeleev's distinction between elements and simple bodies, which played an important role in his discovery of the periodic law,[44] even though Ikeda did not mention the periodic law in this elementary textbook. In his textbook for secondary schools, *Shinpen Chugaku Kagakusho* [A New Secondary school Chemistry Book], published in 1898, Ikeda articulated this distinction more clearly.[45] He explained the periodic law in the last chapter of the inorganic chemistry section.[46] This textbook must have had some influence on the contents and structure of later chemistry textbooks in Japan, including the outline program of instruction for the secondary schools of the Ministry of Education in 1902. It is also noted that after this textbook many textbooks emphasized the distinction between elements and simple bodies, including Takamatsu's revised chemistry textbook.[47]

It seems, however, that Ikeda was not so enthusiastic about the periodic law, because he wrote in one of his textbooks that "this [the periodic law] is only a rule of thumb, but it is very convenient for memorizing the relationships among the elements."[48] While Ikeda published a theoretically refined textbook, he also published rather descriptive textbooks almost simultaneously,[49] where there was no mention of the periodic law. Ikeda's in-depth understanding of the concept of elements and his somewhat cool approach to the periodic law showed that Japanese chemistry was mature enough in the early twentieth century that not all chemists accepted concepts simply because they were recognized in the West.

8. RESEARCH ON THE PERIODIC SYSTEM: OGAWA MASATAKA'S DISCOVERY OF "NIPPONIUM"

Around this same time a Japanese chemist made a discovery based on the periodic law: Ogawa Masataka (1865–1933) claimed in 1908 to have discovered a new element, which he called "nipponium."[50]

Ogawa was born in Edo (Tokyo) in 1865, a few years before the Meiji Restoration, as the son of a samurai family. He received his training in the newly Westernized educational system, graduating from the Chemistry Department of the Imperial University in Tokyo in 1889. After studying one more year under the supervision of Edward Divers (1837–1912), an English chemist who was employed by the Japanese government and taught for more than twenty years (1873–99) in Japanese higher educational institutions, Ogawa became a schoolmaster in a secondary school in the provinces. As one of the first graduates of the Imperial University, he was well paid as a schoolmaster but wanted to follow an academic career. He thus quit his job in 1896 to go to Tokyo, seven years after graduation, again continuing to study under Divers at the Imperial University as an unpaid research fellow and earning a living for his family by teaching in private secondary schools in Tokyo.

In 1899 he obtained a position at the First High School, a three-year liberal arts college, the graduates of which could enter the two Imperial Universities in Tokyo and Kyoto. Even though the teaching burden was heavy, he could maintain a laboratory for his research at the School. He concentrated on teaching from Monday through Thursday and did his research on Fridays, Saturdays, and Sundays.

Because of these efforts, in February 1904 the Ministry of Education decided to send him to England to study at University College, London, under Sir William Ramsay's direction. During his stay in London, he analyzed thorianite, a newly discovered mineral found in Ceylon, given to him by Ramsay. The mineral was found in 1904 in Ceylon, then part of the British Empire, and sent to London by the Indian colonial government to analyze, with various chemists disagreeing on its composition. In the course of analysis, Ogawa identified "an element believed to be new"[51] because of its characteristic line of spectrum. However, during his stay in England, he could not fully establish its nature.

Ramsay himself suggested the name nipponium, based on the name of Ogawa's country Japan, "Nippon," in Japanese, for the new element. It is important to note that the Russo-Japanese War had started in February 1904, just when Ogawa left Japan for England, and Britain was in alliance with Japan because of the

Anglo-Japanese Alliance Agreement concluded in 1902. This was the time when public opinion toward Japan in England was at its most favorable in modern history.

Following his return from London in August 1906, everything went smoothly for him in his country and he was soon appointed professor at the Tokyo Higher Normal School. He continued to study the new element, finding the same element in other minerals, reinite and molybdenite in Japan. The content of nipponium in these minerals, especially in molybdenite, was much larger than in thorianite. In 1908 Ogawa published two preliminary papers on nipponium in *the Journal of the Tokyo Imperial University.*[52] His papers were reprinted in *the Chemical News* in London.[53] He calculated the equivalent weight of the element as approximately fifty and estimated its position in the periodic system as the vacancy between molybdenum and ruthenium, having an atomic weight of approximately one hundred. A chemist soon accepted his claim.[54]

In 1910 he was rewarded with a doctor of science degree for his work on the new element in Japan. The certificate of his degree stated that his research was important not only for science but also for the enhancement of national glory. After fifty years of modernization and consecutive victories in the Sino-Japanese War between 1894 and 1895 and in the Russo-Japanese War between 1904 and 1905, Japan became one of the colonial powers competing for territories and resources in Asia. The discovery of a new element by a Japanese chemist was just the sort of thing needed for the further exaltation of national prestige. In the same year, 1910, Ogawa was awarded the first Sakurai prize,[55] the highest prize of the Tokyo Chemical Society, created just two years previously to commemorate the twenty-fifth year of the professoriate of Sakurai Joji first at Tokyo University and later at Tokyo Imperial University.

In the 1910s Japanese heavy industry began to evolve and consequently, Japanese higher education started to expand greatly. In 1911 Tohoku Imperial University was opened in Sendai, the largest city in northern Japan as the third imperial university. Forty-six-year-old Ogawa was appointed professor in the College of Science at the Imperial University and elected the dean of the College (Fig. 13.3). Highly motivated professors and students gathered to create a research-oriented university after the German model. Even with his seniority, Ogawa was one of the professors who engaged in experimental work until late at night. In 1919 he was named president of the university, the first elected president (all of his predecessors were appointed by the Ministry of Education). He was reelected twice and served as president until his retirement at the age of sixty-three in 1928.

His research proved somewhat difficult, because no one except Ogawa himself could verify the existence of nipponium. Thorianite, in which Ogawa's discovery was first made, contains very little of the desired element. Only Ogawa possessed the technique and patience to isolate the element by classical systematic separations, such as precipitation, solution, evaporation, or extraction. None of Ogawa's students could show even the faintest trace of nipponium. With the benefit of hindsight, it might have been better for them to concentrate on the analysis of molybdenite, which was expected to contain a larger amount of nipponium. Ogawa decided

Figure 13.3 Ogawa Masataka as president of Tohoku Imperial University around 1924. With permission from Tohoku University Archives.

to do the verification himself and spent all of his spare time conducting analyses, even during his presidency at the university.

After his retirement he stayed in the department of Science without salary and continued to work in the laboratory. On July 3, 1930, he collapsed from sudden illness in the laboratory and was hospitalized several days later. He died on July 11, at the age of sixty-five. The cause of his death was cholecystitis, inflammation of the gallbladder.

According to Ogawa, nipponium should be ascribed to z (atomic number) =43. Since this place had been missing for a long time, many researchers tried in vain to find it in minerals, with some even wrongly believing they had discovered it. After Ogawa, Noddack and coworkers analyzed columbite ores and showed the presence of the elements with z = 43 and 75 by X-ray absorption spectra.[56] The authors gave the names masurium and rhenium, respectively, for the newly discovered elements. Although they succeeded in the isolation of rhenium with z = 75 in the amount of 2 mg,[57] they could not obtain a weighable amount of masurium with z = 43. In 1937

Perrier and Segré found the element with z = 43, which was later named techne-tium.[58] The problem of the element with z = 43 was thus considered to be finally settled, Ogawa's work having turned out to be incorrect.

Only recently, a radiochemist and the emeritus professor of Tohoku University, Yoshihara Kenji, persuasively proved that Ogawa's nipponium was actually the el-ement with z = 75 (rhenium), not with z = 43 as Ogawa had supposed.[59] First of all, the spectral line of 4882 Å for nipponium is close to the present value of rhenium at 4889.2 Å. Second, Ogawa's procedure for obtaining a chloride of nipponium MCl_2 instead produces an oxychloride $MOCl_4$ based on present-day knowledge. The recal-culation of its atomic weight based on this result gives 185.2, which is very close to the present value of rhenium 186.2 instead of Ogawa's 100. Third, Ogawa reported that Japanese molybdenite contained a comparatively large amount of nipponium, namely more than 1 percent[60]. It is known that volcanic molybdenite sometimes concentrates rhenium, where nearly 0.4 percent, or in some special cases more, rhenium is found. Unfortunately for Ogawa, it was very difficult to separate rhe-nium from molybdenum by classical procedures. One of the most efficient means of separation is chromatography with an alumina column, which was not known in Ogawa's day.

With hindsight, Ogawa's misplacement of the discovered element resulted from his misunderstanding of the valence of the element, but also came from the exis-tence of two undiscovered elements with z = 43 and z = 75 from the same group, which must have very similar properties. The chances of which element should be discovered looked equal before the discovery of technetium, which showed the non-existence in nature of the element with z =43. The Noddack group's success was due to their claim of discovering both elements (z = 43 and 75) from the same group in the periodic table. The periodic system therefore played an important role in Ogawa's mistake.

Ogawa belonged to the generation of chemists who received the first systematic modern education in the last third of the nineteenth century in Japan. He thor-oughly mastered the classical techniques of analytical chemistry and was fortunate enough to discover an element by the application of his personal classical tech-niques to the fullest extent. Upon his return to Japan, however, even though he was in one of the most privileged positions as a chemist, he worked in a chemistry laboratory that was poorly equipped and far behind the West from the point of view of the development of scientific instrumentation.

Ogawa tried to use more modern technology, albeit somewhat belatedly. In 1927, for example, he asked a researcher at the institute of Materials Research at Tohoku Imperial University to measure his sample by a mass spectroscope, newly introduced there, but the results were inconclusive. Kimura Kenjiro (1896–1988), a geochemist and analytical chemist at the Imperial University of Tokyo, bought a Siegbahn-type X-ray spectrometer in 1927, after returning from his stay at N. Bohr's institute at Copenhagen. The operation of the machine started at the end of the next year. Probably early in 1930 Ogawa sent a sample to Kimura for X-ray spectroscopic analysis. According to a colleague of Kimura's, the sample was found to be very pure rhenium.[61] This fact was not made public, however.

One must point out at the same time that Ogawa did not have the frame of mind to organize a project-type research group, confining himself instead to individual research. This attitude did not help to advance his work fast enough. If he were a researcher of today, it would be unthinkable not to publish anything after his preliminary report in 1908, even though he continued to study his "nippoinium." He should have published something.

These conditions prevented him from further developing his discovery in Japan and consequently, his work ended unfinished. His life and work have shown the success and limitations of his generation of chemists in modern Japan. At the same time, already by the early twentieth century the first generation of Japanese research chemists started to do original work based on the periodic law.

9. CONCLUSION

Japanese scholars of Dutch Learning encountered modern chemistry in Dutch textbooks, many of which were first written in other European languages, translated into Dutch and imported to Japan. One such scholar, Udagawa Yoan, wrote the first introductory textbook of chemistry in the 1830s and 1840s based on those imported Dutch chemistry books, introducing Lavoisier's new chemistry into Japan. Scholars of Western Learning after Udagawa, including Kawamoto Komin and Uyeno Hikoma, went further to introduce Dalton's chemical atomism and the knowledge of modern chemical technology into Japan before the Meiji Restoration in 1868.

The full-fledged modernization of Japan started after the Meiji Restoration. During the early stage of this modernization, some of the scholars of the Dutch Learning continued to prepare books on chemistry in Japanese in the early transitional period. The Meiji government employed many foreign specialists, including in chemistry, from Europe and the United States. Foreign teachers taught chemistry systematically and some of their initial lectures were translated into Japanese. In the early stage of the Meiji period (1870s and 1880s) many popular elementary textbooks of chemistry in English from Great Britain and the United States, such as those of Henry Roscoe and Ira Remsen, were translated into Japanese many times.

In the first stage of the institutionalization of chemistry in Japan, foreign teachers of chemistry like Robert William Atkinson and Edward Divers taught in Japanese higher educational institutions, educated the first generation of Japanese chemists, and showed the possible direction of Japanese chemistry studies by their own studies in Japan. Their students were then sent to Europe or the United States to study further. The institutionalization was completed with the establishment of the Tokyo Chemical Society in 1878 and the foundation of Imperial University in Tokyo in 1886. The discovery of the periodic law between 1869 and 1871 and its dissemination in the 1880s coincided with the institutionalization of chemistry in Japan. This factor helped make the appreciation of the periodic system as a basis for chemistry in Japan easier. Most of the first generation of Japanese chemistry professors accepted the periodic law as one of the recent developments in chemistry in Europe without much doubt.[62]

In the 1890s well-written theoretically oriented chemistry textbooks for higher education were translated into Japanese, such as Lothar Meyer's book and Ira Remsen's textbook. Around that time the first general chemistry professors, such as Takamatsu Toyokichi, Yoshida Hikosaburo, Ikeda Kikunae, and Osaka Yukichi, who were born in the 1850s and 1860s, started to write chemistry textbooks for secondary schools and universities, based on current chemistry textbooks in English and German. Among them, Ikeda was an especially prolific author, who wrote theoretically oriented textbooks, descriptive ones, and purely elementary texts. He was one of the first who paid attention to the difference of concepts of elements and simple bodies. At the same time, his attitude toward the periodic law was somewhat detached and calm, regarding it as a rule of thumb that made it very convenient to memorize relationships among the elements.

The announcement of the outline program of instruction (curricula) for the secondary schools by the Ministry of Education in 1902 marked the completion of the systematization of the secondary education curriculum. The outline program of chemistry included the periodic law and all textbooks contained instruction about the periodic law after that.

Around this time Japanese chemists started to contribute to research related to the periodic system. Ogawa Masataka announced the discovery of a new element, called "nipponium" in 1908, which much later turned out to be rhenium.

The acceptance of the periodic system in Japan looks smooth and swift, but this was largely due to the fact that the discovery of the periodic law coincided with the institutionalization of modern chemistry in Japan, which had started much earlier—from the end of the eighteenth century. This rather lengthy process of the appreciation of modern chemistry from the West helped the smooth acceptance of the periodic system.

NOTES

1. This paper follows the Japanese style of writing family names first, followed by given or first names. Here *Yamawaki* is the family name and *Toyo* is the given name.
2. Sugita Genpaku et al., *Kaitai Shinsho* [A New Book of Anatomy] (Edo: Suharaya Ichibei, 1774). This is the Japanese translation of the Dutch translation, *Ontleedkundige Tafelen*, by Gerard Dicten (1696? –1770), which was in turn the translation of Johann Adam Kulmus's (1689–1745) German *Anatomische Tabellen* (1722).
3. Udagawa Yoan, *Seimi Kaiso* (Edo: Suharaya Ihachi, 1837–1847).
4. Henry's book was translated by J. B. Trommsdorff (1770–1837) into German. This was then translated into Dutch in 1808 by A. Ypey (1749–1820).
5. For his biography, see Kawamoto Yoji and Nakatani Kazumasa, *Kawamoto Komin Den: Kinsei Nihon no Kagaku no Shiso* [The Biography of Kawamoto Komin: The Founder of Chemistry in Early Modern Japan, in Japanese] (Tokyo: Kyoritsu Shuppan, Ltd., 1971).
6. *Kagaku Shisho* (1861) was Kawamoto's translation of Jan Willem Gunning's Dutch translation [*De Scheikunde van het onbewerktuigde en bewerktuigde rijk, bevattelijk voorgesteld en met eenvoudige proeven opgehelderd*] of Julius Adolf Stöckhardt's popular book, *Schule der Chemie* (1846). Even though it was used as a textbook in the *Bansho Shirabesho*, it was not published in those days. It has been published only recently as one of the classics of Japanese science: *Kagaku shinsho* (Tokyo: Kinokuniya shoten, 1998).

7. Uyeno Hikoma, *Seimi-kyoku Hikkei* (Ise: Iseya Jihei, 1862). The reprint edition was published in four volumes in 1976 (Tokyo: Sangyo-Noritsu-Tanki-Daigaku Shuppan). The fourth volume (commentary) contained a short but well-informed biography of Uyeno.

8. See *Shashin no Kaiso, Uyeno Hikoma* [A Founder of Photography in Japan, Uyeno Hikoma] (Tokyo: Sangyo Noritsu-Tanki Daigaku Shuppan, 1975). This book contains a number of photos taken by Uyeno.

9. *Huaxue Chujie*, Canton: Po-chi i-yüan, 1871–75. Japanese translation: *Kagaku Shokai Wakai* [Japanese translation of Huaxue Chujie], transl. by Shishido Itsuro, ed. by Udagawa Kosai (Tokyo: Aoyama-do, Sensho-do, 1873).

10. Jean Pierre Louis Giraldin, *Kagaku Nyumon* [An Introduction to Chemistry], translated by Katsuragawa Hosaku et al. (Edo: Yikkando, 1867–73); John Addison Porter, *Kagaku Tai-i* [Principles of Chemistry], translated and edited by Japanese Naval Academy (Tokyo: Naval Academy, 1873); Thomas Turner Tate, *Mono-wari no hashigo (Seimi no Tebiki)* [An Introduction to Chemistry], translated by Shimizu Usaburo, Kyoto; *Kagaku-tsu* [An Introduction to Chemistry], translated by Kawamoto Komin, edited by Kawamoto Sei-ichi, Tokyo, 1876.

11. Utsunomiya Saburo and Katsuragawa Hosaku, *Kagaku-shinron Mondo* (Tokyo: Ikkando, 1875).

12. Herman Ritter, *Rika Nikki* (Osaka: Itamiya Zenbei, 1870); *Kagaku Nikki* (Tokyo: Ministry of Education, 1876). Ritter's lecture notes were first translated by Ichikawa (Hiraoka) Morisaburo (1852–1882) into Japanese. Ichikawa was a son of a scholar of Western Learning and showed his talent in his early age. He was sent by the Tokugawa government to London to study science in 1866. When he returned due to the collapse of the Tokugawa government in 1868, he became a teacher for higher education for the Ministry of Education. He was sent again to Manchester at the end of 1877 by the Meiji government. He returned because of illness in August 1879 and was appointed as a professor of Tokyo University in 1881 before his premature death at the age of thirty-one. See Sugiura Shigekata, "Ichikawa Morisaburo Kun Ryakuden [A Brief Biography of Mr. Ichikawa Morisaburo]," *Toyo Bungei Zashshi* no.14 (1882): 362–363.

13. Rosuko [Roscoe], *Kagaku-no-hajime*, transl. by Kaneko Sei-ichi, checked by Iinuma Chozo (Akita: Akita ken Taihei Gakko, Dec. 1875); Rosuko [Roscoe], *Rosuko-shi Kagaku*, checked by Hiraoka Morisaburo (Tokyo: Monbu sho Honyaku kyuku [Department of Translation in the Ministry of Education], 1876); Rosuko [Roscoe], *Kagaku-shinsho*, transl. by Kashiwabara Gakuji, checked by Adachi Horoshi (Shizuoka: Kashiwabara Gakuji, 1876–77); Rosuko [Roscoe], *Kagaku-kaitei*, transl. by Udagawa Jun-ichi (Tokyo: Yamanaka Ichibei, Nov. 1881); Rosuko [Roscoe], *Shogaku-Kagaku-sho*, vol.1–3, transl. by Ichikawa Morisaburo (Tokyo and Osaka, 1882).

   Remuzen [Remsen], *Kagaku-sho*, transl. by Yoshioka Tetsujiro and Ueda Toyoketsu (Tokyo: Yoshioka Shoseki-ten, 1888–89); Remuzen [Remsen], *Sho-kagaku-sho*, transl. by Kuhara Mitsuru and Oda Kenjiro (Tokyo: Kengyo sha, 1888–89); Remuzen [Remsen], *Chu-kagaku Uki-hen*, transl. by Hirano Ikkan, checked by Kuhara Mitsuru and Shimoyama Jun-ichiro (Tokyo: Hirano Ikkan, 1893–94); Remuzen [Remsen], *Kagaku-kyokasho*, transl. by Yoshioka Tetsutaro (Tokyo: Kaishin-do, July 1894).

14. It is interesting to note that the same series was translated into Chinese by John Fryer (1839–1928), the famous translator of European texts in nineteenth-century China, and published in 1879–1880 in Shanghai. See Benjamin A. Elman, *A Cultural History of Modern Science in China* (Cambridge, Mass.: Harvard University Press, 2006), 133.

15. Ira Remsen, *The Elements of Chemistry: A Textbook for Beginners* (Macmillan's school class books) (London: Macmillan, 1887).

16. Shiga Taizan, *Kagaku Saishin* (Osaka: Ryushodo, 1877).

17. Shimoyama Jun-ichiro, *Kagaku Shinri* (Tokyo: Kyoeido/ Eirando, 1880).

18. On Sakurai, see Kikuchi Yoshiyuki, "Redefining Academic Chemistry: Joji Sakurai and the Introduction of Physical Chemistry into Meiji Japan," *Historia Scientiarum,* **9** (3), 215–242 (2000) and his unpublished PhD thesis, "The English Model of Chemical Education in Meiji Japan: Transfer and Acculturation," (The Open University, 2006).

19. This higher educational institution was formed in 1869 in Tokyo soon after the Meiji Restoration and changed its name many times with various reforms of the educational system. In 1877 it was combined with the School of Medicine, another higher educational institution under the Ministry of Education, and became Tokyo University.

20. Robert William Atkinson was born in Newcastle in England. After studies at the University College, London, under Alexander Williamson and the Royal School of Mines under Edward Frankland between 1867 and 1872, he became an assistant of Alexander Williamson, professor of chemistry at University College, London. Atkinson came to Japan on September 9th, 1874, and stayed until 1881. After his return, he lived in Cardiff, Wales. See Kikuchi's PhD thesis (note 18), "The English Model of Chemical Education in Meiji Japan."

21. Tokyo Daigaku Rigaku-bu Kagaku-kyoshitsu Zashi-kai, ed., *Tokyo Daigaku Rigaku-bu Kagaku Kyoshitsu no Ayumi* [The History of Department of Chemistry in the University of Tokyo] (Tokyo, 2007) 29–35. The Society was first formed as the Kagaku-kai [Chemical Society] in 1878 and renamed the Tokyo Kagaku-kai [Tokyo Chemical Society] in 1879. It was renamed again the Nihon Kagaku-kai [the Chemical Society of Japan] in 1921.

22. R. W. Atkinson, "Genso no seitai ni kansuru shiso no enkaku ryaku-setsu [*An Outline of the History of the Theory of Elements*]" *Tokyo Kagaku Kaishi* [Journal of the Tokyo Chemical Society] **2** (1): 1–54 (1881), p. 39.

23. Matsui Naokichi, "Genshi-setsu enkaku no Gairyaku [An Outline of the History of Atomic Theory]" *Tokyo Kagaku Kaishi* [Journal of the Tokyo Chemical Society] **3** (1): 35–48 (1883), p. 48.

24. Takamatsu Toyokichi, *Kagaku Kyokasho, Dai ichi hen, Muki kagaku* [Chemistry Textbook, First Part, Inorganic Chemistry] (Tokyo: Ministry of Education, 1890–91).

25. For Yoshida Hikorokuro, see Shiba Tetsuo, "Yoshida Hikorokuro [in Japanese]" *Woko Junyaku Jiho* **70**(3), 2–4.

26. Yoshida Hikorokuro, *Chuto Kagaku Kyokasho* [Chemistry Textbook for Secondary schools] 2 vols, (Tokyo: Kinkodo, 1893).

27. Charles Loudon Bloxam (1831–87). One can find various editions of his *Chemistry: Inorganic and Organic with Experiments*, London: J. & A. Churchill, starting from its second edition, which was published in 1872 (the first edition was published in 1867) in university libraries in Japan.

28. He mentions one more name, "Myua," which is unknown. One possibility would be "William Allen Miller," whose *Elements of Chemistry: Theoretical and Practical* (New York: John Wiley, 1877) was translated into Japanese and published by the Ministry of Education in 1887.

29. The collection of Mori Ogai in the University of Tokyo Library is Mori's former library consisting of 18,800 volumes, including 3,100 Western books. His library was donated by his family to the University of Tokyo Library, when the University lost all of its books due to the Great Kanto Earthquake in 1923 that devastated

Tokyo, Yokohama, and the surrounding area. See http://rarebook.dl.itc.u-tokyo. ac.jp/ogai/ogai.html, accessed November 14, 2013.

30. Lothar Meyer, *Die modernen Theorien der Chemie und ihre Bedeutung für die chemische Mechanik*, 5. Aufl., Breslau: Maruschke & Berendt, 1884 is owned by Tohoku University, University of Tokyo, and Kanazawa University.

31. D. Mendeléeff, *The principles of chemistry*, translated from the Russian (fifth edition) by George Kamensky and edited by A. J. Greenaway (London; New York: Longmans, Green, 1891), is owned by Tohoku University, University of Tokyo, and Kanazawa University. The second edition (1897) is owned by Kyoto University, Doshisha University. The third edition (1905) is owned by Kyoto University, Kyushu University, University of Tokyo, Tokyo Institute of Technology, and Tohoku University.

32. Lothar Meyer, *Kagaku Genron*, transl. by Kondo Kaijiro, Tokyo: Uchida Rokakuho, 1894–95. Meyer's *Outlines of Theoretical Chemistry* is the translation by P. Phillips Bedson and W. Carleton Williams of *Grundzüge der theoretischen Chemie*, which is Meyer's "smaller and less technical work on Chemical Philosophy" (English translators' preface) than his famous *Die modernen Theorien der Chemie*.

33. Remsen, *Chu-Kagaku* [Intermediate Chemistry] in two volumes, transl. by Hirano Ikkan (Tokyo, 1893–94).

34. Ministry of Education ed., *Chuggako Kyoju-yomoku* [The Outline Program of Instruction for the Secondary Schools] (Tokyo: Nihon Keizai-sha, 1902).

35. On the history of science education in Japan, see, for example, Hori Shichizo, *Nihon no Rika Kyoikushi* [The History of Science Education in Japan], 3 volumes (Tokyo: Fukumura Shoten, 1961). Hori Shichizo (1886–1978) was an educator of science in primary and secondary education and had participated in the compilation and editing of various textbooks of science in elementary and secondary schools for the Ministry of Education for a long time before World War II.

36. Here are the major chemistry textbooks for secondary schools from the 1890s to the 1910s: Osaka Yukichi, *Kinsei Kagaku Kyokasho* [Modern Chemistry Textbook] (Tokyo: Toyamabo, 1897); Ikeda Kikunae, *Kagaku Kyokasho* [Chemistry Textbook] (Tokyo: Kinkodo, 1897); Ikeda Kikunae, *Ninpen Chugaku Kagakusho* [A New Secondary school Chemistry Book] (Tokyo: Kinkodo, 1898); Osaka Yukichi, *Kinsei Kagaku Kyokasho* [Modern Chemistry Textbook] 2nd rev. ed. (Tokyo: Toyamabo, 1898); Takamatsu Toyokichi, *Kagaku Kyokasho* [Chemistry Textbook] rev. ed. (Tokyo: DaiNihon-Tosho, 1900); Ikeda Kikunae and Wada Isaburo, *Kagaku Shinpen* [A New Book of Chemistry] (Tokyo: Kinkodo, 1900); Majima Riko, Muki Kagaku [Inorganic Chemistry] (Tokyo: Hakubunkan, 1901); Haneda Seihachi, *Chuto Kagaku Shinpen* [A New Book of Chemistry for Secondary schools] (Tokyo: DaiNihon-Tosho, 1900); Ogawa Masataka, *Kagaku Chu-Kyokasho* [Chemistry Textbook for Secondary schools] (Tokyo: Keigyosha, 1901); Kametaka Tokuhei, *Futsu-Kyoiku Kagaku Kyokasho* [Chemistry Textbook for General Education] (Tokyo: Tokyo Kaiseisha, 1902); Osaka Yukichi, *Kinsei Kagaku Kyokasho* [Modern Chemistry Textbook] rev. ed. (Tokyo: Toyamabo, 1902); Moriya Monoshiro, *Chuto Kagaku Kyokasho* [Chemistry Textbook for Secondary schools] (Tokyo: Fukyukai, 1903); Sawamura Makoto, *Shinpen Kagaku* [A New Book of Chemistry] (Tokyo: Kobunsha, 1905); Sawamura Makoto, *Nogyo Kagaku Kyokasho* [Chemistry Textbook for Agriculture] (Tokyo: Shungyudo, 1911); Muramatsu Shunsuke and Sakurada Ushio, *Chuto Muki-Kagaku Kyokasho* [Inorganic Chemistry for Secondary schools] (Tokyo: Seibido-shoten, 1912).

37. Takamatsu (note 24), *Kagaku Kyokasho*, 1890–91, vol. 1, p. 10.

38. Yoshida (note 26), *Chuto Kagaku Kyokasho*, 1893, vol. 1, pp. 17–18.

RESPONSE TO THE PERIODIC LAW IN JAPAN (303)

39. On Ikeda, see Ikeda Kikunae Tsuioku-kai, *Ikeda Kikunae Hakushi Tsuioku-roku* [Reminiscence on Dr. Ikeda Kikunae, in Japanese], 1956 and Hirota Kozo, *Kagakusha Ikeda Kikunae* [Ikeda Kikunae as a Chemist, in Japanese] (Tokyo: Tokyo Kagaku Dojin, 1994).

40. During most of the Edo period (1635–1862), feudal lords had to move periodically between their domains and Edo, spending every year alternately in each place. Each powerful lord in southern Japan kept a residence in Kyoto, the old capital, located on the way to Edo.

41. After the establishment of Kyoto Imperial University as the second imperial university in 1897, the Imperial University in Tokyo was renamed Tokyo Imperial University.

42. Ikedad Kikunae, *Kagaku* (Tokyo: Bungakusha, 1895).

43. Ibid., pp. 38–39.

44. See my paper, Masanori Kaji, "Mendeleev's Discovery of the Periodic Law: the Origin and the Reception," *Foundations of Chemistry* **5**: 189–214 (2003), pp. 198–201 and the chapter on the Russian case in this book.

45. Ikeda Kikunae, *Ninpen Chugaku Kagakusho* (Tokyo: Kinkodo, 1898), 36–37.

46. Ibid., pp. 273–279.

47. Takamatsu Toyokichi, *Kagaku Kyokasho* [Chemistry Textbook] rev. 3rd ed. (Tokyo: DaiNihon Tosho, 1902), 17–18. Other textbooks (for bibliographical details, see note 37), which distinguished two concepts were Haneda Seihachi, *Chuto Kagaku Shinpen*, 1900, p. 23; Ogawa Masataka, *Kagaku Chu-Kyokasho*, 1901, p. 16; Kametaka Tokuhei, *Futsu-Kyoiku Kagaku Kyokasho*, 1902, pp. 33–34; Osaka Yukichi, *Kinsei Kagaku Kyokasho*, 1902, p. 37; Moriya Monoshiro, *Chuto Kagaku Kyokasho*, 1903, p. 10.

48. Ikeda (note. 45), p. 278.

49. Ikeda Kikunae, *Kinsei Kagaku Kyokasho* [Modern Chemistry Textbook] (Tokyo: Kaiseido, 1903); Ikeda, Chugaku Kagaku Kyokasho [Chemistry Textbook for Secondary schools] (Tokyo: Kinkodo, 1903).

50. This section is based on my paper: Masanori Kaji, "Ogawa's Discovery of a New Chemical Element 'Nipponium': The Emergence of Modern Chemistry Research in Japan and Its Social Background," *Historia Scientiarum*, **12** (3) (2003): 215–218.

51. Ogawa Masataka, "Preliminary Note on a New Element in Thorianite," *Journal of the College of Science, Imperial University*, Tokyo, Japan 25(1908): Article 15, 1–11, on p. 1.

52. Ogawa, ibid.; Ogawa Masataka, "Preliminary Note on a New Element Allied to Molybdenum," *Journal of the College of Science, Imperial University*, Tokyo, Japan 25(1908): Article 16, 1–13.

53. Ogawa Masataka, "Preliminary Note on a New Element in Thorianite," *Chemical News* **98** (1908): 249–251; Ogawa Masataka,"Preliminary Note on a New Element Allied to Molybdenum," *Chemical News* **98** (1908): 261–264.

54. See, for example, F. H. Loring, "The Atomic Weights as Mathematical Functions," *Chemical News* **100** (2611), 281–286 (1909).

55. *Tokyo Kagaku-kai-shi* (Journal of the Tokyo Chemical Society) **31** (1910): 422.

56. W. Noddack, I. Tacke, O. Berg,"Die Ekamangane," *Naturwissenshaften* **13** (1925): 567.

57. I. Noddack, W. Noddack, "Darstellung und einige chemischen Eigenschaften des Rhenium," *Zeitschrift für Physikalische Chemie* **125** (1927): 264–274.

58. C. Perrier, E. Segré, "Some chemical properties of element 43," *Journal of Chemical Physics* **5** (1937): 712–716.

59. Yoshihara H. Kenji, "Nipponium, the Element Ascribable to Rhenium from the Modern Chemical Viewpoint," *Radiochimica Acta* **77** (1997): 9–13; Yoshihara H. Kenji, "Ogawa's Discovery of nipponium and Its Re-evaluation," *Historia Scientiarum* **9** (3) (2000): 257–269.

60. Ogawa (note 52), "Preliminary Note on a New Element in Thorianite," p. 4.
61. Yoshihara (note 59), "Ogawa's Discovery of nipponium and Its Re-evaluation," 266.
62. It might be worthwhile to note the relative smooth acceptance of Darwinism in Meiji Japan. See, for example Osamu Sakura, "Similarities and Varieties: A Brief Sketch on the Reception of Darwinism and Sociobiology in Japan" *Biology and Philosophy* **13** (1998). 341–357.

INDEX